The Future of 24-Hour News

This book is part of the Peter Lang Media and Communication list.
Every volume is peer reviewed and meets
the highest quality standards for content and production.

PETER LANG
New York • Bern • Frankfurt • Berlin
Brussels • Vienna • Oxford • Warsaw

The Future of 24-Hour News

New Directions, New Challenges

Edited by Stephen Cushion and Richard Sambrook

PETER LANG
New York • Bern • Frankfurt • Berlin
Brussels • Vienna • Oxford • Warsaw

Library of Congress Cataloging-in-Publication Data
Names: Cushion, Stephen, editor. | Sambrook, Richard. editor.
Title: The future of 24-hour news: new directions, new challenges /
edited by Stephen Cushion, Richard Sambrook.
Other titles: Future of twenty-four hour news
Description: New York: Peter Lang, 2016.
Includes bibliographical references and index.
Identifiers: LCCN 2016005905 | ISBN 978-1-4331-3047-2 (hardcover: alk. paper)
ISBN 978-1-4331-3046-5 (paperback: alk. paper) | ISBN 978-1-4539-1816-6 (e-book)
Subjects: LCSH: Television broadcasting of news. | Broadcast journalism.
Classification: LCC PN4784.T4 F88 2016 | DDC 070.1/95—dc23
LC record available at http://lccn.loc.gov/2016005905

Bibliographic information published by **Die Deutsche Nationalbibliothek**.
Die Deutsche Nationalbibliothek lists this publication in the "Deutsche
Nationalbibliografie"; detailed bibliographic data are available
on the Internet at http://dnb.d-nb.de/.

Table of Contents

National Contexts and Journalistic Challenges

Introduction: The Future of 24-Hour News: New Directions, New Challenges

STEPHEN CUSHION AND RICHARD SAMBROOK

24-hour news is no longer a new or novel part of journalism. Dedicated television news channels, after all, have been operating for well over thirty years, whilst online news or live rolling blogs are now decades old. More recently, new content and social media platforms—from YouTube to Twitter, Weibo to Facebook, Vine to Periscope—have accelerated the pace of news delivery, instantly publishing at the tap of an app a story, comment, photo, pre-recorded image or live moving pictures. News is now supplied in a more fluid form and faster pace than at any other point in history, radically different to the culture of journalism in 1980 when CNN launched the first-ever 24-hour television news channel. Even at the beginning of the new millennium, the growing army of dedicated television news channels around the world did not face the same level of competition or type of editorial pressures from today's hyper-accelerated news culture.

For news has converged at spectacular speed: from smart phones to radios, television sets to tablets, newspapers to computers, people can—and increasingly do—move seamlessly between an ever-extending menu of media platforms. As internet penetration has increased over recent decades and more people connect online to new portable devices that allow news to be easily consumed *throughout the day*, the role and purpose of dedicated 24-hour television news channels has been brought into sharper focus. Whereas rolling news channels could once claim to bring viewers breaking news first, in most cases today they will be second best to social media or online rolling blogs. For the monopoly of 24-hour news television delivering rolling news as it happens is well and truly over, as increasingly sophisticated and affordable technology allows not just journalists but ordinary citizens the power to instantly share information and observations about what is happening in the world.

In light of these developments, the aim of this edited book is to explore the challenges facing television news channels and to consider the future direction of 24-hour news journalism. While the arrival of always-on, continuous television news channels has made a big impact on the wider culture of journalism over the last thirty years, new digital technology and converged media platforms, as well as changing audience behaviour and demands, prompt not just timely but existential questions about the current and future role of 24-hour news television. We consider four issues that confront television news channels today and in their immediate future, before outlining the structure of the book and introducing each chapter.

First, as new content and social media platforms can publish news far more quickly than television, how have 24-hour news channels responded? Have they adapted to new breaking news pressures? Social media, such as Twitter, connect instantly to a global network of hungry news consumers, whereas many news channels attract tiny audiences most of the time unless a major breaking story erupts. Since many news channels still market themselves as being first with breaking news, will they change editorial direction or try to keep up with the pace of social media? Of course, news channels do not operate in isolation from other media platforms, but will their social media presence soon overtake the importance and resources invested in broadcasting live breaking news?

As it stands, social media do not look likely to imminently replace the television set as the main source of news. According to a recent cross-national survey (Reuters Institute of Journalism, 2015), television news continues to be the dominant source of news for most people in the Western world. Although social media use has increased in recent years, it remains by some distance behind TV, online and, in some cases, even newspapers (as the main source of news, it represents 3%–12% in 12 of the most advanced nations surveyed—Reuters Institute of Journalism, 2015). When younger people are isolated the use of online and social media dramatically increases, but generationally it remains to be seen whether this represents the future of news consumption.

However, it could be technology and changing platforms—not necessarily audience habits—that threaten the lifespan of 24-hour news television channels. A second major issue facing news channels is that the television experience as we know it could soon be coming to an end as online services converge on the same platform. IPTVs (internet television information and online portals) promise to be the future of media consumption in many people's households, with news channels operating side-by-side with the world of online and social media. While many people are already familiar with smart TVs, IPTVs could connect multiple devices and more seamlessly allow news to be consumed between sources and personalized for individual consumer tastes. The old-fashioned view of a TV remote controller would radically change; it would not just act as a means of switching between channels but open up a world of new sources—raw and unedited to polished and packaged—for people to pick from and interact with. Like newspapers transforming their journalism for the online age, 24-hour news channels may need to reconsider how to communicate with their viewers on multiple levels. As the many chapters in this book testify to, this convergence is already well underway. Launching a YouTube Channel in 2007, for example, RT (formerly Russia Today) captures up to a million online viewers each day, allowing users to choose between different genres of news (from uploaded user-generated content to documentaries and extended interviews).

Of course, such changes remain predicated first on technology—how soon and effectively will IPTVs seamlessly guide an ordinary viewer across sources and between different platforms—and secondly on attracting mainly savvy and interactive audiences. While it may appear old-fashioned and costly, anchors fronting a menu of continuous news may still appeal to many viewers content for someone else to be their gatekeeper. Since it has been the dominant way television journalism has been communicated for more than fifty years, the supply of raw footage or live feeds does not presuppose a big demand for it.

But a third issue that may force news channels into abandoning presenters, and producing less polished and packaged material, is the high production costs that go into running a 24-hour rolling news station. After all, the journalistic costs of a news channel—from owning and managing a large studio, maintaining a continuous live presence, paying presenters, operating studio equipment, or dealing with a broadcast licence—far outweigh the outlay of expenditure for most online news and social media platforms. Covering breaking news stories around the world means credible international news channels invest a huge amount of resources into flying journalists, editorial staff and camera crews to far-flung locations.

Of course, how much news channels invest in these resources can differ—CNN, for instance, spends dramatically more on covering international affairs than its more domestically-minded rivals in the US (US State of Media, 2015). By contrast, many smaller, commercially run regional or national regional news channels rely to a great extent on news agencies for images and footage and do not necessarily have a live schedule of continuous news programming 24 hours a day. Needless to say, this is not out of journalistic choice but necessity and explains why—as Robert G. Picard outlines in Chapter 13—many organisations find it difficult to start up a news channel or operate at a loss. Of the international news channels that do operate around the clock, perhaps unsurprisingly many are public service oriented or operate as state broadcasters. In Chapter 12, Rai and Cottle's research demonstrates that 24-hour news channels continue to flourish and expand their reach and influence around the world. But for many, their long-term survival does appear precarious without state backing.

Nevertheless, since the political economy of many rolling news channels has always been built on shaky foundations, their continued existence may rely on the wider take up on new content and social media platforms. While a Putin-backed RT can continue to pump millions into the news channel, for example, at what point will public service broadcasters such as the BBC find it difficult to legitimise the expensive costs of international rolling TV news stations when they could turn to more effective platforms of 24-hour news? So, for example, in 2015 the BBC announced it was considering making its UK television news channel an online only service (Conlan, 2015). It was a proposal put forward to make efficiency savings at the BBC and may not be fully implemented in the short term. But even by considering this option, it suggests television news channels will face more existential threats in the not too distant future.

However, above all, it is the fourth issue facing 24-hour news channels—the editorial values they will champion in the future—that may ultimately play the biggest part in deciding their long-term survival. Rolling television news channels will need to editorially evolve to remain relevant in an increasingly crowded and competitive media environment. As previously acknowledged, 24-hour news channels no longer regularly break news ahead of rival platforms. While putting accuracy above speed has long defined debates about the rolling news genre, it would be difficult to maintain this editorial standard if 24-hour news channels attempted to keep up with the frenetic pace of online news or social media. Indeed, the different levels of trust invested in competing media platforms by news audiences was acknowledged in Reuters Institute of Journalism (2015) cross-national survey: "Social media are not seen as a destination for accurate and reliable journalism but more as a way of getting access to it." Of course, critics will be quick to point to the partisan US cable news stations—notably Fox News and MSNBC (explored in Chapters 20 and 21)—or the highly partial view on the world presented by state-backed channels such as RT (examined in Chapter 4). But these are the exception to the norm, since television remains a closely regulated medium in most countries, and committed to maintaining balance and confidence from its viewers.

If 24-hour news television succumbed to the pressure of delivering breaking news at even greater speed, or attempted to match the personalized and opinionated nature of platforms such as Twitter, many organisations could risk losing their market credibility. Needless to say, there are moments when rolling television news appear to have already adopted this strategy—in the immediate aftermath of the missing Malaysia Airlines flight MH370 in 2014, for instance, when too much coverage reported speculation and wild rumours, rather than establishing the facts or verifying sources. But where television news can be distinctive from

Twitter feeds or rolling blogs is not in regularly beating rival media to a breaking news scoop, but in drawing on a large team of professionally-trained editorial staff to immediately supply the necessary context and analysis to a story. From experienced anchors to expert reporters, a well-resourced 24-hour news channel can make far better sense of fast-moving events than is possible in 140 characters. As a former head of BBC television news put it, "On breaking news stories, a television news channel can guide viewers coherently through a fast-changing situation—and the skill of the presenter is essential for linking and making sense of what is happening. In busy times and quiet times, interviews are at the heart of a news channel: correspondents analysing the latest developments or news-makers popping into the studio to give information or be grilled about what they've done" (Mosey, 2015). Indeed, drawing on the analytical skills of journalists, many news channels now also include specialist programming in their daily schedules. Overall, rolling television news channels have the challenging task of safeguarding these and other editorial values, whilst finding new ways to diversify what is meant by 24-hour news journalism and remaining distinctive from rival online and social media platforms.

This book reflects on these and wider debates surrounding news channels in a global context, considering the future of 24-hour news from a range of industry and academic perspectives. Since *The Rise of 24-Hour News Television: Global Perspectives* (Cushion and Lewis, 2010) was published, there have not been any major book-length assessments of 24-hour news channels or about the changing nature of rolling news television formats. While the rise of social and online media has attracted considerable attention within journalism, media, and communication scholarship, our intention is to put television news back under the spotlight and consider whether and in what ways rolling news channels have been reshaped due to pressure from more instant platforms of news. Our aim is not only to bring together leading academics around the world, but also to ask the heads of some of the biggest news channels or their senior editors and managers to reflect on the present state and future role of 24-hour news television in the context of new competition from online and social media platforms.

The book is split into two parts, with five total subsections. The first part explores the challenges and pressures facing international broadcasters from the perspectives of senior industry figures, whilst the second primarily draws on contributions from academics around the world. Chapter One begins with Richard Sambrook, a former head of BBC news, who helped to launch the BBC News 24 channel in 1997 but is sceptical about the long-term survival of 24-hour news channels. In a provocative chapter designed to elicit debate, he and media strategist Sean McGuire argue that what was transformational 30 years ago is no longer fit for purpose as the internet increasingly meets our information needs.

The second section includes eight chapters from some of the leading editors in the world of rolling news. In different ways, they set out their own views about the current state of 24-hour news television and consider how broadcasters will respond to the many challenges they face in a new media environment.

In Chapter Two, the former president of ABC News in the US, David Westin, argues that the future of video news will be shaped by three forces: the loss of the TV schedule; the expansion of video news producers to include just about everyone; and the transformation of news consumers into editors and of editors into curators. And he considers how the growing partisanship in US 24-hour news is having a detrimental effect on democratic debate.

In Chapter Three, Michael Peters, the Chief Executive of Euronews—the pan-Europe, multilingual news service—argues for the increasing importance of a broad diversity of views being represented in TV news. He explains how this principle, along with maintaining a strong brand, lies at the heart of his strategy for transforming Euronews as, like other TV channels, it seeks to compete with the challenges of the internet and the information deluge offered by digital platforms.

In Chapter Four, the Editor-in-Chief of RT, Margarita Simonyan, argues that the inherent strengths of TV as an information and storytelling medium will ensure 24-hour channels survive in the long term. But she says they need greater differentiation, adopting a wider range of perspectives and deeper more pro-active engagement with audiences via social media.

In Chapter Five, Mark Scott, the former Director General of the ABC in Australia, outlines how TV news can hold its own against fierce competition from a wide range of new digital services—in particular, mobile. By adopting new formats, data journalism, and greater interactivity through the social web he believes TV news can prosper if it retains trust and relevance to audiences.

In Chapter Six, the Head of Sky News in the UK, John Ryley, directly takes on the Sambrook/McGuire argument and says a 24-hour TV channel is essential to provide the core of visual journalism to a successful multiplatform news service. He suggests that technology simply provides new and different tools to achieve the same ends: getting and telling news stories as swiftly and engagingly as possible.

In Chapter Seven, Peter Horrocks, until recently the BBC's Director of the World Service group, recalls the TV news battles between the BBC and Sky and looks at how digital platforms are now challenging in different ways all the core functions of TV news. He argues the channels need to become the gateways to other services, suggesting, like John Ryley, that they can sit at the heart of a multiplatform offer.

In Chapter Eight, Ibrahim Helal from Al Jazeera offers a personal view of the pressures and difficulties faced by news providers in the Middle East. He laments the increased politicisation of media in the region and he sees a growing gap between producers and audiences. He concludes with a call for broadcasters to renew their commitment to strong editorial values in the face of many challenges from competitors and audiences to traditional objective standards.

In Chapter Nine, the former Editor-in-Chief of Reuters News agency, David Schlesinger, recalls the core role of the news agencies in providing TV news and considers how they can renew themselves. He suggests the advent of 24-hour news channels changed the news agencies role forever and now they need to reinvent themselves again by differentiating themselves and moving from business-to-business to business-to-consumer services.

Section 3 puts debates about the present and future role of 24-hour news television into historical perspective and maps out the channels operating around the world. Since CNN launched its news channel in 1980, this section considers the growth and impact of 24-hour news television. In Chapter 10, Stephen Cushion examines how 24-hour news television has evolved over more than three decades. Revisiting a chapter written in 2010 to mark the 30th anniversary of CNN (Cushion, 2010), he asks whether the life-cycle of 24-hour news television could still be defined by three phases as originally argued. After all, he suggests, with new content and social media platforms delivering news instantaneously, it could be that the purpose and character of 24-hour news channels has moved to a fourth phase in order to keep up with the wider pace of news culture. However, in assessing news channels in the social media age, he argues they have not radically changed and remain committed to breaking news instantly regardless of whether they break the story or not. Rather than recent years signalling a departure from the third phase of 24-hour news channels—which marked a period when broadcasters accelerated the speed of news reporting—he suggests not just a reinforcement but an enhancement of it. He suggests a fourth (and possibly final) phase of 24-hour news is likely once the experience of television viewing more fully converges with the online world.

In Chapter 11, Michael Bromley examines the development of news channels by comparing them with how newspapers have evolved in the online era. He suggests that while rolling news television could be considered "new" media towards the end of the 20th century, today they appear old-fashioned just like newspapers. He looks at how newspapers have had to respond to online competition and draws similarities with the challenges faced by rolling news channels. Considering the editorial character of different platforms, he argues that far from "new" media displacing "old" media, each has distinctive value and unique characteristics. Acknowledging that media consumption habits have changed due to new instant news formats and broader changes in people's lifestyles, Bromley concludes that journalism will continue to evolve in communicating news to audiences.

Chapter 12 examines the present landscape of 24-hour news channels on a global level. Revisiting a study carried out in 2009, Mugdha Rai and Simon Cottle empirically explore whether the rise of social media and continued growth of online has reduced the number of rolling news channels operating around the world. Far from 24-hour news channels being in decline, they conclude that more channels have been launched in different regions and languages. They consider different the dynamics at play for local, national and global dedicated news channels, mapping out the continuity and changes compared to their previous study. In concluding the chapter, Rai and Cottle challenge any notion that the 24-hour news landscape represents a "global public sphere" or conveys "global western dominance." Drawing attention to the localization, regionalization and fragmentation of 24-hour news over recent years, they consider the forces that shape not just the continued growth of rolling news channels, but also the factors that influence the diversity of content in different regions around the world.

The six chapters in Section 4 examine the constraints imposed by the political economy of 24-hour news television, explaining how particular journalisms have evolved because of industry pressures and expectations, as well as changes in technology and forces of globalization. In Chapter 13, Robert G. Picard outlines in detail the financial costs associated with running a news channel and the infrastructure involved. He explains how the budget for channels can vary considerably, from international to more localised broadcasters, and according to whether they are state funded, privately owned or commercially driven. Although many commercial channels struggle to deliver an always-on, rolling news service with limited advertising revenue, US cable news outlets are used as an example of delivering high revenues and returning profits. The chapter then examines how particular states fund some of the most widely watched around the world such as Russia Today (RT), BBC World, Al Jazeera and China Central Television (CCTV). Finally, news channels funded by private individuals or enterprises are considered, and the political and economic goals behind them. Picard concludes the survival of 24-hour news will be shaped less by commercial forces, but rather by the governments and private owners funding them.

In Chapter 14, Justin Lewis considers an alternative way of running 24-hour news channels, a departure from the frenetic pace that has characterised the genre for decades. He develops the idea of "slow news," asking what this concept might be and whether it is possible to imagine in the norms and routines associated with much of contemporary journalism. Lewis traces the origins of slow news back to newspapers in the 19th century, when news was part of political and civic education seeking to challenge government or corporate power. But tighter government regulation, new business models and forms of instant communication—live broadcasting or telephones—meant news became increasingly defined by immediacy and bringing the latest or breaking coverage. According to Lewis, the shift away from slow news is a missed opportunity, since it can produce more balanced, contextualised and analytical journalism. Ending on an optimistic note, he argues new forms of storytelling online—hyperlocal blogs, for instance—potentially leave open the possibility for the next generation to popularise the use of slow news sources.

In Chapter 15, Ingrid Volkmer more broadly and conceptually considers these new forms of communication, and how they have changed what is meant and understood by 24-hour news. In an increasingly globalised and transnational media environment, she unpacks concepts such as "crossborder circuits" and "interface" to interpret how digital media is dramatically changing the production and consumption of journalism. She concludes journalism scholars need to pay more attention to interrogating the digital experiences of media consumption and communication. In doing so, Volkmer suggests, it will better represent today's transnational journalism and the interface used to engage and inform people.

Tine Ustad Figenschou's Chapter 16 more empirically considers transnational journalism by examining Al Jazeera and its English-speaking channel, Al Jazeera English (AJE). It begins with a brief outline of the Arab satellite media revolution and the development of the Arab public sphere over recent years, from the mid-1990s into the 2000s. Due to state-sponsored media ownership, censorship practices and authoritarian political regimes, she suggests the new Arab media system can be described as distinctively hybrid. The chapter then considers coverage of the 2011 Arab uprisings, in which Al Jazeera played an important role in explaining the reasons behind the protests to a global audience. Drawing on interviews with journalists from AJE, Figenschou considers the potential and limitations in developing its rolling news journalism online and social media platforms. Overall, she suggests AJE's new digital strategy aims to reflect its wider network agenda of engaging with ordinary people's lives and pursing alternative perspectives to rival Western news organisations.

In Chapter 17, Alan Tomlinson explores the impact of social media on rolling forms of sports news. The use of Twitter amongst sports high-profile celebrities is used as a case study to understand the wider digital transformation of news-sourcing and news-making. Considering some of the most followed sports stars on Twitter—from Christian Ronaldo, LeBron James and Amir Khan—he examines the nature of these sources and how journalists have responded to the new flow of celebrity news and views. Considering the use of Twitter by some sports figures, his analysis examines how social media can be used to enhance their profile, but also create controversy because of impulsive and controversial postings. Drawing on interviews with journalists, Tomlinson reflects on the type of journalism social media encourages, increasing the speed of the news cycle on channels such as Sky Sports News. It is concluded that while Twitter has reinforced if not exacerbated the profiles of the major sports stars, it has not raised the standards of sports journalism.

Mary Angela Bock's Chapter 18 addresses the impact of the 24-hour news cycle on the ever-increasing number of journalists who work with video. Unlike writing, which is largely an abstract, mindful endeavour, she considers video reporting a more material and physical task. Drawing on qualitative research conducted in newsrooms in England and the US, Bock's findings suggest that divergent demands are inducing three trends in the use of video: an increased dependence on sources for access, a preference for softer subjects and a more fragmented production process. By way of conclusion, the chapter argues experimentation with video will continue as media institutions converge, and that journalists who can teach themselves technical skills will have an advantage.

The six chapters in Section 5 consider different challenges faced by 24-hour news channels from either a cross-national or country-specific perspective, including Australia, Canada, the US, UK and China. Colleen Murrell in Chapter 19 examines Australian and Canadian public service broadcasters, considering their international newsgathering practices by conducting interviews with senior journalists and drawing on a content analysis of coverage. Her analysis on international reporting compares coverage on CBS and ABC news channels, along with the evening bulletins broadcast on terrestrial TV. Murrell argues that budget costs on both public service broadcasters have impacted on the quality of international reporting. However, she identifies some notable differences in editorial strategies used in international news output between 24-hour news channels and evening bulletins, most interestingly in the use of live conventions and using journalists as experts. The conclusion suggests international news coverage appears increasingly less reliant on gathering first-hand information from bureaus around the world, and more on senior journalists interpreting major stories.

The next two chapters focus on coverage of politics in US cable news channels, exploring how their journalism has changed in light of competition from new online and social media platforms. In Chapter 20, Jesse Holcomb, an associate director of the Pew Research Centre, argues cable news is today less preoccupied with breaking news coverage because of the presence and increasing influence of social media, notably Twitter and Facebook. Drawing on research conducted at Pew, he shows US cable news has become more partisan, with niche audiences largely made up of people who agree with the political views of particular channels. While acknowledging it is empirically difficult to establish, Holcomb suggests this behaviour is replicated in the way many people use social media to consume news. Questions are raised about the long-term sustainability of 24-hour news channels given the growing use of social media. As a consequence, he concludes the role and relevance of cable news could soon diminish.

Looking at how cable news channels have evolved over recent decades, Alison Dagnes in Chapter 21 suggests their political influence remains strong despite the more competitive media environment. Like Holcomb she argues the presence of social media has actually intensified the level of partisanship in cable news coverage as opinion-based journalism increasingly defines a channel's identity. This, Dagnes points out, is reflected in the promotion of high-profile pundits, who project the ideological views of the channel to identify with left- or right-wing audiences. She also suggests the future of cable news is facing uncertainty as new technologies evolve and people consume media in different ways. This, she concludes, opens up new opportunities for cable news to diversify its content and adapt its opinion-led news coverage for platforms such as Instagram and Snapchat.

Brian McNair, in Chapter 22, examines how Australian news channels have evolved in a news landscape shaped by the commercial market and public service broadcasting. He looks at the editorial differences between ABC News 24, a public service broadcaster, and Sky News, a commercially run channel, identifying how each has developed a distinctive brand of political journalism. He suggests the contrasting ownership model of each broadcaster influences the editorial output of each channel. ABC News's rigorous application of "due impartiality," he argues, results in a more fact-based schedule of news programming. Sky News, on the other hand, does not feel as constrained and has a more opinionated style of journalism. But because Sky News competes with public service broadcasters in Australia, McNair concludes the commercial broadcaster's coverage does not reflect the kind of explicit partisanship found on US cable news. The national news market, in other words, remains critical to safeguarding editorial standards in Australian journalism.

Similarly, Daya Thussu's analysis of Indian news channels in Chapter 23 focuses on the implications of media deregulation in the late 1990s and early 2000s. Because of the enhanced competition between commercial broadcasters and light touch rules about cross-media ownership, he argues the quality of news has diminished in the pursuit of audience ratings. Thussu traces the huge growth in Indian news channels, which he estimates at 800 as of 2015, and their relationship with reality and entertainment-based programming. In doing so, Thussu considers much of Indian 24-hour news journalism "infotainment" because the focus is on sensationalising content and entertaining audiences, rather than accurately informing people about the serious issues facing the largest democracy in the world. To conclude, the chapter suggests Indian news channels could make more effective use of the country's global profile, producing journalism that has international appeal, but which also addresses some of the nation's underlying issues about poverty and inequality which currently receives scant news coverage.

In Chapter 24, Yunya Song, Yin Lu and Tsan-kuo Chang examine CCTV's 24-hour news service, considering how its editorial strategies and output have evolved from offline to online media. It first considers the tension between the Chinese government's control of the broadcaster and the market position of CCTV, which they consider crucial to understanding how its journalism has evolved in the era of 24-hour news. Like other news channels, Song, Lu and Chang argue CCTV has adopted new editorial strategies to compete with rival outlets and new online platforms, but these polices are also subject to strict state control. They outline how CCTV has responded to the digital era, developing a multiplatform service, but one that remains centralised with limited opportunities for people to interact and participate in. They also conclude CCTV's editorial strategy remains largely national in focus, and its institutional practices could be improved to engage with audiences around the world.

Overall, this chapter—and the final section—act as a useful reminder that while some rolling news channels compete to set the agenda globally, the organisational structure and editorial aims behind many continue to be influenced by national pressures and anxieties. At the same time, the 24 chapters in this volume also demonstrate that news channels face many similar challenges in a new digital landscape, and provide fresh insights into some of the ways different 24-hour news channels have either redefined rolling news journalism—or potentially could do—in order to remain relevant and effective in supplying continuous news for 21st century audiences.

REFERENCES

Conlan, Tara (2015). BBC considering move to make news channel online only, *Media Guardian*, July 7, http://www.theguardian.com/media/2015/jul/07/bbc-considering-move-news-channel-online-only

Cushion, Stephen (2010). The Three Phases of 24-Hour News Television. In Stephen Cushion and Justin Lewis (2010), *The Rise of 24-Hour News Television: Global Perspectives*. New York: Peter Lang

Cushion, Stephen and Lewis, Justin (2010). *The Rise of 24-Hour News Television: Global Perspectives*. New York: Peter Lang

Mosey, Roger (2015). The BBC News Channel must be saved, *Media Guardian*, July 9, http://www.theguardian.com/media/2015/jul/09/the-bbc-news-channel-must-be-saved

Reuters Institute for Study of Journalism (2015). Digital News Report 2015. https://reutersinstitute.politics.ox.ac.uk/publication/digital-news-report-2015

PART ONE

Industry Challenges and Pressures: International Perspectives

Setting the Scene and Provoking Debate

CHAPTER ONE

Have 24-Hour TV News Channels Had Their Day?

RICHARD SAMBROOK AND SEAN MCGUIRE

In 2014, we wrote two articles for The Guardian website on the future of 24-hour TV News channels.[1] As a former Director of BBC News and former Head of Strategy for BBC News, we had been instrumental in the launching of a number of news channels, but felt there was now little recognition of the imminent impact of digital and mobile platforms on their function and performance. The articles were polemical—intended to provoke debate in an industry we felt had been insufficiently innovative in response to digital technologies and which had fallen into some complacency.

The reaction was decidedly mixed. Some editors and executives welcomed and joined the debate in an open fashion. Others told us they privately agreed but publicly would have to reject our arguments. Some simply disagreed. A number of editors initially rejected the arguments, but later said they recognized some of the points we made. We were invited to talk to several broadcasters in Europe and take part in internal discussions on the future of TV news. In the UK, the Head of Sky News, John Ryley took on the arguments in public at a speech at the Royal Society of Arts.[2] (See Chapter Six.)

As the future of BBC funding began to be publicly debated speculation grew about whether the BBC News Channel could be closed to bridge the gap in finances and become online only.[3] And, in turn, there were calls from former executives for the TV channel to be saved.[4] We are pleased our thoughts provoked a discussion that had not taken place until then and recognize, not as a result of our articles, that new options for 24-hour TV news are starting to be considered. We do not have the answers but remain convinced that the combination of factors we outline make the current 24-hour news channel proposition, adopted by broadcasters around the world, unsustainable in the long term.

Fundamentally our argument can be summarized as follows:

- 24-hour TV news is expensive, soaks up resources, distorts editorial judgements and now fails to deliver breaking news first—which is its principal market proposition.
- The internet provides greater speed and depth of news provision.
- Internet-connected TV is about to alter viewer behaviour in a way which will further diminish news channel audiences to a point of economic unsustainability.
- Channels therefore need to radically re-invent their purpose, structures and delivery fit for an age of interactive digital technology and greater consumer choice.
- There is greater justification for pan-national channels than national ones, but even in the sphere of soft power, social media is the new forum of influence.
- Channels are a symptom of the age of satellite broadcasting which has determined their formats and proposition. The satellite age is now all but over and the internet age offers very different formats and opportunities.

In saying this, we should acknowledge that 24-hour TV news has been transformative in the last thirty years: breaking the news cycle, changing consumer expectations, allowing live and real time experience of events around the world. TV remains the single most important platform for news and information around the world. This is why it is important it continues to adapt to new technologies and consumer expectations, finding its place at the heart of visual storytelling in the digital age. The *Guardian* articles unpacked the argument as follows.

The past two decades have seen a revolution in every aspect of the media industry—technological change has enabled consumers to develop sophisticated and subtle patterns of behaviour, constantly being updated from a variety of sources. Cable news established the 24-hour news habit, but today social media and mobile phones fulfil the instant news needs of consumers better than any TV channel can. Yet, around the world hundreds of millions of dollars continue to be invested each year in news networks. Is this money well spent?

24-hour TV news broke the audience away from the daily news cycle, focused on a flagship primetime newscast. But should linear satellite channels still be the focus of so much attention in the interactive internet age? They don't quite give us news when we want it—we often have to wait 15 or 30 minutes for the story to come around—so it's news-not-quite-on-demand. If we want it now, we will go online and get it instantly.

Twitter—and increasingly live blogs of breaking news events—consistently beat 24-hour TV channels. And on those defining moments that bring the nation together, the multichannel broadcasters will, and regularly do, clear their main mass-audience channels. So that makes a news channel perfect for those quite big, but not really big, stories for people who want information quite fast, but not immediately. By anyone's judgment, that's a small (and slightly weird) segment. Beyond that, it's great for people stranded in hotel rooms, office foyers or trading floors. But even that doesn't provide a huge audience—and probably not one in need of an entire network.

Rolling news imposes too many costs on the system. The infrastructure behind a 24-hour news channel is impressive—and formidably expensive. A studio, with two anchors and a steady stream of contributors and guests who all have to be booked and taxied to the studio. Behind them a shift system of producers, graphics designers, crews and editors. Reporters and camera crews around the clock. And underpinning them, continually open satellite links, transponders and digital terrestrial TV channels. All in all, that's an incremental cost of £50m to £70m each year.

The biggest cost comes from having created a machine that has to be fed. Every 15 minutes we go back to our reporters in the field for an update on what's happened since the last time we visited them. Most of the time the answer is "nothing." But even if something had happened, the chances are we wouldn't get to hear about it—as all they've done is stand in front of the camera waiting to go live; "Dish monkeys" as they are unflatteringly known. To actually go and get the news they'll need to send—and pay for—a second crew.

Newsgathering becomes a sausage machine, dedicated to filling airtime. Hours a day are spent on live feeds waiting for something, anything, to happen—"vamping" it's called in the business. A correspondent talking to fill empty airtime until the press conference or event begins. The editor can't risk broadcasting a different report or going live somewhere else in case he misses the start and a rival channel can claim to be "first."

24-hour channels warp editorial judgments. The need to fill airtime—and particularly the need to be seen to be live—means that in the heat of the moment questionable editorial judgments can be made. Everything seems to be "breaking news." For example, we saw the BBC showing live pictures of an empty courtroom in the US, eagerly anticipating the sentencing of already convicted kidnapper Ariel Castro—a story of interest to few if any in the UK.

In the US itself, we had terrible misjudgments in the aftermath of both the Boston Marathon bombing and the Navy Yard shooting. Al Jazeera America—recently shut down but once keen to make an impact in the US market—followed the lead of the other news channels and would "vamp" for 20 minutes or more until the president's press conference began. When a presenter feels compelled to say "Plenty more to come, none of it news. But that won't stop us" (BBC News's Simon McCoy, waiting for the royal birth[5]), then there really is a problem.

The genesis behind the news channel was the advent of global satellite links. News could be transmitted from anywhere, repackaged and then delivered to people's homes. When CNN launched in the 1980s, the live capability of a satellite network was breathtaking and transformative.

Now, technological developments mean that for the most part the internet has replaced satellite links for capturing and distributing the news. At the same time, consumers increasingly have broadband links to home, office, tablet and phone. Yet the industry remains wedded to the idea of a single, linear channel. Audiences have never been convinced. Viewing figures for news channels have always been low—spiking when a big event happens. The justification for broadcasters was to have a rolling spine of coverage that could be turned to at moments of need. Increasingly, however, we turn to the internet.

News channels prize being first, but increasingly that is a race that they can't win against the immediacy of online, and the audience, only watching one channel at a time, cares less than the professionals. "Did we beat CNN?" is a phrase often heard in a newsroom. But in the digital age, social media will always win the race to be first (if not always the race to be right). And who, other than the inhabitants of newsrooms, is watching enough news channels simultaneously to know who was first anyway? Those 30 seconds might be important for commodity traders—but for news audiences?

In today's media environment any broadcaster is first for minutes at most, by which time Twitter or the competition will have caught up. Being first—the primary criterion for 24-hour news channels—is increasingly the least interesting and effective value they offer.

What is "live" anyway? What do we mean—and what do consumers expect—from "live"? Some news events are clearly reported truly live—the second plane hitting WTC2 on 9/11 or Sky News's Alex Crawford broadcasting live as she entered Tripoli with the rebels. But beyond this, very few news events are covered as they happen. Press conferences are edited and reported; two-ways with reporters often cut away to pre-recorded package where the real storytelling takes place. News editors have conflated on-demand with live, and in doing so, have added costs for very little audience benefit.

Television news can be powerful, moving and informative. It can, in the space of a few minutes, change the outlook of an entire nation—Walter Cronkite on Vietnam; Michael Buerk on Ethiopia. Yet the number of stories that are conveyed by live, as-it-happens pictures is vanishingly small. Many stories—the economy, climate change—aren't best served by pictures; others (inside Syria, Iraq, Afghanistan or Zimbabwe) often don't have pictures available until days after the event; many more work better with a well-crafted, tightly edited package rather than a live feed.

There is some great journalism on the news networks, but seldom live and often in spite of the platform, not because of it. News channels have their own narrow agenda outside of big breaking news, one of the lost opportunities of all that airtime is coverage of under-reported places or issues or providing more analysis or depth. The reason is that all the resources are tied up, waiting to "go live" on the same narrow agenda as everyone else.

Global news channels have their own parallel world of timeless, rootless programmes that work as well at 2 a.m. in an airport as at 2 p.m. in a hotel suite. Their agenda strains to find common ground for a global audience, so is full of pictures of middle-aged men getting in and out of cars at international summits. Plus the live correspondent two-way confirming that although it's a very important event, nothing much has happened.

Global news channels are the old "new imperialism." Most of the rash of global news channels that have been launched in the past 10 years are in some way state-backed and—although this is frequently denied—are there to reflect a particular set of values to the world.

These channels seldom if ever make money—they are not commercial propositions. It's about "soft power," which at least is a purpose. China has invested more than $7bn in international broadcasting and talks of laying "cornerstones to underpin a de-Americanised world."

Exposing the world to our political and social values may be the strongest justification for global news channels. But in the meantime, much of the audience, including in developing countries, are looking at their mobile phones and posting to Facebook. Those are the new arenas for global influence.

News channels feed partisanship and the echo chamber. This is particularly true in the US—where TV is unregulated—and a consequence of the undeniable success of Fox News. Talk shows and argument fill the airtime more cheaply than on-the-ground newsgathering. To create impact and get noticed, both hosts and argument become more partisan and more extreme. People choose the channel that agrees with their views and become less exposed to other viewpoints—encouraging partisanship, political polarisation and a political echo chamber that ill serves open democratic debate.

The problem isn't the consumer. At heart, the problem is a closed, linear technology failing to keep pace with the growing on-demand, interactive expectations of the public. News channels suffer from low audiences—at times vanishingly small. These audiences were boosted by a switch to multichannel and digital TV; now they are at best flat and in many cases declining. This isn't a sign of a lack of interest in news. In every major market, well over 80% of consumers read, watch or listen to the news each day. But they are becoming increasingly discerning, using multiple sources to create their own news agenda, many of them online.

So what's the answer? The BBC's director of news, James Harding, has acknowledged the need for more research and development (R&D) by creating a "Newslabs" team looking at data and visual journalism and launching a debate about the Future of News.[6] But perhaps the industry needs a bolder vision. What might a reconfigured on-demand news service look like? Integrating TV feeds into the web (and remember all TVs will soon be internet connected) could save cost, free resources and provide improved speed and depth of coverage. No need for a channel, or satellite space, or a DTT slot. More journalists gathering news, fewer filling space.

Give consumers what they want, as much as they want, when they want it. A menu of on-demand packages that can be assembled into a personalised bulletin, with the ability to go into as much depth as you want, accessing comment, pulling in charts, data and analysis from specialist sources as part of the experience. The bulletin waiting for you on any device, learning from you as you go, or interrupting you with the things you really need to know about right now. Look, for example, at Watchup TV aggregating and curating news video.

Return newsgathering to what is says on the tin—a service that goes out to speak to people, investigates, considers and then files packages as needed, with updates and commentary, freed of the need to fill empty space. When something happens, or new information comes to light, a new story can be generated. A package can be updated and be ready to go as soon as the consumer needs it.

Spend money on what matters and ignore what doesn't. It's not two bodies in a studio waiting, hoping for something, anything to happen, or a miserable guy under an umbrella filling empty time. It's both far more, and far less, than that.

Satellite news channels have played a hugely important role in the development of 24-hour news and information over the last 30 years. But technology, and consumers, have moved on. Is it time we recognised that, like the emperor's new clothes in Hans Christian Andersen's tale, news channels are not all they pretend to be?

As consumers increasingly demand immediacy and relevance, as technology demands investment, it seems unlikely that spending upwards of £60m a year on a closed, satellite infrastructure that delivers the news—late—to 1% or 2% of the available audience will remain a sensible use of resources. Managing the transition away from that expensive infrastructure to a low-capital digital future is the hardest task. Starting from scratch with minimal physical overheads is easier than weaning an organisation off printing presses or TV transmitters. However, as the Reuters Institute's Digital News Report 2015 highlighted,[7] the change in consumer behaviour, driven by the exponential rise of smartphones and tablets to access news, is already with us. Consumers are already deciding how they want their news, and a broadcast stream figures low in their preferences. John Ryley, head of Sky News, has said that "if (the news channel) didn't exist, we'd need to invent it."[8] His argument was that the TV channels generated the content they could then exploit across multiple other platforms, and always would.

But in an age of digital interactivity it seems perverse to insist a linear channel will always be the core of what you offer. Ericsson Consumer Labs has estimated that by 2020, fixed broadband connections will exceed 1bn homes globally;[9] there will be more than 50bn connected devices of which 15bn will be video-enabled and reliant on mobile internet networks. This will transform TV and open opportunities for innovation, new business models and new consumer services. There will be an expectation of permanent connectivity, cloud storage, and services freed of location, time, or platform. On-demand and time-shifted viewing, it is estimated, will grow to at least 50% of consumption. Immediacy and relevance—personalisation—will become core expectations.

One of the lessons from the bruising experience of print news organisations is that adapting to new consumer expectations driven by technology is more than a matter of rearranging the desks in the newsroom or adopting a "digital-first" slogan. It takes a fundamental rethinking of processes, production and even purpose. In other words, in confronting the reality of the task, some sacred cows have to be sacrificed. Dispense with the long-treasured methods of production and focus relentlessly on what the consumer needs, which isn't the strictures of a fixed schedule, or the choreography of the studio and satellite links. Serving the channel first, with digital platforms second, is unlikely to work for much longer.

In all areas of news, as recently identified by Kevin Delaney of Quartz,[10] the premium is now on tight, timely information or on longer, in-depth specialism. The "middle" is dead. But of course that "middle" content is where a lot of broadcasters (and newspapers) have traditionally focused. Newsrooms are configured to deliver the 800-word article or the two-minute video package, which increasingly doesn't work in a digital environment.

So what are the options available to channel operators? Thankfully for them, brand still matters and loyalty persists for many, which provides time to innovate. But it won't last forever. One option is closure and reinvestment in on-demand services. However that is not—nor should it be—the solution for most.

But a solution needs to be found. The American news networks continue to suffer,[11] with declining and ageing audiences for even the most successful.

Another option is to find a new purpose and different content to appeal to a wider group. CNN's Jeff Zucker has said while CNN will roll on the biggest stories, it is seeking to develop more "informational based, educational, entertaining programs."[12] In other words, if news channels can't win with breaking news, then let them offer greater depth of programming. It might, however, take them closer to channels like Discovery or National Geographic.

Another option is impact through partisanship. It has worked for Fox News, although less well for MSNBC which set out to be the yin to Fox's yang.

But these approaches are wedded to the past—a world of fixed, closed, linear broadcasting.

Clearly the future will involve some means of combining the advantages of TV news channels with the speed, flexibility and openness of the internet. A newsgathering operation which can deliver to the screen the advantages of live blogging—with speed, flexibility and transparency—with the weight and power of great visual reporting. A service focused on the story, rather than the choreography of delivery or the traditional formats of the past. They could develop an online video stream, at lower cost than a permanent satellite broadcast. It's a strategy which is growing the audience of CBS.com faster than many of its competitors.[13]

When, as is likely the case by 2020, the TV in the corner of the room is a screen which can take any internet-delivered service, the difference between an online video stream and a continuous broadcast channel disappears for the consumer. Not for the producer, however, who has to measure the cost of satellite and terrestrial distribution infrastructure against delivery via the internet.

This was the logic which drove the BBC to make BBC3 a web-only, iPlayer-led service. It seems unlikely, as the use of on-demand media players grows, that it will be the only channel to do so. For video news content, how about an app on a mobile device or TV set that aggregates the latest packages filed by newsgathering—personalised to your taste, but with the three things that you really need to know about, ready in a bulletin whenever and wherever you want it. Not waiting for the news-wheel to turn to the top or bottom of the hour.

Broadcasters might, as Al Jazeera announced recently, integrate online and TV. AJ+, they say, is "for the generation connected to the real world"[14]—an online network of short videos aimed at a younger audience to rival Vice News. They can drive their online video through relentless interactions with users which has made RT (Russia Today) the most successful news channel on YouTube,[15] clocking a million views a day. (See Chapter Four.)

News consumers are turning to mobile devices—for short clips or text rather than live TV streams—and the age profile of those wedded to their TVs continues to grow older. Which makes it odd to think a linear broadcast channel should be the heart of your news service in the digital age.

In TV news, if you didn't already have a 24-hour channel, you'd probably invent something else.

(This chapter is largely taken from two articles written by the authors for The Guardian in February and August 2014.)

NOTES

1 http://www.theguardian.com/media/2014/feb/03/tv-24-hour-news-channels-bbc-rolling
http://www.theguardian.com/media/media-blog/2014/aug/28/tv-news-channel-phone-digital
(Accessed August 10th 2015)

2 https://www.thersa.org/events/2014/06/rolling-news--the-backbone-of-a-digital-future/
(Accessed August 10th 2015)

3 http://www.theguardian.com/media/2015/jul/07/bbc-considering-move-news-channel-on-
line-only (Accessed August 10th 2015)

4 http://www.theguardian.com/media/2015/jul/09/the-bbc-news-channel-must-be-saved
(Accessed August 15th 2015)

5 https://www.youtube.com/watch?v=agWlKIiVXRA (Accessed August 12th 2015)

6 http://www.bbc.co.uk/news/uk-30999914 (Accessed August 12th 2015)

7 http://www.digitalnewsreport.org (Accessed August 12th 2015)

8 http://www.theguardian.com/media/2014/jun/12/sky-news-chief-alain-de-botton-john-ryley
(Accessed August 12th 2015)

9 http://www.ericsson.com/news/140619-what-is-tv-going-to-be-like-in-
2020_244099437_c?query=networked (Accessed August 12th 2015)

10 http://blog.newswhip.com/index.php/2013/12/article-length (Accessed August 12th 2015)

11 http://www.nytimes.com/2014/05/29/business/media/may-brings-a-loss-of-viewership-for-
cable-news-channels.html?module=Search&mabReward=relbias%3Ar%2C%5B%22RI%3A5%
22%2C%22RI%3A14%22 (Accessed August 12th 2015)

12 http://www.capitalnewyork.com/article/media/2014/05/8545669/jeff-zucker-talks-cnns-post-
plane-plans (Accessed August 12th 2015)

13 http://www.businessinsider.com/cbs-news-grows-online-audience-with-distributed-live-
streaming-strategy-video-2011-2?IR=T (Accessed August 12th 2015)

14 http://ajplus.net (Accessed August 12th 2015)

15 https://www.journalism.co.uk/news/how-russia-today-reached-one-billion-views-on-youtube/
s2/a553152/ (Accessed August 12th 2015)

The View from the Control Room:
Executives and Editors on the Future
of 24-Hour Television News

CHAPTER TWO

The View from the United States: Three Forces Shaping the Future of Video News

DAVID L. WESTIN, BLOOMBERG NEWS AND FORMER PRESI-
DENT, ABC NEWS

Questions about the future of television news in the United States are easily answered: there is none. The days when we got most of our news on television are either gone or rapidly going. Today, we routinely watch video news reports throughout the day and night on all sorts of devices, only some of which can be called "television sets." Increasingly even our TV sets are fully connected to the internet, so at any given time, we're not even using them to watch what we've traditionally considered "television news." Those producing video news reports aren't necessarily named "ABC" or "CNN," but often now are companies like Yahoo!, BuzzFeed, Vice, and the *New York Times*.

What's more, consumers don't much care. We want our news accessible, compelling, and convenient—convenient to us, not necessarily convenient to those producing it. If that means watching it on television, fine, but convenience now means watching it also on a smartphone or a tablet or some new device soon to come on the scene. Televisions may not disappear altogether, but television news will no longer be a genre onto itself.

Questions about the future of video news in the United States, on the other hand, are more difficult to answer and much more interesting. To get a glimpse of the future of video news in the United States, one needs to consider how we've gotten to where we are. Most trends tend to continue until there's a truly disruptive force—and the United States is still in the midst of adapting to the seismic disruption that's come with the shift from analog to digital technology.

Digital has created the three powerful forces that will shape video news in the United States for the foreseeable future: the loss of the TV schedule; the expansion of video news producers to include just about everyone; and the transformation of news consumers into editors and of editors into curators.

LOSING TRACK OF TIME

The move from analog to digital video news in the United States did away with the tyranny of time. When television became widely adopted in the United States in the 1950s, people tried to plan their days and evenings around the broadcast schedule. Children came home from school in the afternoon and sat in front of enormous cabinets surrounding small black and white picture tubes to consume series like *Superman* and *Zorro* and the *Mickey Mouse Club*.

Parents made sure that they watched the "evening news," which started out as a 15-minute broadcast in the early evening each day and later expanded to a full half hour, preceded and sometimes surrounded by local news programming. Later at night, families knew when their favorite shows would be on, whether it was *I Love Lucy* or *Gunsmoke* or *The Flintstones*. *TV Guide* was a hugely profitable business built simply on publishing and selling a list of all the TV shows and when they'd be on each week.

It all gave us predictability and a structure to our days, but it also meant that we had to adjust our schedules to meet those set by programmers in New York, rather than making them come to us when we wanted them.

Those days now seem about as distant as ancient Rome. Today, we can still watch the evening news when it first airs, if we want. But we can also record it on our digital video recorders or go to a convenient app that allows us to browse through archived content that's already been broadcast. More and more Americans no longer turn to the evening news at all because there are a host of video news alternatives presented to us from the moment we awaken in the morning straight through the day.

In the last 20 years, the average number of Americans watching one of the three traditional evening news broadcasts has gone from 40 million people to 24 million, and at this point there's every reason to believe it will continue to decline into the future. The imminent death of the broadcast evening news programs has been confidently predicted for all of that 20 years, but there's no sign that they're in mortal danger at least over the next few years. Reaching 24 million people a night, they remain the single largest source of national video news in the United States. But their role and influence are diminished, and they long ago gave up any hope of being the broadcast "news of record" for the country.

There is one part of video news in the United States that remains tethered to time (although often not tied to a schedule known in advance): the truly live news event. When a 9/11 happens or there's a tsunami or a plane crash, most people want to follow it as it's happening. Where once people would turn to the broadcast news for "special reports" of live, breaking news, they've come increasingly to rely on the 24-hour cable news channels, as they know that they're always on with the news.

Cable news is nothing new in the United States. CNN was founded back in 1980 (although it came into its own a decade later with the Gulf War). Fox News and MSNBC came along a bit later, in 1996, and introduced into the United States video news mingled with partisan opinion. What's new is that cable is being overshadowed by the new digital upstarts.

Even with the convenience of cable news, viewers still have to wait until the story they're most interested in comes around in the rotation set by the cable news producers. Digital technology, however, makes available the complete array of video news reports as they are completed and leaves it up to each individual news consumer to choose which reports to watch and in what order. Today we can go on the website of any video news provider and instantaneously see a range of news videos covering all the current news topics.

These non-linear providers of video news include the traditional TV news outlets, both broadcasting and cable. But increasingly print news organizations like newspapers and magazines are moving to provide video on their websites and apps. And, of course, there is a wide and growing array of new providers that have grown up specifically to provide news—including video news—over computers and smart phones and tablets.

Ironically, we are now gaining access to this instant news video, not just over computers and smartphones and tablets, but over televisions themselves. Our television sets are now internet ready, so that we can see these video news reports on TV on whatever topic we want whenever we like. And programmers, both news and non-news, are rushing to create "over the top" channels and programming services delivered over the internet, rather than over the air or over cable.

In short, cable news channels may be hoist with their own petard of viewer convenience and instant gratification. Now we will all get to watch the news when we want to watch it.

TOO MUCH NEWS?

Just as digital technology has freed us from the constraints of time on our seeing video news, it has opened the gates to a flood of new video news suppliers. Television news in the United States originally developed based on the scarcity of spectrum and the government licenses required to broadcast in any location nationwide. Initially, television was largely limited to the "Very High Frequency" band of spectrum; with time came the technically inferior licensing of TV channels in the "Ultra High Frequency" band. To start a television station, one had to apply to the Federal Communications Commission in Washington, which granted limited duration licenses that had to be renewed periodically with a showing that the broadcaster operated "in the public interest." This gave rise to a host of regulations and regulatory proceedings consuming great time and money (and keeping many Washington lawyers well compensated).

It also led, as a practical matter, to a system limited to only three national broadcast networks affiliated with local television stations living off of the expensive, high-quality news, entertainment, and sports programming the networks provide. The resulting product was a combination of high-quality national content with locally generated programming, largely given over to news, traffic, and weather.

What began as a business with incredibly high barriers to entry (and the exceptional profits that went with them) began to erode by the 1970s. First came the expansion of licenses for UHF TV stations. By the 1980s, cable television was coming into its own, first as an antenna service to improve reception in remote locations and then as a provider of programming in its own right. By the mid-1990s several local or regional news channels were begun, principally in major metropolitan areas like Washington, D.C.; New York; and Boston.

Cable programmers by and large did not have to contend with the barriers to entry posed by the FCC and its broadcast license regime. They faced a different challenge: getting the cable companies (who'd invested to install the cable) to put them on their systems and pay them for their programming. But over time, a handful of cable programming services—news and otherwise—gained a foothold and were joined by dozens of others. Some of these were spin-offs of the first cable program providers, as ESPN built ESPN 2 and then several sister outlets; Discovery created many related channels; and CNN created Headline News and an array of international varieties for its news. Other cable programmers grew up on their own, providing everything from wrestling to recipes and home improvement tips.

The advent of cable news posed a challenge for broadcast news operations and made them less profitable, but looking back on it now, even as the new millennium began, news providers in the United States remained members of a relatively small and exclusive club. The low-quality, unreliable streaming video that began to be delivered to mobile devices early in the new century seemed at first to be just a novelty. But soon streaming video from a multitude of producers and sources created an environment that let just about anyone join the news club.

Today, we all can gather video, edit it, and provide it to the world for very little cost—and an awful lot of people do. The trend began with digital first companies such as the Huffington Post (part of AOL), Vice, and Yahoo! But it now includes countless individuals who happen upon news events as they occur and post them to their own social media sites and/or to YouTube and other video aggregators.

Nor is this rapidly expanding field of video news providers limited to local, regional, or even national borders. Where once we could get video news only from our local television stations and a handful of national news organizations, now we have instant access to the BBC and the Guardian and Le Monde—and videos posted by terrorists shooting high-resolution video of hideous acts and streaming them to us from unknown locations halfway around the globe.

Today, any self-respecting mainstream newsroom counts its live Twitter feed as an essential part of its newsgathering operation. The traditional news bureau is either being re-thought completely or eliminated. New services are beginning to search social media outlets and identify locally shot video as it is posted to Facebook and Instagram and Twitter, and they direct traditional news organizations to it. Other services are forming loose relationships with individuals around the world who are willing to step out their door and shoot news video for a set, low fee to be fed to the large news organizations for their use. Mainstay news agencies such as the Associated Press and Reuters and Agence Presse Francais will find themselves competing more and more with freelancers and citizen journalists around the world when it comes to breaking news video.

We're rapidly approaching a world where there will be video that gives us news from everywhere all the time. Some of it we will access directly; some of it we'll find because others recommended it to us by providing us links on social media; and some of it will be picked up by mainstream media outlets and incorporated into their own news gathering and reporting.

So, as we go forward we will not only get to watch video news whenever we want to, but we'll be able to watch the stories and the sources we prefer, rather than being compelled to settle for what's available. The question is what we will do with this nearly unlimited access.

EVERYONE IS AN EDITOR

There was a day when we had editors to sort through all the information and mis-information that was available around us, put it all into a digestible form, and present it to us in compelling ways. We thought that the reason we trusted the broadcast networks or national newspapers like the *New York Times* was that the editors were so very good at their jobs. But the last few years have brought a rude awakening: The authority of the relative handful of powerful editors came as much from scarcity as it did from excellence.

Today, that scarcity is rapidly disappearing. There is no hierarchy built on the foundation of having only a few sources for our news, and most of us see little reason to let a claimed authority slice and dice the material before we get to look at it ourselves. The position on the "dial" of the television set is gone; even the position in a lower "tier" of a cable package is rapidly disappearing.

There are so many news websites (and sites that include news) that the concentration of reporting on a given website doesn't necessarily command our attention, either. Increasingly we don't even go to websites as a destination in the first place. We come to video news through the "side door," with a link from another site or a social media outlet. Or, most recently, without even a link as social media applications seek to host news content directly on their platforms. We turn to the applications that we've chosen to reflect for the moment our favorite places on our smartphone or tablet for news and information.

Each day the news consumer finds another way to gain access to video news, some of it raw and some of it produced. Not all of these new video news services will survive, but the next few years will offer those interested in video news an ever-richer array of alternatives.

This gives each of us unprecedented power over the video news we turn to and consume. Now, it's up to us to decide what's important and what's not—regardless of whether it's on the front page of the *New York Times* or leads the evening newscast. It's up to us to decide what we believe is true or not. It's up to us to sift through the sources, weigh the evidence, and come to our own conclusions.

We're not alone in this endeavor. We have our "friends" and others we've come to trust to direct our attention to things they regard as important. And we have the strength of the crowd to help us vet reports that come in that we may question. For example, when the CBS evening news anchor, Dan Rather, first had his report on the war record of President George W. Bush questioned, it came not from any of the established news organization we knew, but from an internet source most of us had never heard of. We watched as a range of participants on the internet vetted this report for us.

Where are Americans to turn when they face an unlimited number of news reports from all sorts of sources? To the "tried-and-true" mainstream media of old? To the large, new "digital first" sources who are investing in collecting and reporting the news? To our friends who have placed their faith in a particular piece of video news? To the latest, hottest gossip regardless of the source?

The answer is simple: all of the above. As we move forward we will rely on all these sources for our news—and no doubt sources yet to come. We may trust some sources more than others (at least for a time), but for each of us the palette will differ. Each of us will ultimately trust only ourselves to sort through all the news and data and information that are flooding us and draw our own conclusions about what is important and what is truthful.

Each of us will become his or her own editor. Given the abundant wealth of news video, the issue will not be whether we're getting enough video, but whether we may get too much. In a world where we have instant access to everything, we can quickly find that we have effective, meaningful access to nothing. There may be so much of it and it may be so disorganized that we may struggle to get to the parts that we truly need.

WHERE'S THE BUSINESS?

Digital technology hasn't just disrupted time, news providers, and editors; it's also upended the traditional business model for how companies pay for gathering and reporting and producing the news. Scarcity once provided healthy revenues for news organizations—whether it came from the advertising in the news programs themselves or from the wealth of income from the entertainment programs that surrounded them.

As competitive sources of news proliferated and audiences fragmented, video news organizations had to look to new sources of revenue. Cable news organizations, which were competing to take the audiences away from broadcast news, also were among the first to realize the benefits of adding subscription to advertising revenues. Viewers paid their monthly bill for cable service, and a healthy piece of those revenues went straight to the programming services, including the news channels.

Fox took this to a whole new level, realizing that the size of its audience often mattered less than how dedicated, even passionate, that audience was. No cable operator can afford to lose many of its subscribers, so Fox gained a powerful position in extracting high subscription fees from cable companies because a small but significant portion of cable subscribers demand that Fox be available to them. Fox News may not reach more people than CNN in a given month, but it has a much larger audience at any given time because the people it does reach come often and stay long. By following this business plan, Fox News has become the most profitable video news provider in the United States, perhaps the world.

This is the single greatest change to the US business model for video news in the last 20 years: The shift from emphasizing how many people watch a network to focusing on how often they watch it (and for how long). Depth of engagement has come to replace mass.

More recently, we've seen this approach carried over into the world of the internet and mobile devices. Glenn Beck started on Fox News, but then he took his narrowly focused, dedicated audience with him to a subscription video news service that was profitable almost from day one. Where Fox News made a million viewers more profitable than the several million the broadcast networks were reaching, Glenn Beck made a few hundred thousand viewers very profitable by having them pay him directly every month.

This quest for deep engagement (even instead of large audiences) will be an important part of the future of the video news business in the United States. Not everyone will be able to engage with a defined audience through political orientation. But there are other ways of developing strong relationships with specific subsets of the news audience. National Public Radio has done this with a relatively well-educated, sophisticated audience. (It's ironic that of all the broadcast news media in the United States, the one that is built on the oldest platform—radio—may ultimately survive the longest because it has captured a small but devoted audience.) Vice Media is doing it today with video produced for young men interested in kinetic international news.

The business model based on deep engagement works well for attracting subscription revenue, but we should expect that it will affect the way that advertising revenue is raised as well. Increasingly, news providers are seeking ways to redefine how they measure their audiences to reflect levels of engagement. For example, the blogging site Medium has pioneered the measurement of Total Time Read ("TTR") as a step toward selling advertisers the attention of their audience in blocks of time, not just how many eyeballs they can collect.

The search for new ways to raise revenue for video news will include a variety of approaches to increasing audience engagement, but we can expect it will go beyond engagement. It's not just how engaged people are with a particular outlet or piece of video; it's also who is engaged. We are in the early stages of a world where big data are being applied to video news to make it possible for video news advertisers to target their products and services to the people most likely to respond. The quest is on for increasing the "return on investment" for video news advertisers so that those paying the bills are assured that their dollars are well spent.

Finally, it's only a matter of time before video news follows the trend toward "native advertising." What began with text and stills at BuzzFeed and others will soon spread to video, as advertisers produce (or have video news providers and others produce for them) video that is informative, entertaining, and enhances the appetite of the consumer for the product being promoted.

Those of us from "old media" may have some hesitation about this development, as it's long been a hallmark of serious news organizations that the line between advertising and editorial never be breached. But we're already beginning to see consumer goods companies producing their own video that they make available over the internet and mobile and that looks a great deal like what more traditional video outlets are providing. As people come to video through social media and other routes that don't require going to the news source itself, we should expect that the distinction will be harder and harder to discern. And, there's every reason to believe that the news organizations themselves will become open to native advertising as part of their arsenal as they fight to keep their businesses healthy.

Increasingly, it will be a question of how native advertising in video news is done, not whether. The goal should be to make sure that the viewer is always aware of the difference between what is being presented purely because the editors believe it is important and/or interesting and what, although perhaps important and interesting, is there first and foremost because someone is paying for it to be there. Then it will be up to each of us to decide how much we care about the benefits of what the traditional craft of journalism brings to news reports.

VIDEO NEWS IN THE PUBLIC INTEREST

In many ways, video news going into the future will be healthier and more robust in the United States than it has ever been. There will be more outlets producing more video and covering more news stories. It will always be available when you want it. It will be more global with more competition and more innovation. Some of it may not be well reported or produced, but there has always been a wide range of quality and importance to video news, and there's no reason to believe that will be all that different in the future. We'll just have more of it.

What remains to be determined is what we will do with all this rich and varied video news available to us everywhere, all the time. And, what will it all mean for the country overall? In short, what's to become of that "public interest" that was for many years a mainstay of television news in the US? Will it become an anachronism left over from an out-of-date government licensing system, or will it find some new basis in what people need and want?

We already have access to far more information than our parents or grandparents dreamed of. We can go deeper faster than ever before into any particular subject or issue that interests us. In the end, however, it will be up to us to determine whether all this information makes us more informed.

To be more informed, we will need to ensure that we can make some sense out of all the competing voices coming at us. This will be difficult, if the fracturing of the sources for video news continues indefinitely instead of there being some re-grouping that will begin to return some sense of order to what could become chaotic.

We, the news consumers, may demand this. As the amount and sources of video news continue to increase, we simply lack the time and ability to sort through it all. Life is short, and most of us have better things to do with our time than spend it endlessly on our computers and smart phones and tablets trying to make sure we're not missing video news reports we care about (or would, if we knew about them). It makes sense that the cycle at some point will turn, and we'll begin to see some generally accepted ways of consolidating all these video news reports in ways that make sense and that address our needs and interests.

This doesn't mean that we'll return to a day when a handful of major news organizations invest hundreds of millions of dollars doing bespoke news reporting from bureaus around the world and expecting people to accept that they're getting everything they need. There's simply too much out there to cover, there are too many potential sources for video news, and no one can cover it all. We'll find it increasingly necessary to turn to computers and algorithms to sort through it all, identify our individual interests, and provide each of us with our own, personalized version of "the news."

But will the algorithms and "big data" be enough to sort and edit the news? At least for the foreseeable future, experienced, smart editors and reporters will likely continue to have an important role in making editorial decisions about which stories are true and which are not; which will interest people or not; and which are important or not. They may not be the all-powerful editors of old, but they can be important, even necessary, curators of what's out there. The ideal would be for someone to find a compelling way to combine the best of the smart editors and reporters with the best of what algorithms make possible and give us all a curated—not just aggregated—version of the news in video. We have yet to see this approach made successful and taken to scale. But in success, it could help return some of what is being lost as we move away from television news and toward undifferentiated video news.

The public interest requires more than that we, as individuals, just make sense of our particular corner of the world. It requires that we learn together about the important issues we face as citizens and what our alternatives are. When anyone and everyone can provide news video, anyone and everyone can find a version that addresses their particular interests and needs and, yes, prejudices.

From the beginning of television, there has been a close relationship between the news and the political process. The 1960 debates between Richard Nixon and John F. Kennedy are often identified as the moment when television came to politics. But in truth, it began in the 1950s, as the political conventions were televised for the first time and a chastened Vice President Nixon went on television to give his famous "Checkers" speech to put behind him a growing scandal. The FCC for many years has required television stations to sell advertising time to political candidates at discounted rates. Despite the discounts, political advertising has been a driving force behind the astronomical amounts now required for political candidates to run for office in the United States.

Political divisions in the United States have grown dramatically over the last 20 years, and the changes in television news have been a large factor in that polarization. The divisions we've seen include the growing partisanship among those most engaged in the political process, but even more important has been the growing division between those who participate—by contributing, by volunteering, or by just showing up to vote—and those who do not. Political science research shows that increasingly the electorate in the United States is being made up of the partisans, and that those in the political center are not bothering to vote. What's more, part of that shift is because there are so many alternatives to video news to attract and engage those in the middle. When there was nothing else on the television to watch, a good number of us stuck around to watch the evening news. And, when we learned more about the issues and the political races, we were more inclined to vote.

But now that we all have so many choices, there's a growing separation between those who want to watch the news no matter what and those who don't. It turns out that the ones preferring the news are the partisans; the ones preferring entertainment or sports are those most likely to be in the center.[1] As the partisans on the left and the right watch more and more video news confirming their points of view, they're even more likely to show up on election day and vote. As the centrists wander off to spend their time on things other than news, they're more likely to leave it up to the partisans to decide on who is to lead us.

We're not going back to a United States where there are only a handful of sources of video news that we have little choice but to watch. Nor should we. The new digital world gives us a richness of information and news that we couldn't have imagined 20 years ago. But the most important question going forward is whether there will be new approaches and new innovations—both in the media and elsewhere—that come to restore some of what we've lost. Whether new programming and institutions and customs rise up to take the place of what television news once was. Whether something serves to gather us together to learn about and to care about the issues we face as a nation.

The answer to that last, all important question is very much open.

NOTE

1   See Prior, M. *Post-Broadcast Democracy* (Cambridge University Press, 2007).

CHAPTER THREE

The View from Europe: "All Views" First

MICHAEL PETERS, CEO, EURONEWS

The news media environment has changed and keeps changing—it is a journey and ephemerality is the timeframe. New platforms, new social networks constantly appear and the audiences are keen to discover new watercooler spaces in which to interact.

What is the future of TV news? To say it straight: I believe that traditional 24-hour TV news channels have had their day. Yet, I am nonetheless very optimistic about information media's future, TV included.

TODAY'S NEWS MEDIA PARADOXES

The news media industry is evolving at an incredible pace. It evolves so fast that several major paradoxes have emerged. I believe that solving these paradoxes is the key to surviving the next era of news media.

Let me step back for a moment and look at the world we live in. Over the past 50 years, our world has dramatically changed. We live in an uneven and unrestful age. We see, in places, sagging enthusiasm for democracy, polarisation of opinion, disengagement from society and a crisis of citizenship. Irredentism explodes and, as geographical borders lose importance in this digital world, we observe increased nationalist movements and societies getting more and more fragmented. Over the past 20 years, the number of traditional media has been amplified with the rise of new media and purely digital players. Information is now available everywhere. Information is so overflowing, that people don't need to access it anymore, it comes to them!

Nowadays, as a result, we live in a world of noise. Traditionally it was the role of media to educate and help people understand the context of things and explain complex situations in order to help them build informed opinions and to stabilize a world of peace. This, as well, has changed.

It seems that most of these "new" media outlets have little interest in any public interest mission and tend to increase social divisions instead of bringing people together. One of the reasons behind this is that the global news channels launched in the past decade are primarily produced for political reasons with clearly related agendas. It is not a surprise as international news media are hardly very profitable in the first place. Many are directly or indirectly state-backed and their unspoken objective is to show events from a national perspective. There are new players, such as Russia, Qatar and China, the latter having invested more than $7bn in its international broadcasting, to drive their reach up to 220m foreign households.

These new global news channels are made in a unified manner, each promoting its specific ideologies. Some have even become the tools of a "new imperialism." However this highlights another downside of the current media environment: the logical consequence of the above described "political agendas" is that on these channels some major themes of modern life and global issues stay unreported or at least under-reported. Instead of counter-balancing this political approach, traditional networks with commercial intentions have made matters worse, as their main remit is to capture the audience's attention rather than responding to its needs and genuine interests. They have therefore focused on the creation of celebrity anchors, heavy infotainment and emotional storytelling. As a result, news shows have become less relevant and news has become less and less serious or balanced, both on a national and on an international level.

Technology has also given birth to many new platforms, including "social media." As information is an intangible service and no longer strictly linked to a defined distribution tool, the media sector has been one of the first to be shaken by the "digital revolution."

This is the first paradox I want to point out: Today there is more information, more immediately, in more formats, on more devices and to many hundreds of millions more people than ever before. But this "info-besity" does not help the consumer to be better informed.

Yet, this revolution is undoubtedly offering unparalleled opportunities for media to spread information.

It has become a multiscreen world and the areas for global influence have become blurred between linear and non-linear. A few figures: in 2015, mobile phones outnumbered people for the first time. By 2020, 3G and 4G connections will be available everywhere and there will be roughly 10 connected devices for every human being on earth. Ericsson Consumer Labs has estimated that by 2020 fixed broadband connections will exceed 1bn homes globally; there will be more than 50bn connected devices of which 15bn will be video-enabled and reliant on mobile internet networks. By 2030, everyone in Europe will get television over the internet.[1] The media environment has changed from an economic model of supply to a model driven by demand. There will be an expectation of permanent connectivity and services freed of location, time, or platform. Immediacy and relevance—personalisation, experience and scarcity—will be core expectations.

But whilst the barriers get blurred, information finds itself spread everywhere. This raises another interesting question: what information? Who is creating the news? Is it the blogger, the consumer or the journalist? Maybe all but the journalist. Today, anyone with an internet connection and a social media account can make the news, find, tell and share stories. Internet is bypassing the professional reporter. People in power, celebrities, and politics are finding they can speak directly to the public. The journalist's competitor is no longer another journalist. Often, it is the subject of the story.

Even computers can do what journalists used to, namely compile the sports results, write up keynote reports and financial results coverage. The services that used to be essential parts of the news are increasingly automated and available separately online. All aspects of news activities are being disrupted: robots write articles at Bloomberg and USA Today, computers produce videos, news is distributed through aggregators.

In the meantime, more than 5,000 editorial jobs were cut in a decade across the press in the UK[2] and over the same period international reporters working for US newspapers declined by a third.[3] This thinning out of reporters is happening everywhere. Yet, we need professional journalists more than ever. It is not by cutting workforce that we have to adapt. This would be a mistake. Traditional news media must modify their offer, workflows and behaviour, but stay true to their values. And while it seems obvious that consumers are tired of a "top-down" journalistic approach, consumers still need the "top," even if media should try to become more interactive.

This is the second paradox: information is everywhere, but news media and journalists are less and less the source of this information, and less and less the ones that spread it. So a star political journalist from CNN moves to Snapchat.[4] It is part of the paradox.

The third paradox is that we have entered a new era, but with the same actors and sometimes with the same processes. Technology has enabled so many new ways of consuming news that it has changed consumer expectations. Yet, many networks have not changed their way of working. Some media tend to do what they used to do and just replicate it on new platforms. They add teams dedicated to create content for new media, or simply copy-and-paste their traditional content to new platforms. This will not work and traditional news networks will continue to suffer from declining and ageing audiences. We are seeing historical news brands closing down (Newsweek, LCI), big leaders being shaken (NYT, CNN), start-ups taking close to monopoly positions on some content verticals (BuzzFeed) or services (Flipboard).

Is this sustainable? The answer is obviously no. This paradox may solve itself. The theory of evolution is equally operating in the news media industry. Hyper-competition is clearly calling for changes. Not only has the number of media dramatically risen, but barriers between traditional media have fallen apart. On digital platforms, TVs compete with radios, newspapers, and so many others. And as users have an increasingly fragmented mode of consuming news, you cannot be your traditional self anymore. You cannot be "just" a TV, or "just" a newspaper. You need to reach consumers when and where they want to consume your news. Experienced brands find themselves shaken. Their processes are not aligned with the era, and their "industry" competitive advantage is not sufficient anymore. The paradox for these media is that their model is no longer relevant, but their brand still is—brand-advantaged but model-disadvantaged traditional media as opposed to agile new comers with little credibility.

Therefore, while some see the fall of some traditional media, I see the opportunities of having a strong brand: credible and renowned. Through all this noise, thousands of stories on the same issue, which media are you going to consume? The one that you know better, the brand that you like the most. It is not content that media organisations need to copy and paste to new platforms, but rather brand credibility.

There is more information—but people are not better informed. More news—but fewer journalists. It's a new era—but with the same actors. The changes undergone by the news media industry call for a transition. But to what? In the words of Christine Rosen, I believe news media must manage the transition to the age of Egocasting,[5] the era of customized and personalized content (as opposed to the era of broadcasting [early 1980s], where content was "imposed" on the general public, targeting the largest audience, and the age of NarrowCasting [1990s] with selected content targeting specific audiences).

HOW TO MANAGE THE TRANSITION

The transition needs to take place on two levels. First internally, where a new spirit of flexibility needs to be implemented and understood by all. This is probably the most important, yet the most difficult part of the transformation process. Editors and technical teams need to change their workflow and Human Resources professionals need to search for new talents, combining traditional and new digital professional capabilities. Then, the transition needs to take place externally, by the channel's capacity to target and attract new audiences without losing its established support.

When the tape recorder was invented, this did not kill the radio, neither did CDs. Radio does still exist for a specific usage; passive listening to editorial programs comes as a complement to active uploading and consumption of personalized playlists. Whether on television, the web or mobile, millions of viewers will continue to rely on the broadcast offer for original reporting on the important and relevant topics of the day.

Systems will co-exist for a while and serve both a passive consumption, via editorialized content, and an active consumption, where people expect to build their own media experience with social networks and blogs. What does this co-existence imply? That there is a generational change, becoming one of the key elements when it comes to implementing a transition away from TV and towards a complete media offer.

In Sweden, the average age of the nightly TV news bulletin audience on the public service broadcaster is 66. Meanwhile, a recent survey found that 26% of two-year-olds in Sweden are online at least once a day. TV News in the UK reached 92% of over 55-year-old in 2014, but among 16–34s, this falls to 52%, down from 69% in 2004. The younger generation wants to own and share news and become media actors. If we do not understand this, and don't manage to be flexible, to adopt workflow and talent management, we will not survive.

So what is the future of TV? The future of news media? The answer to this question may not be as complicated as it seems. One should definitely not underestimate the power of the porting effect (the fact that as people age, they don't abandon the media they grew up with), which can clearly be seen in our Swedish example: those who were raised with TV still consume news on TV, and younger generations use internet and mobile devices.

It is a simple truth: the future lies with youth. You therefore need to develop a structure and an appropriate mindset to cater to the needs of your target audience, without missing the consumption patterns of the next wave. You have to become an "Egocaster."

At Euronews, this is what we aim for, and have therefore translated our belief into three main pillars:

1) PRODUCT. Understanding that content and context need to be combined and require on one side the best editorial team and on the other the best user-experience team. We must differentiate ourselves through the quality, the standards and the uniqueness of our content. Content that is easily adaptable (and not simply replicated) to multiple products and services.

2) ECOSYSTEM. Considering all platforms as elements of an ecosystem that contribute putting Euronews content wherever and whenever is relevant to the user and to us. Mobile-first definitely (already 40% of our views on YouTube), if not mobile-only in the near future. We're not using

digital to drive people to TV or the website. We're using digital to offer our content in the most relevant way to our audience. We also need to engage with our external ecosystem, interact with our audience and manage communities.

3) INNOVATION. We're not super big, but we're nimble and agile and able to move quickly—we were the first on Vine, first on Sony Smartwatch, first EU media on Google Glass, first on BlackBerry Messenger, soon first on Ubuntu, among best of class on Google+.

In fact, at Euronews, we talk more about a *journey* than a transition. And we definitely believe ephemerality is here to stay.

We are proud to collect the first fruits of our capability to adapt to multiple digital platforms, notably mobile, through our mobile website and mobile apps delivering a customized user experience throughout most mobile operating systems, but also applications for wearable devices, connected cars and smart TVs. Our content is well exposed on video sharing platforms, with channels dedicated to news in multiple languages and to specific interest themes on YouTube and Dailymotion. The journey is in full swing with our website. We are changing the front, the back, the engine and the journalist workflow. Editorially, too, we need to change. And although our newsroom was integrated from the start, changing the focus of 600 people takes time, requiring digital breaking news first, speed with enriched coverage, etc.

With the right structure and mindset, we are confident that we will improve our audience, even as the market suffers from hyper-competition. But of course, can the new 360-degree media model be sustainable? We believe that there are many ways to make the news media of tomorrow economically viable. As a media company, we create highly valuable content to highly valued segments of population. Targeting, therefore, is a huge asset when it comes to monetization. But branded content is also a fantastic option for brands to deliver value and relevant services. Advertisers are keen to create value through content, we see the major brands becoming media centres and they need our expertise to produce high-quality content: whether curated business news, themed magazines, or industry reports. Native Advertising is another example, of course. This approach isn't purely a commercial matter; it also involves the product and the editorial divisions, all working side by side. It is all the more important to adopt the right structure and mindset to make this journey profitable.

NEWS, NOT NOISE

Adapting to new consumer expectations driven by technology is more than a matter of rearranging the desks in the newsroom or adopting a "mobile-first" slogan.

Serving the channel first, with digital platforms second, is unlikely to work for much longer.

It definitely takes a fundamental rethinking of processes, multiplatform-production, competencies (to serve new needs such as data journalism) and even purpose. People don't necessarily buy what you do, but *why* you do it.

So before concluding, I wanted to come back to a key element, I want to talk about what matters to people, what they want, not only about what news media should do. And I believe that people are eager, in a world of info-besity, to find serious, reliable and independent information. And this is why I believe that we are living in the most exciting times for journalism since the advent of television.

Social media will always win the race to be first on breaking news. Social media will convey a bunch of inaccurate breaking news, but also accurate reports. Even though people say it's easier to get the news, people are increasingly unsure of the facts and unclear what they mean. A media that gives clarity to this noise is what people are waiting for.

Moreover, people are tired of mainstream news, of hearing about the same big headlines. On social networks, they like to share under-reported stories, which will have something special (hence why BuzzFeed has so much success). People are waiting for a news media to cover these under-reported stories.

Last, but not least, as consumers are more and more becoming prisoners of their personal information bubble made of their personalized Twitter feeds and networks, they tend to limit themselves to a unique viewpoint. The internet is, in fact, as an "information bubble," magnifying information inequality, misinformation, polarisation and disengagement. People are looking for a media that conveys a diversity of points of view.

If you look again at what people are seeking, they are asking for quality journalism. Having a strong newsroom is more important than ever. Journalists need to find under-reported issues, send the signal in all that noise, and tell a well-researched story with authority, perspectives, and with context that helps all of us to understand what is actually going on.

At Euronews, we want to satisfy the consumers' newfound needs, but maintain our top-notch editorial standards. In the world of a click, the most spectacular headline will always win. If you care what comes after the click, the best quality content will win. Bring in different people, perspectives and backgrounds to discuss and better understand news and what is important to all. News media are here to make sense of the world, to cut through the noise, whether on linear or non-linear media.

Perspectives. Diversity of point of views. This is what gives sense to stories. This is what people long for; they are tired of being told what to think. It is also the, perhaps forgotten, essence of journalism, the quintessence of news media: "All views" first. This is our belief. We mean it to be our identity and our *modus vivendi*.

To cut through the noise, this "all views" approach must be more than a claim, more than a strategy. It must be a game changer, and disrupt the news media landscape. It can allow us to get out of the bias of current news media. I often like to joke by saying "Tell me what you think, and I will tell you what media corresponds to you." If you are conservative in the US, there is a channel for you. If you are pro-Russian, or pro-Ukrainian, chances are some media outlets share your point of view. Liberal? Patriot? Same thing. Telling people what they need to hear, and not what they want to hear; this is "all views." Never imposing a point of view, an ideology, nor siding politically; this, too, is "all views." Allowing journalists and protagonists to tell the story, without ever forgetting the viewer or user; this is "all views." As a matter of fact, the viewer or user must interact and participate to give "all views" its full meaning. I would even go as far as to say that "all views" would have no meaning without the user's view.

Today, as well as in the future, we want Euronews to be the place people come to for the real story. What really matters. What's really going on. What it really means. Empower people to form their own, informed opinion. Change the world of media, to enable us to be better citizens, equipped with what we need to know. How many conflicts, wars, and deceits have been fuelled by media? No wonder the first thing that happens in wartime or a coup d'état is the taking over of media. Media own a real power, that some call "soft power," the power to influence people. "All views" media could be the only media to bring foes to watch the same stories and make their own informed opinion about it, helping them to understand each other and rely on facts more than on fear.

We believe that serious news will always prevail—regardless of the medium—and can only be made by professional journalists, driven by strong values, open to all views and opinions. At Euronews, 600 journalists from 30 nationalities work in a common newsroom, confront their ideas constantly, discuss global priorities and share a common sense of public service.

The Euronews model of diversified views to create reliable news has proven its effectiveness over the past 20 years and is, in today's world, more relevant than ever. The desire to share and represent "all views" has been defined as part of our DNA, acknowledged as a core value by all employees of the group. Recently, the same approach has been formalized by six major European dailies (*Die Welt, El Pais, La Repubblica, Le Figaro, Le Soir, Tages-Anzeiger*, and *La Tribune de Genève*) in creating the Leading European Newspaper Alliance (LENA). The objective of the LENA is to sensitise public opinion in Europe on the important mission to deliver high-quality journalism.

No one can really estimate the impact of technological development in the coming decade. It will be huge and we need to be alert. Artificial intelligence will serve big data, new outlets will be created, new usages linked to the generalisation of wearable devices and new reporting and viewing techniques—such as 3D virtual reality—will continue to shape the news media of the future. But high-quality journalism will persist and Euronews will be part of it.

NOTES

1 http://www.ericsson.com/res/docs/2014/consumerlab/tv-media-2014-ericsson-consumerlab.pdf (Accessed 28th July 2015)
2 https://www.journalism.co.uk/uploads/laidoffreport.pdf p.8 (Accessed 28th July 2015)
3 http://www.journalism.org/2014/03/26/the-losses-in-legacy/ (Accessed 28th July 2015)
4 http://www.politico.com/blogs/media/2015/04/peter-hamby-leaving-cnn-for-snap-chat-206178.html (Accessed 28th July 2015)
5 http://www.thenewatlantis.com/publications/the-age-of-egocasting (Accessed 28th July 2015)

CHAPTER FOUR

The View from Russia: "Your News Channel" Is Here to Stay

MARGARITA SIMONYAN, EDITOR-IN-CHIEF, RT
(RUSSIA TODAY)

The resources, expertise and shared social experience of TV news will keep the medium alive, but your 6 p.m. big-screen line-up may soon look like your Facebook feed.

The demise of TV news at the hands of the all-mighty internet has been repeatedly anticipated over the last two decades.[1] Authoritative broadcast anchors will be replaced by popular bloggers. Social media will deliver news straight from the source, forgoing professional intermediaries. Mobile technology will turn viewers away from their TVs, and user-generated footage will eliminate the need for professional video crews. A grim outlook indeed for the TV news industry.

Yet, it turns out reports of the death of TV news have been greatly exaggerated. On the contrary, the exact same period during which the internet was going to render TV news obsolete saw an explosion of TV news channels around the world. CNN International, launched in 1985, really hit its stride during the 1990s, which is when BBC World News and Deutsche Welle kicked off their own international TV broadcasting. Euronews and Japan's NHK followed suit. Fox News and Al Jazeera went on air in 1996 and became the powerhouses of their respective regional news ecosystems.

The next decade brought even more diversity to the global news scene. Russia entered the field in 2005 with the round-the-clock English-language news channel Russia Today, now RT. China's CCTV went global, as did Al Jazeera, with their English platforms. Saudi Arabia launched the pan-Arab Al Arabiya News. And now even American network stalwart CBS has jumped into the dedicated

TV news game by launching CBSN via a range of TV-connected platforms, such as Roku, Amazon Fire TV and Apple TV. If the internet, mobile technology and social media were supposed to retire TV news, so far they have failed spectacularly.[2-5]

A CHALLENGE, NOT A THREAT

Yet it is undeniable that all this new technology has had, and continues to have, an immense impact on all aspects of the TV news industry: how news is produced, distributed and consumed. The challenges that traditional news providers face today are manifold. A news editor no longer sets the news agenda. With increasing frequency, news breaks not on TV or in newspapers, but on social media, especially Twitter.[6] You might watch the first footage of a developing story on YouTube. The event is then actively discussed, debated and fought over on Facebook.[7]

These platforms present real competition to TV news, and are increasingly determining what *becomes* news. Instead of leading the audience, conventional media—due to a longer production cycle—sometimes has no choice but to follow it. The ascent of digital also erased competition barriers between professional news providers, both in terms of geography and medium. Regional and national channels find themselves competing with international broadcasters, which are available via satellite and cable platforms. TV is pitted against newspapers, radio and independent websites—domestic and foreign alike.

Digital feeds have become de facto news channels. Customization options on social media platforms like Twitter and Facebook, and reader applications such as Feedly and Flipboard, enable news consumers to easily combine various mediums from anywhere in the world into one convenient text and video package. Instead of sitting through a fixed 30-minute bulletin arranged for you and millions of others by a news editor, you have the choice to consume only the information that you want, and, thanks to mobile technology, when and where you want to.

This applies in equal measure to subject matter (Are you a fan of European politics, royal scandals or climate change?) and ideological leaning (Are you left, right, communist, or libertarian?). We've got news for all of you! News feeds are feeding us exactly what we want. Out of those personalized news menus, consumers create tailored "news boxes" for themselves—and stay in them for most of the time.[8] Breaking into those boxes is a challenge for any media player, news or otherwise.

But if old-school broadcasters treat these challenges as threats to their influence, or even existence, they risk missing out on opportunities to extend both.

ADAPT, DON'T DIE

High-quality, professional news is expensive to produce—and even more expensive to distribute. RT entered the international news media scene when it was already crowded, dominated by lavishly financed juggernauts from the US and UK. Competing on their turf on their terms was beyond the realms of possibility. So the RT team turned its efforts and much more limited resources toward the opportunities provided by the new media.

Streaming RT newscasts and shows online allowed our young channel to reach millions of viewers around the world before we could enter their countries' cable and satellite networks. Our reporters tweeted constantly from the events they were covering, which added a dimension of authenticity and immediacy to the coverage, raised the journalists' profiles and fostered a closer bond with their audience. Listening to our viewers, and engaging with them via our website and social media platforms, was how we learned which stories resonated the most. It turned out that what they were craving from their TV news was a different perspective on current affairs—something that has become RT's calling card ever since.

Back-and-forth communication with its viewer community made RT the go-to news source for Occupy Wall Street from the very first day of the protests, which were not found anywhere in the mainstream media for weeks. Carefully sourcing, curating and diversifying engaging video content catapulted RT to the #1 TV news channel spot on YouTube in a few short years. Embracing new media opportunities helped RT to establish itself in the old media realm.

BREAKING NEWS BREAKS THE BOX

Still, even if a TV network wholeheartedly and competently adapts all the vanguard tools of the new media trade, this does not guarantee admission into an individual's carefully crafted "news box." What does? The answer is simple: the news itself.

In 2014, the Maidan protests in Ukraine turn violent. Boko Haram kidnaps hundreds of girls in Nigeria. Ebola comes to the United States. Scots vote to stay in the United Kingdom. There was hardly a person in the civilized world that remained ignorant of these stories, no matter their usual interests.

There will always be big news stories like these that force people to open up their "news boxes" to a wider range of media. During these events, conventional news providers will not just be relevant—they will rule the news scene for the foreseeable future.

Despite the growing likelihood of someone discovering a news story online, particularly on a mobile device, TV still rules as the source of breaking news. "Half of news consumers recalling a breaking news story they followed recently said they first heard about it on TV," says the American Press Institute (API).[9] This finding echoes the results of the international study commissioned by the BBC, which surveyed news consumers in Australia, Singapore, India, UAE, South Africa, Poland, Germany, France and the US.[10]

Other platforms then enter consumers' news cycle: you might turn to a search engine for more story details, then go on Facebook to chew it over with your friends. The next day you might read an in-depth newspaper article about it by your favorite reporter or blogger. But TV remains an essential part of the news mix and the most efficient way to get comprehensive coverage of an event. Be it RT, CNN, BBC, or Al Jazeera—it's hard to tear yourself away from the TV screen during a rolling news event.

LEAVE IT TO THE PROS?

During a big news event there is no substitute for professional reporting. Proper journalistic organizations, be they news agencies, newspapers or TV channels, are the ones that have human, financial, intellectual, technological and logistical capital to bring together all the components of high-quality, comprehensive coverage from virtually any corner of the world in one place.

Such coverage is made up of on-the-spot information gathering, regular updates, high-definition video footage, eyewitness accounts, verified facts, larger context, in-depth analysis and a spectrum of opinions. This makes up the essence of news. And reporters, editors and producers act as high-level aggregators—collecting raw information, establishing credibility, sourcing experts, and filtering out the noise.

Political analysis requires the kind of access that a citizen journalist cannot provide. Journalists are able to virtually, or even sometimes literally, "parachute" in to get to areas where a natural disaster has paralyzed the infrastructure, where most people are more concerned with getting away from the epicenter, rather than getting in for a scoop. Warzone reporting demands expert crews and vigilant security measures.

Even when news takes place "in plain sight" and breaks on Twitter or another platform that enables a live-blogging experience by an eyewitness, it takes either a large enough crowd of "average Joes" to share it to generate a "trending" momentum, or for a professional news organization to pick it up and pass it on to its millions of followers.

Of course, news editors have to accept the reality that today some consumers are willing to partially sacrifice professionalism for high-speed delivery, to which they have grown accustomed on social media. Bringing the speed of professional news output as close as possible to that of social media will remain a challenge, but the professionals' other virtues secure their indispensability.

And professional journalists are naturally part of the social media action themselves. In 2011, RT's Lucy Kafanov made The Huffington Post's top 20 list of most important Twitter accounts to follow for Occupy Wall Street coverage for regular reporting from the scene of events.[11]

PRO NEWS: BETTER ON TV

When it comes to covering the big news events, TV news is a strong marriage of content and medium. Television delivers news in a way that has not yet been attained by any other single platform: it is at once professional, comprehensive, visual and often live. "The 24-hour cable channels…are the source most often cited for four of the topics probed: politics, international news, business and the economy, and social issues," says the API.[12] This is true for the UK market, too: according to Ofcom, the UK broadcasting regulator, "television has a 42% share of [news] references,"[13] more than any other news source.

While different types of news outlets can produce quality reporting, television's major advantage over newspapers, blogs, radio or social platforms is that it is the most visual of news media. Yes, a picture is worth a thousand words, but moving pictures are worth a great deal more.

On top of, rather literally, completing the picture of an event, visual storytelling is one of the cornerstones of audience engagement.[14] Even Twitter isn't just about the 140 characters, but the photo or video that makes it go viral.[15] Being a visual medium at its very core makes TV a highly compelling form of news delivery.

TV news is also engaging in a way that is becoming a rarity in the increasingly individualized digital age—by enabling a shared social experience. How often have we heard that "the whole country is *watching*" a story unfold? A breaking news event can galvanize an audience in front of their TV screens in a way of a thrilling sports game or a new episode of a hit show. People feel that if they don't tune in, they're missing out.

You tweet to your select group, read a magazine article whenever you want, browse Facebook at your own pace—maybe on the beach or in line at the store. But when you watch breaking news on television you know millions of others are doing it, too. TV enables consumers to share that zeitgeist moment with a world of unseen others.

So while we get our news from an ever-growing number of devices and information sources,[16] it's still a fact that more than 70% of Americans still watch news on TV,[17] most of them particularly engaged by breaking news stories.[18]

BACK IN THE BOX

The operational challenge for any news outlet is that "breaking news" is by definition an exceptional event. Those electrifying stories punctuate days (or even weeks) of comparatively mundane bulletins. Even sensational events that take on any sort of continuity eventually generate viewer, or even compassion, fatigue.[19] Once a breaking news event is over, news consumers retreat to their preferred media content. To become part of a person's regular media mix, a news provider has to follow the audience, and engage viewers where they want to be engaged.

And yet, at a time when our choice of sources is virtually unlimited, the resources available to news organizations are anything but. And since viewers' interests run the widest gamut imaginable, no one news provider can be all things to all people.

The key to success is finding the golden mean among a) offering a unique product that informs your brand (such as geographic expertise, an original perspective, or unique presentational style); b) providing diverse content that goes beyond traditional news, as to be interesting to a wide range of consumers; and c) determining optimal platforms for delivery of each type of news product. This is the essence of "curating content" and the way for a news brand to find its audience and keep it engaged.

ENGAGING WITH DIFFERENCE: THE PRODUCT

Since entering the crowded global media landscape in late 2005, RT has established itself as the preeminent alternative to the mainstream news. RT has become a welcome voice of dissent on a wide range of current issues, the place to get the stories overlooked or ignored by the majority of the Western mainstream media (MSM), as well as a go-to source for the Russian perspective on the matters relevant to Russia. From Occupy Wall Street to the Arab Spring, from fears over global surveillance to the crisis in Ukraine, people from around the world turn to RT to hear the other side of the story: to Question More, as our tagline goes. This is the core of RT's identity, its unique offering.

This was not always the case. At the outset, RT—originally styled Russia Today—was primarily trying to tell the world about Russia. This was a noble goal. Most people either knew little about this country or their image of it was limited to clichés and Cold War stereotypes. International press coverage wasn't much help; a mix of sporadic reports on corruption, lucrative investment opportunities and the perennial myth of the "dying bear" (demographically speaking)[20] painted a highly skewed picture of the modern Russia. It was time for us to tell our story, ourselves.

RT correspondents traveled far and wide across Russia, reporting on our nation's diverse culture and complicated history, impressive technological advances and buoyant economic development, ongoing challenges and efforts to address them, complex domestic politics and broader geopolitical context.

There was, of course, a segment of curious and grateful viewers who enjoyed learning new things about a country which, though taking up the biggest landmass area on the planet's surface, remained a mystery for the majority of the world. Yet how many Russia fanatics were out there, really? How big was the audience that wanted to watch news from Russia, about Russia, 24 hours a day? Not that big.

The trouble was, in those early days of Russia Today, we were disproportionately focused on the stories we wanted to show—instead of what our international audience wanted to see.

The shift toward what RT is today was partly organic, partly deliberate, but first and foremost catalyzed by our audience. Although a great share of RT content was made up of Russia-focused stories, the channel still defined itself as a global news outlet. When a big international story broke, our journalists often found themselves well positioned to offer a different point of view, especially when the news in any way related to Russia or the region at large. From the 2008 South Ossetian conflict to the NATO ballistic missile defense system, RT gave a voice to the unheard, and paid special attention to the facts that had escaped the interest of the establishment media.

Audience response—via websites' comments, social media and thousands of emails—was an overwhelming and unequivocal "More please!" Switching RT's focus from local to international, paying attention to the underrepresented voices from the US and Europe, and reporting on untold stories from the Middle East to Latin America has earned RT a loyal, engaged, and ever-growing following from viewers and readers worldwide.

RT has really taken off since embracing this approach. We were the sole TV channel on the scene of the Occupy Wall Street protests for nearly three weeks. The movement that was originally dismissed by the MSM as a non-story went on to become a global phenomenon, and earned RT an International Emmy nomination—its second of three.

In 2011–2012, RT was the voice of caution during the mass media exuberance over the Arab Spring, warning about the likelihood of sectarian violence in Libya and the dangers of arming Syrian rebels. By 2014, many of those rebels had gone on to join the terror organization Islamic State, and Libya has been engulfed by tribal strife.

In Europe, RT was among the first to pay attention to the growing Eurosceptic sentiment, which has since become strong enough to redefine the European Union as we know it. In Ukraine, RT brought to global screens the humanitarian tragedy that has unfolded in the country's eastern regions while being largely ignored by the Western press.

The demand for an alternative perspective on news is far from having been met. On a weekly basis, RT receives requests from viewers to launch local-language versions of the channel—in French, Romanian, Slovak, Serbian, Portuguese, Farsi, Pashto, and Hungarian, and the list goes on. In 2014, a German-language petition for RT to launch RT Deutsche gathered more than 30,000 signatures in just a few months. Several Latin American countries, including Argentina, Venezuela and Cuba, have made RT or its programming an integral part of their national TV grids. In Iraq, 44% of those who watch TV news turn in to RT on a daily basis.[21]

This is the reality of today's news landscape. Instead of dry facts, people want opinionated news and engaging stories, and they know that, in most cases, each story has many sides. They are drawn to strong, original perspectives, be they from RT, Fox News, or Al Jazeera, which made a splash by being a voice of dissent during the Iraq War.

Some consumers want to stay in their comfort zone, and if what they are craving matches your unique product, then you've got yourself a permanent place in their "news box." Many others crave variety, however. When they scroll through channel after channel and see the same lead stories, talking points and expert commentators, it is only something new and different that will grab their attention. Originality earns you a spot in a diversified news mix.

There is no longer any single voice that is in a position to dominate the conversation. Though the balance still heavily favors the mainstream US, UK and European news media, new players are growing stronger, connecting with the audience and disrupting entrenched narratives. In the process they are also stirring up more than a modicum of panic among the media establishment that for so long enjoyed being unchallenged.[22, 23] Thanks to new technology and diversification of media, we now know that there are many more voices eager to be heard. Will we soon see a global news channel emerging out of Argentina or Brazil, giving a platform to Latin American perspectives? Or perhaps a South African or Kenyan channel, which will change how the world sees sub-Saharan Africa? If we do, it will only work to benefit international news consumers at large, and create new opportunities for the countries themselves.

ENGAGING WITH DIFFERENCE: THE PLATFORMS

RT, just like every other TV channel out there, no longer exists solely on TV. Websites, social media, and mobile apps all have become inextricable components of a modern television network as we adapt to new consumption patterns. These media are especially critical for keeping your audience engaged during the downtime from the big news events. The key to success is curating varied content, concurrently yet differently, for various platforms.

Despite being a relative newcomer to the news landscape, RT has achieved record-breaking success across a variety of new media platforms. We don't treat all social media with a "one-size-fits-all" approach. Consumption of online news, YouTube, Facebook and Twitter all differ in their demographic profiles, content preference, and engagement patterns. Taking the same piece of content, such as a comprehensive broadcast news bulletin, and plugging it into every single platform will not do much to attract or excite your audience. Customization is key.

As soon as YouTube launched in Russia in 2007, RT was one of the first to jump onboard. By 2013, RT's flagship English-language channel had become the top TV news channel on YouTube, with over a billion views and a million subscribers. It surpassed all its competitors, including CNN, BBC News and Al Jazeera, all of whom had the opportunity to join the platform before 2007. By 2015, with more than 2 billion views and 2.5 million subscribers combined across five channels,[24] RT has firmly established itself as the #1 TV news network on YouTube.

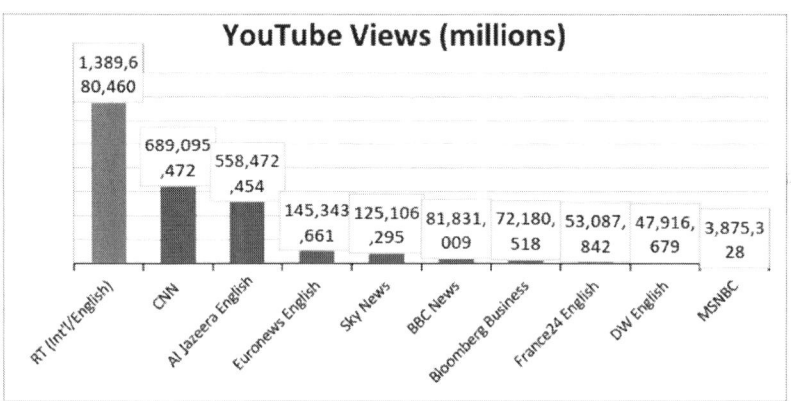

	YouTube Subscribers	YouTube Views
RT (Int'l /English)	1,488,357	1,389,680,460
CNN	758,756	689,095,472
BBC News	404,202	81,831,009
MSNBC	11,554	3,875,328
Euronews English	126,904	145,343,661
France24 English	46,714	53,087,842
Bloomberg Business	174,963	72,180,518
Sky News	94,884	125,106,295
DW English	64,988	47,916,679
Al Jazeera English	675,931	558,472,454

Source: YouTube, April 2015

RT's YouTube success is a result of tuning into what online video news consumers want: strong raw footage—without studio packaging, and without being told what to think. YouTube is also RT's go-to platform for uninterrupted, long-form and exclusive broadcasts of live events rarely accessible to the foreign-language media, such as press conferences by senior government figures or important cultural events. These streams, dubbed into English, Arabic and Spanish, are then picked up by the Western press—such as the *Guardian*—strengthening RT's news brand and reach.

RT's online platform, the multilingual RT.com, gives our on-air channels much broader broadcast reach. Anyone with internet access can watch a live RT broadcast, commentary shows, or exclusive documentaries. Online broadcast is an equally strong proposition on RT.com for regular news consumers and specialists who will not find broadcasts of key Russian political events anywhere else.

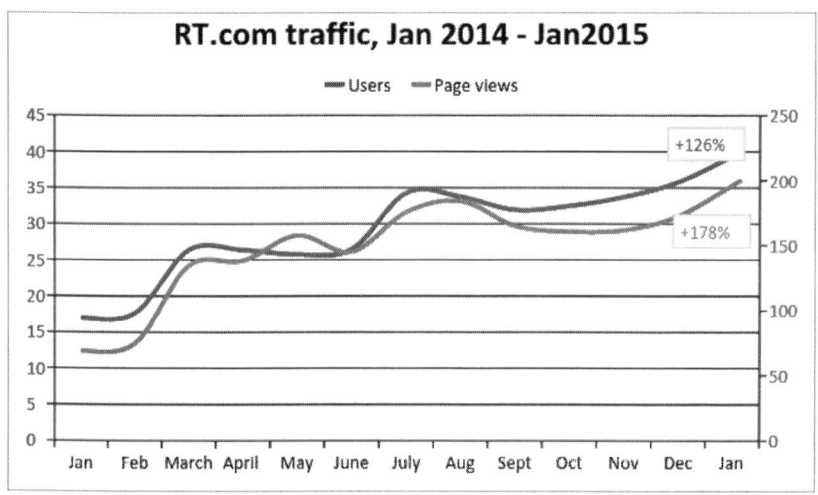

Month, 2014–2015	Users	Page views
Jan	17,009,321	68,742,930
Feb	17,635,691	74,580,907
March	26,374,596	134,402,984
April	26,376,384	138,318,226
May	25,766,858	157,778,088
June	26,504,385	145,065,117
July	34,284,877	175,651,740
Aug	33,728,961	183,977,915
Sept	31,902,961	164,875,069

Oct	32,466,484	160,559,250
Nov	33,702,888	161,709,619
Dec	35,798,406	172,627,960
Jan	38,409,402	191,348,842

Source: Google Analytics

Twitter is a prime tool for gathering and quickly disseminating information. It's the fastest way to attract attention to a story or issue, and to keep your audience updated on developing stories in real time. Effective engagement has to be concise and to the point—no burying the lede here. Many RT reporters compulsively tweet live updates and images from the scene of events, giving their followers a chance to compare how a particular event is being perceived by journalists from different countries, as well as by civilian witnesses.

By contrast, Facebook is the go-to platform for in-depth discussion and substantive feedback. Sharing and commenting on a story generates a community around it, with a much longer engagement period. Long-form articles and op-ed pieces particularly benefit from that kind of exposure.

By carefully matching varied content to appropriate online outlets, all of which are accessible on mobile devices, RT's flagship channel achieved 80% audience growth for both its Twitter and Facebook platforms in 2014 alone.

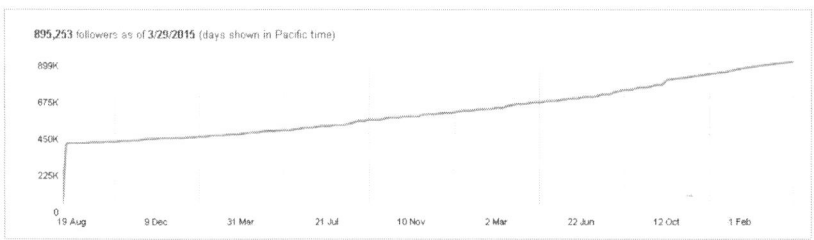

Source: Twitter, 2012–2015

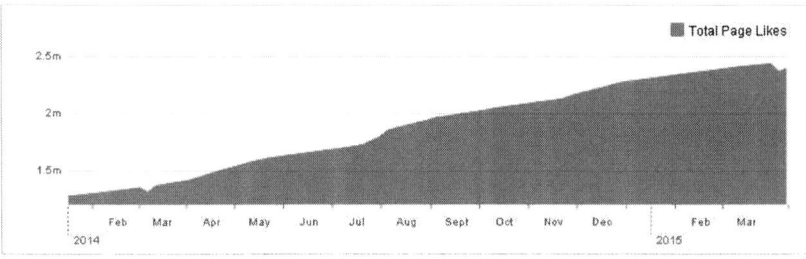

Source: Facebook, 2014–2015

Even more remarkably, RT's audience engagement level on Facebook, as represented by the "People are talking about it" metric, is far ahead of virtually all other international news channels.

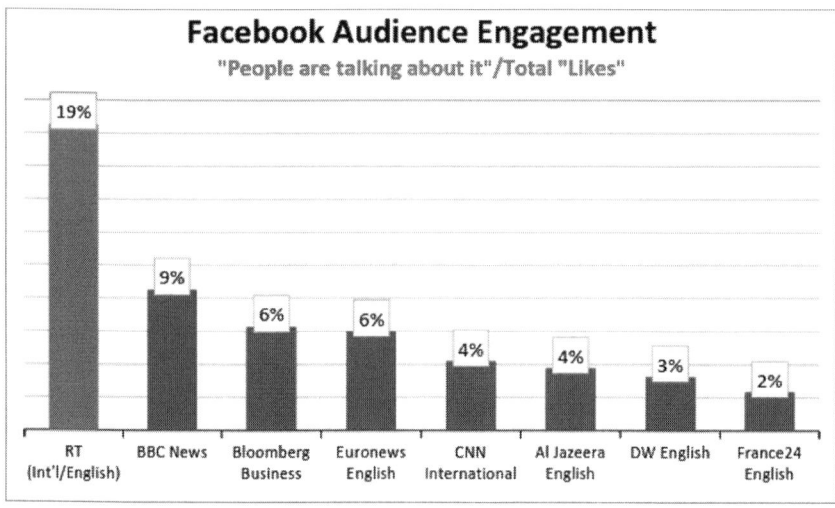

	"People talking about it"
RT (Int'l / English)	19%
BBC News	9%
Bloomberg Business	6%
Euronews English	6%
CNN International	4%
Al Jazeera English	4%
DW English	3%
France24 English	2%

Source: Facebook, April 2015

These active purveyors of social media content are one of the key elements that drive RT's robust growth, particularly by enabling RT to engage a younger generation of news consumers. Although television remains a dominant force with this demographic in many markets, including the US, where over 70% of young adults get their news from TV,[25] 33% of Millennials surveyed say that they receive news from Twitter regularly; that number jumps to a whopping 88% for Facebook. Overall, 85% of Millennials say that staying up on news is important to them, and 69% said they get news daily.[26]

THE FUTURE OF TV NEWS: YOUR CUSTOM TV STREAM?

By offering unique content and properly leveraging multiple platforms, professional television news will remain a powerful force on the news landscape. And despite the growing proliferation of mobile devices, which enable news consumption on demand, live news broadcasts will dominate breaking news events by galvanizing broad swathes of audience in front of big screens.

In the meantime, our definition of what is a TV screen is changing. Internet connectivity has been turning flat-screen televisions into giant monitors for the last decade. Wide-screen computer monitors can now carry high-quality live video feeds no worse than most traditional TV sets. TV screens and computer screens are close to becoming fully interchangeable. As these devices continue to converge and adapt to the insatiable user appetite for customization, the TV itself might become the new Facebook—a highly personalized stream of video news reports from multiple sources.

For now, how exactly "TV news on demand" will look is anyone's guess. Perhaps you will turn on "Your News Channel" at 6 o'clock in the evening and will see, side by side, the first and second lead stories on RT, CNN and Fox, and will choose accordingly, as several breaking news tickers run at the bottom of the giant screen. Perhaps your interest in European financial markets will pull that day's related video packages from the BBC, Deutsche Welle and RT—and supplement them with an audio and graphics feed from the Wall Street Journal online edition. Or you might passively listen to the stories your TV has recommended for you based on your previous consumption patterns, while doing the dishes. The possibilities are endless.

Yet one thing is certain. Whatever hardware and algorithm we might use two, five, or even 10 years from now, done the right way, the TV news experience is here to stay.

NOTES

(All accessed July 30th 2015)

1 http://thepomoblog.com/papers/pomo35.htm

2 http://stakeholders.ofcom.org.uk/binaries/research/tv-research/news/2014/News_Report_2014.pdf

3 http://www.americanpressinstitute.org/publications/reports/survey-research/social-demographic-differences-news-habits-attitudes

4 http://www.bbg.gov/wp-content/media/2014/02/Russia-research-brief.pdf

5 http://www.bbc.co.uk/mediacentre/worldnews/news-consumption.html

6 http://www.journalism.org/2013/11/14/news-use-across-social-media-platforms/

7 http://www.pewresearch.org/fact-tank/2014/09/24/how-social-media-is-reshaping-news/

8   http://www.pewinternet.org/2014/02/20/mapping-twitter-topic-networks-from-polarized-crowds-to-community-clusters/

9   http://www.americanpressinstitute.org/publications/reports/survey-research/personal-news-cycle/

10 http://www.bbc.co.uk/mediacentre/worldnews/news-consumption.html

11 http://www.huffingtonpost.com/2011/10/19/occupy-wall-street-essential-twitter-follows_n_1020215.html

12 http://www.americanpressinstitute.org/publications/reports/survey-research/personal-news-cycle/

13 http://stakeholders.ofcom.org.uk/binaries/research/tv-research/news/2014/News_Report_2014.pdf

14 http://thenextweb.com/dd/2014/05/21/importance-visual-content-deliver-effectively/

15 https://blog.bufferapp.com/how-buffer-ab-tests?utm_content=bufferae77c&utm_medium=social&utm_source=twitter.com&utm_campaign=buffer

16 http://www.americanpressinstitute.org/publications/reports/survey-research/how-americans-get-news/

17 http://www.journalism.org/2013/10/11/how-americans-get-tv-news-at-home/

18 http://www.americanpressinstitute.org/publications/reports/survey-research/social-demographic-differences-news-habits-attitudes

19 Compassion Fatigue: How the Media Sell Disease, Famine, War and Death By Susan D. Moelle - http://www.amazon.com/Compassion-Fatigue-Media-Disease-Famine/dp/0415920981

20 http://www.forbes.com/sites/markadomanis/2014/11/03/russias-demography-continues-to-improve/

21 http://rt.com/about-us/press-releases/rt-arabic-top-3/

22 http://www.theguardian.com/world/2009/jul/26/china-arabic-tv-channel

23 http://www.theglobeandmail.com/news/world/worldview/as-western-media-contract-the-china-daily-expands/article4367720/

24 English-language RT International, RT Arabic, RT Spanish, RT Russian and RT America

25 http://www.americanpressinstitute.org/publications/reports/survey-research/social-demographic-differences-news-habits-attitudes/

26 http://www.americanpressinstitute.org/publications/reports/survey-research/millennials-news/single-page/

CHAPTER FIVE

The View from Australia: How Will We Be Heard?

MARK SCOTT, FORMER DIRECTOR GENERAL, ABC

After noting historian George Rudé's estimate that Robespierre had made 900 speeches, Hilary Mantel made a telling observation: "He had spoken, of course; but had he been heard?"[1] Questions of a similar nature keep me awake at night. What impact can news make, what is its future audience and how will news be heard and seen?

The ABC currently produces thousands of hours of television news and current affairs each year. And, in the context of existing audiences for television in Australia, audiences for TV news are exceptionally good. The evening news bulletin of the ABC, along with those of the two most successful commercial TV services, Seven and Nine, are fixtures in the Top 10 of Australia's most watched programs. Audiences for the ABC's rolling news service *ABC News24*, the home of breaking news on ABC TV, continue to grow.

TV news is in great shape in Australia—whatever happens to it in future, it will have the advantage of coming off a strong base. However, it's difficult to imagine how this degree of audience interest might be sustained, both now as transformation is underway and later, in 10 to 15 years' time, when most television will be delivered via the internet.

Over the next five years in Australia, both standard and high-definition channels will become increasingly available via a high-speed National Broadband Network that is now being rolled out. This will be the most disruptive development Australian television has had to contend with in more than 60 years.

We need to avoid the temptation to look for the single replacement for existing TV news as all signs now suggest that TV news will exist, but there will be many kinds of TV news. Marshall McLuhan noted that:

> When faced with a totally new situation we tend always to attach ourselves to the objects, to the flavour of the most recent past. We look at the present through a rear-view mirror. We march backwards into the future.[2]

We may not know what news services will look like but some journalistic practices from the past will still be part of that future.

When news bulletins first appeared on ABC television in the 1950s, they conformed to the format that had previously been established in radio. In fact ABC Commissioners, equivalent to today's Board directors, had insisted that bulletins have no more than two minutes of pictures—on the basis that images would distract viewers from the content of the news.

This limit did not last long. While the historic format of TV news—newsreader, script and links—has proven remarkably durable, definitions of the news itself have continued to evolve and expand. The immediacy of online and mobile news means that by the time an evening news bulletin rolls around, the day has already been peppered with headlines, news snapshots, and fragments of the day's stories for most people.

Yet, a bulletin that wraps the news of the day, adding context and explanation to reporting in one hit, still seems potentially valuable. Which might explain why an evening bulletin has remained so popular in Australia. It explains too some of the cautious optimism that prevails among Australia's TV news services. But is this simply a holding pattern before the inevitable decline? No one who's seen a toddler with a tablet is going to suggest that television is disappearing into cultural irrelevance. More television is being produced and consumed—time-shifted, binge watched and in so many other settings than the television itself.

And while TV news will continue to exist, it will come to us in different formats, in different places. It matters less where the news originates than where it ends up, as new technologies and devices continue to experience exponential growth. As Ben Thompson, the technology analyst, has noted:

> Every single person on the Internet has the same addressable market—the entire world— as *The New York Times*. 'News' can come from anywhere.[3]

The competition TV news faces will vastly increase once television experiences are fully assimilated into the more crowded, open environment of the web.

While the National Broadband Network will increase competition, it will also create opportunities, granting all news services—commercial and public—a means of providing, via the internet, more personalised, interactive and better-targeted content to multiple audiences. Even without a national high-speed broadband, the ABC's internet TV service, iview, has already experienced substantial audience growth with average monthly program plays rising by almost 30% over the past year.

Significantly, 70% of *all* plays came from mobile with just 16% via the iview website and 14% through connected TVs, a possible sign of things to come. This new means of distribution, however, also has financial implications for the ABC. For while there is no marginal cost in extra TV viewers or radio listeners, each extra *online* user means an additional cost. The ABC will have to first compare distribution costs—over existing poles and wires against over the internet—to ensure it achieves the best, most efficient arrangement to reach audiences.

Meanwhile, the lessons of failure for the business of news journalism are abundant and the few success stories difficult to emulate. Australian commercial TV is still finding its way through the fallout of a fragmentation of audiences that will not stop. Sustaining the present performance of TV news will become even more difficult.

The pattern of digital decline for many long-established print players is a familiar one. There are more readers and less revenue. The print advertising dollar, once decimated by the online environment, is decimated once more now in mobile. Many heritage news providers have been outrun if not outclassed by new players with a winning combination of digital DNA and lower overheads. This is one reason to keep the future of *journalism*, which is blooming with promise and opportunity, distinct from the *financial* future of journalism, where the search for solutions continues.

This distinction proves useful when, for instance, BuzzFeed[4] is being described as "the most important news organization in the world,"[5] a claim that has more to do with BuzzFeed's business model—and its powerhouse performance in social, which drives 75% of its traffic—than its journalism.

Yet BuzzFeed's success is one of the clearest ways of illustrating how dynamic and volatile conditions are in the digital space, and the extent of change now underway. Measures of success are shifting. Story shares are becoming more important and clicks, page visits and even the homepage itself become less. Digital natives like Huffington Post and BuzzFeed, both of which have Australian branches now, first found financial success through a package of curation, aggregation and innovative storytelling.

These fresh takes, voices and formats have innovatively drawn audiences to news. They are casebook studies in how to create stories that people want to share. Each, as it has become more successful, has invested in more original reporting.

As part of the epic shift in power from media providers to audiences, along with the rise of social media, audiences now influence and broaden news story selection, bringing more depth to research and access to a broader range of sources. Interaction with these additional voices will continue to enrich the news mix.

Stories that begin with an ABC reporter and are then developed and strengthened through audience input will continue to grow.

New sources of big data have also opened paths to new research possibilities, bringing a new dimension to both investigative journalism and storytelling in a digital environment.

Australian TV news services have taken advantage of these developments with data and greater interaction with audiences, embedding them in contemporary news practices. The conditions in which we now operate have never seemed more promising nor, paradoxically, more perilous.

FUTURE FORMATS

To secure a place in the crowded, converged environment in which news must compete, news formats will continue to change. Elements of audio, video and text must be remixed and repackaged to suit particular platforms and increasingly particular audiences. News services no longer compete simply with each other.

The ABC is one of the choices available in the same space as Australian news services, and global competitors like the BBC, the *Guardian* and the *New York Times*. Its news must also be discovered in the same space and on the same device as Clash of Clans and Candy Crush. Instagram and Snapchat are but a swipe away, also gaining attention that might otherwise be directed to news, as do the 100 hours of video being loaded up every minute to YouTube and 4 billion Facebook shares and 500 million tweets flowing through social streams each day.

So how are we to present stories, ensure that they reach audiences inundated with choices about how to spend time and attention? Smart reporting, smart curating, smart sharing and distribution will continue to be crucial. It's become very clear that without access to curation that accords with interests, the condition of being information-rich often comes at the cost of being time-poor—which makes the future of the news bulletin seem secure. In a crowded, converged digital space where attention is harder to gain and often fleeting, curatorial skills will have to be even sharper.

Public opinion polling in recent years has consistently confirmed the ABC as Australia's most trusted news service, directly due to the ABC's publicly recognised news values of impartiality and independence. Much around the provision of news may change, but not these values. That trust will also continue to be critical to future news provision. The infinite variety of news sources now available includes an infinite variety of standards, quality and integrity. And in a whirlwind news cycle with so much focus on immediacy, a race to break news fastest and first, news that can be trusted will be an even more vital point of differentiation.

The challenge is a greater one for the ABC. As Australia's primary national public broadcaster, it has a budget fixed by Government appropriation and obligations to be universal, to sustain democracy. It can't pursue populist stories to boost audiences or, like a commercial news source like the *Economist*, pick and choose its audience. As a public broadcaster, it must instead provide an accessible news service for all Australians. The challenge for the ABC is to pursue the innovations necessary to keep providing that service, and do it within budget.

SOCIAL AUDIENCES, SOCIAL NEWS

A TV news story, distributed first through the ABC and its range of platforms, swiftly makes its way into the various social network streams. Broadcast and online networks are merely the first point in an increasingly mixed digital distribution chain.

The primary source for news, for almost 4 in 10 people under the age of 35, is social media.[6] As The Media Insight Project's "How Millennials Get News: Habits of America's First Digital Generation" reported:

> Simply put, social media is no longer simply social…It has become a way of being connected to the world generally—to…follow channels of interest, get news, share news, talk about it, be entertained, stay in touch, and to check in and see what's new in the world.

Given the prominence of social media in the lives of these audiences, it's in our interests to ensure that the stories we tell are not just good, but travel. To distribute and connect to make news and information go further, with as many chances as possible to be watched, heard or read.

As a public broadcaster, the ABC is also obliged to ensure we deliver the best possible return on the public investment in news. To engage as many citizens as possible. This obliges us to address news to audiences beyond those with an established interest in news, to reach the casual user as well, the more difficult and arguably more important target.

By absorbing lessons of the social web, rendering stories in a way that makes sharing a seamless experience, TV news stories will continue to appear in those significant social streams.

Distribution within social media does, however, place news organisations at the mercy of an algorithm—as any news provider that's woken up on the wrong side of a sudden switch in Google or Facebook rankings knows. At 1.1 million likes, ABC News is the most followed Australian news service and our page administrators are provided some very valuable audience insights by Facebook.

However, Facebook's business model relies on having as much activity as possible happening behind its walls. Clarity around the interaction with news stories, beyond a news service's own pages and accounts, is limited. Facebook remains the most powerful and significant player in social media, and its possession of vital data on its users' relationships with our stories means that, to a certain degree, we're flying blind behind its walls. Alex Madrigal, writing for the *Atlantic* identified this vulnerability and the risk it entails:

> The takeaway is this: if you're a media company, you are almost certainly underestimating your Facebook traffic…The bad news is that, if you didn't know before, it should be even more clear now: Facebook owns web media distribution.

A further negative about online activity being kept within Facebook is that it inevitably results in fewer visits to news services own sites where gateways to other stories from the service are featured, resulting in a reduced return. When writing about the value of the incoming traffic and figures for "page views by source," Frederic Filloux, the tech and media analyst, estimated that while direct access to a site will typically yield 5 to 6 page views and Google search results in 2 or 3 views, social referrals tend to result in just 1 page view.[7]

Few news services would be able to emulate Netflix's mastery of audience data to build a recommendation engine that sustains audience commitment and drives growth. Yet, access to some of the audience data collected by Facebook would significantly improve understanding of audiences and their relationship to news stories.

Meanwhile, we rely on other means to find out more about the audience expectation of news. In May 2015, the ABC is piloting a GfK Audience Appreciation panel system to provide insights into the quality and distinctiveness of its content and that of other media outlets. The panel will consist of 2,000 Australian adults, in each of the mainland State capitals.

During the month-long trial, the panel will be asked about TV news on a daily basis with a specific weekly focus on audiences using digital news services. There will also be a short series of questions every few days about television programs, radio stations and online activities. This system is also used by the BBC in the UK, RTE in Ireland, NPO in the Netherlands, VRT in Belgium and TV3 in Spain.

Qualitative research can give us insights into the impact of a news story: did it lead conversation, alter public thinking in a way that led to changes in public policy—as so many ABC stories continue to do?

While ratings figures are integral to commercial broadcasting business models, metrics on the *impact* of news stories, the ways in which the civic purpose of the news was fulfilled, will be increasingly significant in the fragmented audience space.

The Reuters Institute's Digital News Report of 2014—based on a survey of over 18,000 online news consumers in the UK, US, Germany, France, Italy, Spain, Brazil, Japan, Denmark and Finland—notes that smartphone users access "a narrower range of news sources on a smartphone than they do on a desktop or tablet," with 37% on a single source each week. On laptop or desktop, it's 30%. The differences are more marked in the UK, with 55% of smartphone users and 45% of computer users relying on one news source each week.[8]

In the cable news TV space, Fox has proven the effectiveness, as a business model, of valuing confirmation over information. Facebook's algorithm, for the moment, offers a variation on this, with past use determining the future, giving priority in users' Facebook feeds to the sources, types of media and points of view with which they have interacted previously. As a public news service, dedicated to impartiality and providing a diversity of voices, the full range of opinion, this narrowing effect is emerging as a significant challenge.

There are other implications for stories and how we tell them brought about as stories are increasingly shared. While figures vary, it's clear that up to 50% of Facebook activity and more than 60% of Twitter activity is happening within mobile. The ways in which we both tell and format our stories must fit the demands of the scroll-and-swipe mobile and social environments and behaviour and consumption patterns within them.

More than 50% of Facebook activity takes place within the Facebook app. Mobile users already interact with a homepage far less frequently than desktop users, and for more than 80% of Australians using smartphone, mobile is integral to media use.

THE CIVIC PURPOSE OF NEWS

There are not just infinitely more competitors in the battle for attention, but television itself is far less a living room experience than it once was. The *habit* of television watching, developed through that experience created over several past generations, has not been handed down to the present. New generations have new and different habits, reflecting the shift from an age of scarcity to one of plenty. Is this a demographic time bomb waiting for television networks, including rolling news networks? Television commands attention a long way from the living room now, in so many places, through many other platforms and devices, distributed through social as well as broadcast networks.

We have to ensure not just that the news stories are found, but that they are worth searching for, worth sharing. While Australians under the age of 30 spend less time watching TV than on their phones, simply taking the news to the phone and into the social streams will not be enough. By holding those in power to account, making institutions and businesses more responsive, and informing debate around policy decisions that shape Australian life, news services contribute to a more meaningful citizenship. While technology and audience behaviour continue to change, this underlying objective and civic purpose of news will not.

But what of the disconnected? Engaged citizenship is conditional upon citizens actually connecting with news and information, not just on its availability. And no matter how necessary news is to civic life, younger audiences engage far less with it in its traditional formats than with alternatives that speak more directly to them in different ways. What is a public broadcaster without a public?

With so many news services to choose from, how is it that Jon Stewart, a satirist, became "arguably the most trusted man in news," as *New York Times* media and technology writer David Carr described him? And what is to be learned from the way in which BuzzFeed or ABC Australia's the *Chaser's Media Circus* made themselves interesting, particularly to younger audiences, as many other news services have not?

Debate around news success often resembles an old debate, that around values of tabloids and broadsheets, where populist and quality are regarded as mutually exclusive, rather than essential elements to any successful news service.

For ABC, the news "sensibility" itself will change just as it has previously as the influence of a new generation of journalists now entering newsrooms is felt. In recent years, the ABC has recruited cadet journalists from a broad range of academic disciplines—law, economics, environment, science—many of whom have been online since their teens. Uploading, posting, tweeting and sharing text, images and video is second nature to them—an activity that becomes part of the workflow of their journalism just as it's part of their lives. Onto this foundation of digital skills, the ABC builds training in writing for specific media, in ethics, law, verification, and editorial standards. A new voice for ABC news will emerge from this combination of digital and journalistic skills, and with it, as generational change feeds into the culture, a new journalism. The cultural project of connecting younger audiences with news is essential to securing future relevance for our news services.

AUDIENCES AT THE CENTRE

> The days of desktop dominance are over. Mobile has swiftly risen to become the leading digital platform, with total activity on smartphones and tablets accounting for an astounding 60 percent of digital media time spent in the U.S. (The U.S. Mobile App Report comScore Whitepaper, August 2014)

With the migration to mobile, the greatest audience shift in the history of the ABC is now underway. In focusing on news on mobile and online, the ABC is not leaving audiences behind, but going with them.

Just as the music industry was reinvented by iTunes, the unbundling of the album, TV news, in being unbundled and disaggregated, is being reinvented. Stories now travel individually, often independent of one another and the context of trust associated with the source, to reach different fragmented audiences.

We have been able to base digital news around both location and community interests, providing news and information that is uniquely matched to unique audiences. This has led to one of the most significant recent breakthroughs for ABC News, an extensive process of localisation, with stories reaching a far greater total audience through a wider range of distinct local audiences. Opportunities for further individualisation continue to increase. ABC Spoke—a project to understand how people in regional and rural Australia get their news and information—has provided some significant insights into how fluid and complex audience notions of "local content" are, influenced as much by social interests as geography. Communities prefer to see "local content" not as a separate category, but to be embedded alongside national and international content, adjustable for topic preferences and personal interests as part of a whole.

Today, realigning resources from platforms, prioritising online and mobile, where the ABC's present audiences already are and future audiences will be, has a sense of urgency. The majority of visits to ABC digital news are coming through our flagship app and mobile website, with the usual peaks around the working day: 7 to 9 a.m. and 7 to 11 p.m.

We are examining the editorial product, content management systems, remixing story elements for the digital space, tailoring news from commissioning point to the specific needs—such as headline length and image load speeds—for the app and devices, attuned to the behaviour and interests of the mobile audience. As digital business designer John Oswald has noted:

> Ten years ago Apple knew what consumers wanted, before they wanted it—but perhaps it's the consumers who will dictate what comes next.[9]

TV news services are in the midst of a revolution that hasn't finished being born. As Australia's national public media organisation, audiences have expectations about both the values of ABC news, its impartiality and independence and of the ABC itself—that it is universally available to all Australians. While technology and behaviour must inevitably change, those values will not. Since we are not able to predict the future of TV news, our best efforts are going into inventing a future for it by being vigilant to public expectations of the ABC and its values, and to keeping up with changes that continue to affect news services and how they are used and delivered.

NOTES

1 http://www.lrb.co.uk/v28/n08/hilary-mantel/if-youd-seen-his-green-eyes
2 McLuhan, M. and Fiore, Q. (1967). *The Medium Is the Massage: An Inventory of Effects*. Coordinated by J. Agel. New York: Bantam Books (p. 74).

3 https://stratechery.com/2015/buzzfeed-important-news-organization-world/
4 http://stratechery.com/2015/buzzfeed-important-news-organization-world/
5 ibid.
6 In terms of demographics, social media are far more important for younger groups, with 18–24s more than twice as likely to find news that way (38% vs. 17% for over 55s). Reuters Institute Digital News Report 2014 (p. 64).
7 http://www.mondaynote.com/2015/04/06/jumping-in-bed-with-facebook-smart-or-desperate/
8 https://reutersinstitute.politics.ox.ac.uk/sites/default/files/Reuters%20Institute%20Digital%20News%20Report%202014.pdf
9 http://www.digitalstrategyconsulting.com/intelligence/2013/04/apples_itunes_turns_10_can_it_keep_its_lead.php

CHAPTER SIX

The View from the UK: Sky News

JOHN RYLEY, HEAD OF SKY NEWS

The first television broadcast I remember was Sir Winston Churchill's funeral on January 30, 1965. I was three years old. My parents and I huddled around the tiny flickering black-and-white screen as the great man's coffin was borne on a gun carriage from Westminster to St Paul's. It was watched on television by 25 million people in Britain and an estimated 350 million around the globe. Such was the power of a shared event, even in those early, pioneering days of television.

In the years that followed there have been many more moments when TV brought us together: the moon landings, the splashdown of the crippled *Apollo* 13, Princess Diana's funeral, the 9/11 and 7/7 terrorist attacks, the "shock and awe" bombardment of Iraq. We often refer to them as "where were you?" moments, though perhaps "who were you with?" moments is better because so often they were shared experiences. The arrival of Sky News, in 1989, and the other rolling news channels cemented in our collective consciousness the idea of witnessing events live, as they happened.

Half a century after Churchill's funeral, almost every home in Britain has multiple TVs, yet an audience of 25 million for a single live event is very rare indeed. In recent times only the opening and closing ceremonies of the Olympics came close.[1] The news media landscape has been fragmented and transformed by the digital revolution to such an extent, it is argued, that shared, live events have had their day.

Marx said revolutions are the locomotives of history,[2] yet they are slippery things. They promise dramatic, disruptive upheaval, but the eventual outcomes do not always align with the initial promise. Did Robespierre anticipate Napoleon after executing Louis XVI? Did Lenin foresee Putin when he overthrew the tsar? As Jean-Baptiste Alphonse Karr, a 19th century editor of *Le Figaro* famously claimed: "Plus ça change, plus c'est la même chose."[3]

Unarguably, however, the digital revolution *has* transformed the news industry. As the outgoing head of GCHQ said recently, the growth of the internet has been the biggest migration in human history.[4] Technology has opened up previously undreamed of frontiers in the way news is distributed and delivered. Yet, as with all upheavals, the digital revolution in news has brought with it battalions of fellow travelers, in the shape of academics, pundits, consultants and self-styled futurologists, all eager to write off the old world without entirely understanding where the new one will take us. The late Roy Amara of Stanford University summed it up succinctly: "We tend to overestimate the effect of a technology in the short run and underestimate the effect in the long run."[5]

I was appointed Head of Sky News in 2006, the year the digital revolution really took hold as the roll-out of broadband in Britain took off. It was also, coincidentally, the first year the Office for National Statistics recorded Britain's internet usage, on annual basis, and its initial set of figures was startling. Over the preceding eight years the number of British households with internet access had grown from 10% to nearly 60%.[6]

In 2006, Sky News was a successful rolling news channel which, like most other major news organisations, regarded its website and mobile applications as add-ons, not core parts of its proposition. Although few of us in the news industry really appreciated the sheer scale of change which was underway, it was clear that fast broadband internet would bring new opportunities for Sky News, whose main business was being "First for Breaking News." I realised we had no choice but to transform ourselves from a TV channel to a multiplatform, nonstop 24/7 news organisation.

Today, you can get Sky News, not just on TV, but on fourteen different digital screens, everything from phones and tablets to connected boxes and outdoor displays, and new platforms are being added all the time. It means that all news is now available nonstop, live and on-demand.

In Britain, 70% of adults access the internet while on the move every day, according to the latest ONS statistics. News is the web's fastest growing content type: today 55% of adults get news online, compared with 20% in 2007.[7] So for those of us in the business of news, it is a time to look on the bright side. As Martin Baron, the Executive Editor of the *Washington Post* said recently, optimism is an inescapable requirement for those at the helm in the digital revolution: "The upheaval means unparalleled opportunities for those who are willing to learn what they need—and know that the learning must never stop."[8]

For that reason, if no other, there has never been a better time to be a journalist. The pessimists who look backwards to some mythical golden age of journalism are mistaken. The golden age is now, and in the future. The range of technology available for gathering news, as well as distributing it, continues to grow at a breathtaking pace. The advent of 4G has transformed the smartphone into a miniature live camera-cum-satellite truck. Our reporters have even edited their own packages on them. Often, the first pictures of breaking news events are taken by the public on their mobiles. All news organisations now have drones to shoot aerials, where permitted, at a fraction of the cost of putting up a helicopter. Our graphics department has some of the world's most sophisticated equipment of its type. It means that we have more tools at our disposal than ever, not just to report events as they happen, but to analyse them, explain them and contextualise them. Multimedia journalism is no longer some text, a few stills and perhaps a bit of video. It is an intricately woven tapestry using the most suitable media to tell each aspect of the story.

The opportunities for news organisations generated by the mobile internet have been highlighted by Marc Andreessen, a Silicon Valley veteran who co-founded Netscape. Andreessen argues the smartphone revolution has been under-hyped, citing the sobering fact that more people now have access to phones than to running water. The analysts Gartner estimates total global sales of smartphones in 2014 alone were 1.2 billion, with the strongest growth in emerging markets.[9] Andreessen predicts this rocketing ownership of internet-connected mobile devices could grow the global addressable market for news a hundredfold over the next two decades.[10]

By then we can assume that more or less everyone on the planet will be connected to the internet. That will be around 8.7 billion people, according to the UN's mid-range prediction on global population.[11] We can also assume, as Nic Newman of the Reuters Institute for Journalism notes, that future Wi-Fi internet connections will be ubiquitous and fast and video streaming will be in ultra-high quality, not 4k but 40k, with no irritating buffering. Battery life will no longer be a problem: they will either constantly recharge wirelessly, on the move, or new technology will keep them going more or less indefinitely.

This relentless technological march presents a wealth of opportunities for news organisations. Markets no longer have boundaries. The rise of mobile has driven spectacular growth in the levels of audience engagement on our digital platforms with time spent by users doubling in two years. Downloads of our mobile applications have exceeded 10 million. Our content is watched and read by more than six million fans and subscribers in social media, everywhere from Facebook and Twitter to Vines and Snapchat. Sky News, live and video-on-demand, is available on all the main connected TV applications including Apple, Roku, Amazon Fire and Xbox.[12] Around half a million people a month, many in the United States are seeing Sky News on them, often for the first time. In Britain, Sky's own Catch-Up service supplies regularly updated news, in-depth coverage and special programmes on demand.

Back in 2006 when I took over Sky News, I wrote in the *Independent*: "Some media watchers believe the internet's speed and immediacy will destroy 24-hour news channels. They are missing a trick. 24-hour news and the internet are not mutually exclusive. Both rolling news and on-demand news will grow…we create masses of material everyday which is currently used and seen often only once: but in news, anything can be made into anything else and reused on new platforms."[13]

Where, perhaps, I got things slightly wrong in my 2006 *Independent* article was the assertion that everything could be repurposed on multiple devices. Today it is clear that different platforms sometimes demand different content formats: TV is better for live event coverage and longer-form video reports. On mobiles, short-form, bite-sized news works best. People mainly tend to look at news on their tablets in the evening and at weekends, while using their PCs mainly at work, viewing news in their downtime. Every significant news organisation knows it must allow its audience to consume the news not just when they want it and where they want it, but *how* they want it, while maintaining consistent core editorial values across all its journalism.

This has challenged the structures and workflows of newsrooms in ways we never foresaw. In the proto-digital world (which some news organisations still inhabit) we ran separate teams on each platform: TV, web and iPad. The old media platforms (mainly printed newspapers, but also TV) got the resources and dictated the agenda, while digital teams repurposed the content as best they could. Many modernising newsrooms have adopted "digital first" strategy, trying to turn on its head old ways of doing things. This may be the only way to drag some newsrooms into the digital age, but at Sky News, our policy is that all platforms come first.

Today our journalists must be multi-skilled and work across all our outputs. Reporters and correspondents have long been expected to contribute to all our platforms equally. Lately, we have also trained each of our one hundred producers to work on digital, in all its forms, as well as TV. Meanwhile, we (and other progressive news organisations like the *Financial Times*) are analysing the patterns of consumption by the audiences to work out where best to deploy our resources at different times of the day. In the future, I expect demarcations between news teams will disappear completely and job titles instead of relating to platforms as they were in the past (TV reporter, newspaper photographer, broadcast journalist) will simply reflect functions.

I cite this as powerful evidence against the argument made by Professor Richard Sambrook and Sean Maguire that the legacy of 24-hour news channels is holding back broadcasters in adapting to the potential of the digital age.[14] On the contrary, the strength of our journalistic legacy is providing the building blocks to construct a viable mixed economy of mainstream journalism delivered in every way the audience wants it.

The Sambrook argument runs roughly as follows: in the digital age of instant on-demand news delivered via the internet, specialist *rolling* news channels no longer have a place. Instead you should stick the whole lot online, spend the savings on more journalism and therefore serve the audience better.

Their case depends on a central premise that there is a finite demand for nonstop news: as it increases on digital platforms, it decreases on old media including TV. The evidence is otherwise: the Reuters Institute for the Study of Journalism, in its 2014 digital report, found that in the ten developed countries it surveyed, including the UK, US and Japan, an average of 86% of those who access news on a tablet say *they also watch TV news*, with similar patterns for smartphone users. Overall, the survey found that more people are accessing news through a greater number of devices than ever before.[15] This is certainly our experience at Sky: in March 2014, a remarkable month for news when flight MH370 went missing, Russian annexed Crimea and Oscar Pistorius went on trial, our digital platforms recorded their best-ever consumption figures to date while the TV channel's consumption was the third best overall for a decade. As the Reuters Institute report says, although there are clear generation differences with younger people preferring digital sources for news, traditional media platforms are far from being replaced by digital platforms.

A striking and recurring theme in the digital transformation of the news industry is the recognition that a nonstop 24/7 video news service needs a production backbone. Many of the major non-television news services have effectively created rolling news channels, but streamed online not on TV: for example, the *Wall Street Journal*[16] and *Huffington Post Live*.[17] Interestingly, the US television network CBS, which does not have its own rolling news channel, has created an online one which is also available on connected TVs.[18]

Hence, I argue that our rolling news channel is a necessary component of a nonstop news service, though as I have stressed, not sufficient on its own. It is its spinal cord, its vital and fundamental source of strength and as I told the Royal Society of Arts in 2014, if it didn't exist, we would have had to invent it.[19]

Live video news does something which is hard to replace: it reports events *as they happen*. In the same way that the early 20th-century invention—the cinema—remains a good place to watch films, so news on big TV-size screens will still be there in the future. Crucially, live TV news also offers something which on-demand cannot do: it delivers arresting TV moments where you realise you are witnessing something momentous, or terrible or simply remarkable. It unfolds right in front of you, and you don't know how it will end. In 2011, Sky's Alex Crawford and her team reported live from a rebel truck as it drove into the Libyan capital, Tripoli. It was a remarkable scoop; using power from the truck's cigarette lighter, Crawford's producer pointed the broadband satellite dish at the sky, tweaking it a few millimetres at a time to keep the signal steady.

Before Sky News, the ability to witness momentous events live was a novelty. Arguably, we gave birth to the 24-hour cycle of nonstop news. As things happen, we explain what we know, assess the credibility of our sources and analyse what it all means. We take viewers into our confidence and as new details emerge and the story changes, we update it there and then.

During the Charlie Hebdo killings, the world was gripped by an unfolding series of terrorist atrocities in France. The story produced huge spikes in audience numbers on all Sky News platforms. On TV, every dramatic twist and turn has been played out, in real time, before our eyes. The desire to know what is happening, as it happens, on news stories like this is what makes rolling television news so gripping, and continues to drive viewer numbers. It even moved a rival editor, Ben de Pear of Channel Four News, to tweet "#skynews absolutely compelling watching."[20]

There is another reason to believe that 24-hour news channels have plenty of life left in them. As Douglas McCabe of the media consultancy Enders Analysis has pointed out, in the future more people than ever will be at home, and thanks to technology will have more leisure time to watch TV, as well as browse their internet connected devices. In 30 years' time, it is estimated that roughly a quarter of the UK population will be over 65, yet since most will be very fit and likely to live to their nineties, will barely count as middle-aged.[21]

Moreover, television is profitable. Recently, the high priest of digital news, Professor Jeff Jarvis of the City University of New York, tackled Jason Mojica, editor-in-chief of Vice News, one of the new disrupters in online video journalism. Why, demanded Jarvis, did Vice also put out its content on television (on HBO)? "You've got the f***** Internet! Why would you even dance with the old models?" Mojica's reply was telling: "Why do people rob banks? That's where the money is."[22]

Nevertheless, start-ups like Vice, Buzzfeed and Vox have shown the established news media that there *are* different ways of doing things, and we are all scrutinising them hard to see what we can learn. But it is important to note that none of them make any pretence at providing a nonstop news service, covering all the main global stories, in the way that organisations like Sky News do. Big news can come and go and they will report it, or not, as they please. They certainly have value for their audiences, but they offer something different and, it must be said, something which is considerably cheaper to produce than the mainstream coverage.

Interestingly, the disrupters, like Vice, are in many ways very traditional, especially when it comes to editorial curation. Their journalists put the news in order, according to their own subjective judgements, much as it has been done for a century or more. Our parents and grandparents tended to get their news from a very small range of sources, perhaps one newspaper and one TV channel, which they chose largely on the basis of trust or habit. They were content to accept the editor's judgement on content and hierarchy.

Today, thanks to the digital revolution, the fragmentation of news supply has demolished that model. Your Twitter feed and Facebook page have become your own personalised information stream, delivering the news brands you have chosen to follow because you are interested in their content and trust them. You can mash up mainstream, professional news sources with an eclectic range of bloggers, mavericks and rumour-mongers. It's your choice, your news.

Large news organisations have been philosophically nervous of the concept of personalisation, because traditionally they think it's their job to curate the news, deciding what to put in and what to leave out, what will be the lead, and so on. Meanwhile, news consumers set about personalising their *own* news, on their own. As well as using social media, they have turned to news aggregators like Flipboard, Circa and Pulse, which pull in feeds from your chosen sources and deliver them as a single coherent service. Some mainstream news organisations have already responded to this by allowing people to personalise their news or are doing the personalising for them with fiendish algorithms that understand your personal preferences for news.

But this trend, which looks set to grow, will not spell the end of curated, edited news. On the contrary, apart from delivering live event coverage the great strength that a mainstream news brand like Sky News offers is a wealth of expertise to analyse and contextualise what is going on. As the industry continues to fragment the value of that expertise is likely to get even greater.

A couple of years ago, Demos reported this rather obvious truism: on the internet you find "…unprecedented amounts of mistakes, half-truths, mistruths, propaganda, misinformation, disinformation….and general nonsense."[23] As the digital world grows, it will become even truer, but the value of trusted mainstream journalism and journalistic brands will become commensurately more valuable in the information ecosystem. In many ways, our specialist reporters will define our brands. This trend is already evident in the emergence of the individual journalist as trusted brands in their own right. In the United States, some have even broken away from their employers to do that. They include Kara Swisher and Walt Mossberg, formerly of the *Wall Street Journal*, Glenn Greenwald who broke the Snowden story for the *Guardian* and Ezra Klein of the *Washington Post*.[24] What is more, venture capitalists are keen to invest in these individuals; as *Forbes* said, there are billionaires "who realise the value of star talent in the new journalism economy."[25] In the UK, only a handful of journalists' names resonate well beyond their own news organisations—people like Robert Peston at the BBC and Alex Crawford at Sky—but it seems probable that in the future the phenomenon of the individual journalist brand will take root here as it has in the United States.

In 2009, on Sky's 20th anniversary, I wrote that "technology is the engine of change in how news is consumed—and the ambition of Sky News is to be first for breaking news on every new platform, just as it is on television."[26] That remains as true now as it was then, but with a caveat. In the future, it seems likely there will be much less distinction between televisions, mobiles, tablets and the rest. They will all just be screens, of varying sizes, but all ultra-smart and designed to suit your individual needs at a particular time and in a particular place.

Journalists must also change to reflect the growing emphasis on technology. When I left university in the summer of 1984, my mother sent me on a typing course. Today that idea seems absurd: typing is a ubiquitous skill which everyone needs to engage in almost every form of digital communication. In the same way, it seems likely the barriers between editorial and software engineering will be eroded; journalists of the future will need to be able to write whatever code is necessary to tell stories effectively. Already, one of the biggest growth areas in newsrooms is audience development through social media and search, and this trend can only expand.

As is evident, the challenges of the digital revolution have only heightened my optimism, but I am not blinkered either. The financial model for news remains a hard puzzle. News is much less profitable than it was because people have gotten used to the idea that it is free. Some niche services like the *Financial Times*, the *Times* and the *Wall Street Journal* are making a success from subscriptions and state news organisations like the BBC continue to be funded by some form of license fee or taxation. But advertising on digital platforms remains stubbornly cheap because the supply of inventory outstrips demand. Much of this is due to the growth of Facebook and Google, which soak up an increasingly large share of the global advertising spend.

Here again, technology may help provide the solution. There is an ancient advertising adage, attributed to everyone from Henry Ford to William Lever, founder of Unilever, which says that "Half the money I spend on advertising is wasted. The problem is that I don't know which half." That may have been true in a pre-digital era, but today it's possible to target advertising by demographics so you only spend the right half of your advertising budget. Sky has introduced a system called AdSmart,[27] which uses data like age, location and life-stage to show the right TV adverts to the right people. For example, if you know a household doesn't have children there's not much point showing them ads for nappies. When all media are digital, that process will have become extremely sophisticated and showing people the adverts which suit them, and measuring their success, will be the norm. Increasingly, much of this advertising will be "content marketing" or "native adverts," a growing trend which according to Lewis DVorkin of *Forbes* will shake up a century of journalism.[28] Undoubtedly, news purists will resist any form of advertising which blurs the boundaries between editorial and commercial space, but if the private sector is to continue funding journalism, we must develop sustainable business models to support it.

As I wrote this contribution, I watched the barge which carried Winston Churchill's coffin along the River Thames make the journey once again—all part of a series of events to mark the 50th anniversary of his state funeral. Today, at a click, you can watch the 1965 funeral coverage presented by Richard Dimbleby, listen to the war hero's speeches and read the funeral service. It's a reminder that the digital revolution is the most exciting, disruptive and challenging force I have experienced in my journalistic career.

It has forced all of us who work in the news business to re-examine our assumptions, systems and working practices. Our traditional criteria when recruiting young journalists have gone out of the window; we need people with skills we hardly knew existed a few years ago.

Last week's new technology is routinely superseded by something else. Our software development teams do their best to satisfy our persistent demands for more and better features on our digital platforms. In the field, our news-gathering teams astonish us daily, producing the latest gadgets to conjure up live pictures from the remotest of locations. But, at heart, nothing has changed. We still work to the same philosophy that our business is telling stories as fast as we can and as well as we can. We get as close as possible to the action, and provide as much detail as we can gather. We use all the technology we can muster to analyse and contextualise stories, but in ways that are accessible as well as authoritative.

Winning in this complex business will be all about offering world-class visual journalism to a global audience across a variety of products, known and unknown.

NOTES

1   http://www.barb.co.uk/resources/tv-facts/tv-since-1981/2012/top10

2   https://www.marxists.org/archive/marx/works/subject/hist-mat/class-sf/ch03.htm

3   http://www.oxfordreference.com/search?siteToSearch=aup&q=%E2%80%9CP1 us+%C3%A7a+change%2C+plus+c%27est+la+m%C3%AAme+chose.%E2%80 %9D++&searchBtn=Search&isQuickSearch=true

4   http://www.zdnet.com/article/gchq-boss-defends-internet-trawl-but-rejects-mass- surveillance-claims/

5   http://boingboing.net/2008/01/03/roy-amara-forecaster.html

6   http://www.ons.gov.uk/ons/rel/rdit2/internet-access-households-and-individuals/2014/ stb-ia-2014.html

7   http://www.ons.gov.uk/ons/rel/rdit2/internet-access-households-and-individuals/2014/ stb-ia-2014.html

8   http://www.washingtonpost.com/pr/wp/2014/11/19/martin-baron-speaks-at-florida- international-on-the-future-of-news/

9   http://www.gartner.com/newsroom/id/2944819

10 http://a16z.com/2014/02/25/future-of-news-business/

11 http://esa.un.org/wpp/unpp/p2k0data.asp

12 January 2015 figure.

13 http://www.independent.co.uk/news/media/never-been-a-better-time-to-be-in-tv-news-414538.html

14 http://www.theguardian.com/media/2014/feb/03/tv-24-hour-news-channels-bbc-rolling

15 https://reutersinstitute.politics.ox.ac.uk/sites/default/files/Reuters%20Institute%20Digital%20News%20Report%202014.pdf

16 http://www.wsj.com/video/

17 http://live.huffingtonpost.com/

18 http://cbsn.cbsnews.com/

19 https://corporate.sky.com/documents/speeches/john_ryley_at_rsa.pdf

20 https://twitter.com/bendepear/status/553543183045386240?s=03

21 *www.parliament.uk/briefing-papers/sn03228.pdf*

22 http://www.poynter.org/news/mediawire/270781/q-why-is-vice-on-tv-a-why-do-people-rob-banks/

23 http://www.demos.co.uk/files/Truth_-_web.pdf

24 http://www.theguardian.com/commentisfree/2014/jan/06/ezra-klein-leave-washington-post-personal-brand

25 http://www.forbes.com/sites/maxrobins/2013/10/22/billionaires-want-to-pay-millions-for-journalists-who-make-trouble/

26 Ryley, J (2009) Introduction in 20 Years of Breaking News, London, Harper Collins.

27 http://www.skymedia.co.uk/sky-adsmart/

28 http://www.forbes.com/sites/lewisdvorkin/2012/08/06/inside-forbes-the-advertising-trend-that-will-shake-up-100-years-of-journalism/

CHAPTER SEVEN

The View from the UK: The BBC—"Channel Wars, Streaming Wars"

PETER HORROCKS, FORMER DIRECTOR BBC WORLD SERVICE GROUP

I remember the good old days of the TV news channel wars in the UK. It felt like we were battling for victory one remote handset at a time. How the BBC won that war holds some clues for the digital battles now being waged for the aftermath of 24-hour TV news.

Back in 2005, I was tasked by BBC bosses with ensuring that BBC News 24 (now the BBC News channel) overtake Sky News in the UK. The first battleground was not in viewers' homes, but in the newsrooms of London and the political offices of Westminster and Whitehall.

Firstly, I persuaded Kevin Bakhurst, foremost producer and editor of BBC TV news bulletins at that time, to apply for, and then become the controller of News 24. (The fact that the respected Kevin Bakhurst might demean himself by getting involved in "rolling news" in itself sent waves through the BBC newsroom and its correspondents around the world). Kevin and I then embarked on a comprehensive plan—of PR. We toured the news desks of London, pointing out exclusive stories and interviews that the BBC was beginning to break as fast as Sky News.

Journalists and politicians had used Sky News as their default, always-on news source ever since it had launched in 1989. The BBC's belated launch of News 24 in 1997 didn't shift their habits. The initial News 24 was bedevilled with initial production errors that led to its being the butt of jokes from journalists. Its reputation in 2005 was still low, despite enormous improvements in its quality, largely under the tireless leadership of Rachel Attwell.

What was needed was to match opinion to the reality of the improved BBC product. And key to that was shoe leather from Kevin and me, persuading news-room editors and the custodians of the TV remote in ministerial waiting rooms to turn over to the BBC and sample the channel. And we needed to create a noise, so we appointed new editors. We lured the BBC's main TV news presenter Huw Edwards to present nightly at 5 p.m. on News 24, as well as fronting the Ten O'clock News and launched "a war of words," as the papers helpfully dubbed it, on Sky News.

This was the fun bit (and Sky to their credit, enjoyed it too). Every time BARB audience research figures showed BBC News 24 creeping ahead ("BBC News nosed ahead last month in the vital 11 a.m. channel slot") or when we broke a story ahead of Sky, we would press release it with a fanfare.

But of course Sky would retaliate with their equivalent of the neutron bomb—the breaking news banner. Sky News took to labelling almost everything (except the weather, or maybe also the weather) as breaking news. Any predictable diary event would do: "Breaking: Blair begins PMQs." In turn, we had our own counter-attack. Cardiff University published a helpful "authoritative analysis" of the channels' criteria for breaking news and who was fastest. This was sufficiently positive about the BBC's previously slow performance and Sky's highly flexible definition of breaking news to allow us to be able to describe Sky's breaking news policy as "simply a marketing trick." Heady days.

So why am I wandering down memory lane for an essay on the future of TV news channels? The focus then on achieving reputational recognition was as important to the BBC as the accompanying genuine improvements in editorial quality. We had backing from the very top. (The BBC even relaxed its previous strictures about using news presenters in dramas when we pointed out how often Sky News was appearing in BBC1 contemporary dramas as they had no such in-hibitions on dramatising news.)

It mattered hugely to the BBC that it shed the damaging baggage of the stumbling early launch of News 24. For the BBC, news is first and foremost and it could not miss out on the news format of the moment. And, despite advances on other platforms, the TV news channel in the corner of the office or the hotel bar is still a formidable brand indicator for any media organisation.

Over the last six years, I was responsible for the BBC World Service Group, including BBC World News, the BBC's international 24-hour TV channel. In the global market, the importance of that presence in airports and offices throughout the world is enormous. The music, the spinning logos, the sweeping shots over the studio, are power symbols that few want to give up.

So media organisations are not going to give up on the brand building (and national reputation-building) impact of TV channels any time soon. But there is another lesson I would draw from my experience of the battle when BBC News took on and beat Sky News. If you look at the audience figures of the time when News 24 gradually overhauled and then went well ahead of Sky, there is one stand out reason for it.

Much as I would like to put it down to Kevin's brilliant editorial leadership, or Huw's presenting, or the stories leading correspondents like Nick Robinson or Robert Peston broke first on the channel, I know that's not the real reason.

If you plot the increase in News 24 against the sales in that period of "Freeview" Digital Terrestrial Television boxes, there is an uncanny correlation. Put simply, every time a retired guy came home from the High Street with a cheap Freeview box under his arm, that was another potential long-term BBC News 24 viewer.

And I draw from that a clear lesson for the debate this book is about—when a new technology or platform emerges that provides news in a more relevant, convenient and cost-effective fashion, that platform will win out. The brand that is on that platform will thereby win too, as Sky News did first on satellite TV and as the BBC did when its platform, Freeview, came to dominate in the UK.

So the coming battle for the future of 24-hour TV news channels is one between the heart, that feels the importance of the splendour and drama of a TV news channel in its pomp, and the head, saying that news can be most easily consumed on the convenient and functional form of a smartphone. But where is the smartphone's grandeur? For news organisations that are potentially as strong in digital news as they are in rolling TV news (and there are vanishingly few of those), there are some uncomfortable choices to be made, or some very smart thinking to be done that avoids stark decisions needing to be made.

When one of the distinguished editors of this book (and my former boss), Richard Sambrook, first produced his "TV news channels are dead" thesis, I remember the magnificent Paul Mason marching over to see me. The then Newsnight economics editor waved a *Guardian* triumphantly in the air and asked "Have you seen what Sambrook is saying? So, just close the channels down. You'd save a fortune. There's the future of journalism. Digital only." Paul is probably correct for the kind of journalism he believes in—trenchant, diverse, in depth. The average news channel mixed schedule is unlikely to contain the specificity of content that he and his classy kind stand for. The TV news channel has to do too many things—being all things to all news people—while digital services can meet specific needs according to users and circumstances.

Let's consider the various roles a news channel fulfils and then move on to consider which of those are still required, and how they can be delivered, in a fully digital era. There are three broad functions TV news channels play for the audience.

1. What is going on right now around the world? (The desire to witness/ participate in the daily drama of global events.)

2. What has been happening in the past 24 hours or so? What do I need to know and is what I am hearing right or wrong? (The desire to be well informed and up to date.)
3. Making sense of what has been happening. (The desire to learn, reflect and be enlightened, especially through hearing others' views.)

Each of these roles is challenged by digital news, but to different extents. In "what is happening now," well-resourced TV channels can still have an edge over digital. When something literally unfolds before our eyes, channels can perform an invaluable function. (Examples might be the Mumbai hotel siege, the attack on the Sydney cafe, and the assault on the Paris Hebdo attackers.)

But even with these right-in-front-of your-eyes events, digital intrudes. When I watched the culmination of the Paris attacks, I was both monitoring TV and keeping watch across social media. Despite the TV channels handling events speedily and smoothly, because some events happened beyond the sights of the cameras, information inevitably appeared piecemeal, through a variety of sources, most of which appeared on a social media feed before they got to the screen, no matter how fast the channels checked and processed information. How often does one now find oneself shouting to the TV screen, literally or metaphorically, "Twitter told me that ten minutes ago"?

In the second category, the look back over the day's events, TV news channels can still benefit from inheriting quality TV packages—the compendious items that pull together and make sense of the day's material on any given story. But if the stories of the day don't lend themselves to TV package treatment, or if that day's stories are simply dull, a linear TV channel is a less satisfactory experience for catch-up than digital.

In the third role, explanation, analysis, reportage and debate, long-form TV can still grip and illuminate, but inevitably it is patchy. The most vibrant timely discussions can be as important as the news itself. But too often, time spent watching a sterile TV discussion might be better spent reading or watching short video clips of the world's most interesting speakers, provided by one's social media stream. Of course, much of that video material might be originated by high-quality, topical TV-channel production. The question is whether the primary channel should be giving over significant amounts of time to long-form documentary and discussion when that takes up valuable screen time that detracts from more topical material.

So all three of the audience-functions are being eroded, with increasing intensity through the three categories. In other words "right now" is less threatened than long form.

But the funders of TV news channels don't run them simply for the immediate news needs of their TV viewers. The bosses of news channels (and the bosses' bosses) provide them increasingly for reasons of power and promotion. They want to feel the prestige of their news and their brand, displayed on the worlds' notice boards—those airport, hotel and office public screens.

Curiously, despite the technological and audience shifts to digital, the desire to build TV news channels shows no signs of abating. Just look at the extraordinary investment by the Chinese into CCTV in many languages, running the risk of services like BBC World News and BBC World Service being outgunned. Other global examples: the Kremlin's investment in RT (Russia Today) which, after an interesting start as a provocative leftish alternative channel, has now turned into largely a Putin mouthpiece, at least on stories of relevance to Russia. Al Jazeera poured extraordinary sums into Al Jazeera America for scant audience and commercial return. AJA was clearly and fundamentally a soft power play. And, not to be outdone by Putin, Erdogan is planning a Turkish-run and funded global TV news channel. Whether the Turkish leader will get value for money from his proposed channel is another thing entirely. One wise old bird in the channels game quipped that it might be cheaper for the Turkish authorities to offer each potential audience member a free holiday to Turkey than to run the channel.

That's just a snapshot of global news channel developments. In individual country markets, channels continue to grow, or at least there is little sign of scaling back. Take India, with its dozens of news channels. Any politician and business leader who wants to get noticed wants to own a channel. In Africa, local TV news channels are proliferating.

And in advanced markets, channel numbers are growing or at least not declining. Despite the audience and economic challenges in the US, there has been no loss of major news channels. In the UK, since the sad demise of the ITV news channel there have been no signs of the BBC and Sky reducing their fundamental commitment to the channels, or at least the TV real estate they represent. (It still matters hugely to the BBC and Sky whose news is seen in key public spaces.)

So, we have a paradox here. At a time when technological change is reducing the utility of news channels for audiences, their perceived value to their creators is increasing. Some of this is due to natural time lags—maybe world leaders who want political virility symbols haven't noticed the digital news revolution. But I would argue that this dysfunction between promotional value and audience utility will continue. The battle of the airport lounges will persist. It really would be significant if in Africa the Chinese, or indeed, a locally run pan-Africa TV channel were to take the default place as a public news source now played by CNN or the BBC. And, crucially, by withdrawing from the TV battle, brands would thereby weaken their brands in digital.

Indeed, in a converged, multiplatform market with broadcast news organisations competing with digital newspapers, digital natives and platform aggregators, the advantage of video, both on demand and live, is considerable. The barriers to entry for the effective presentation of video and live video are higher than for most other parts of the news value chain. (For instance, HuffPost Live is a worthy effort, but has little of the distinctiveness of its text-based mother ship.) The irony is that live video may be watched less than before on TV, but its promotional value across screens in a digital environment will only increase.

How can the cost and production contradictions between utility and promotion be resolved? I believe the answer is to treat the channel as a promotional product for the audience as well as the owner/publisher.

I can hear my many TV news presenter and producer friends recoil at the thought they might be merely brand or soft power promotional functionaries. Rather, the role of the channel should be as an interface, an entry point or navigational device to the best that a news organisation has to offer, both linear and non-linear. The advantages an always-available live stream has as a promotional/sign-posting tool would be considerable on all screen formats, not just conventional TV screens. The stream should be at the heart of an "omnichannel" approach.

To carry out this function in a way that works for viewers, TV channels will need to become comfortable with acknowledging, on air, that they are no longer primary news destinations but now play an introductory or secondary role. This will be uncomfortable for many, but unless channels understand how the digital world has changed the way they are used by the audience, they will become increasingly anachronistic.

The shift towards channel-as-interface requires a fundamental overhaul of channels' content and technological proposition. As the content proposition will depend on the technology available in each market, we need to assess platform first.

In advanced countries with plentiful bandwidth, high levels of device ownership and growing penetration of IPTV, there are exciting possibilities around the corner. We have seen a variety of clever approaches to delivering on-demand video to various screen sizes, for example in the UK, the BBC has produced the BBC News video app for connected TVs.

What has not yet happened is the full integration between a linear TV channel and on-demand video or text experience. That is because in most news organisations, the leadership of TV and digital has been separate and there has not been sufficient mutual understanding between the relevant platform editorial and technical teams. And, to be fair, the technology has only recently become available to deliver these experiences. As an ordinary news consumer now, rather than a manager who is responsible for devising and producing the solutions, these are the use-cases that I would like news organisations to fulfil. They could be delivered either through the primary (usually TV) screen or via a second screen device.

For users with connected IPTV (Internet Protocol TV), the linear TV channel could deliver a comprehensive video menu showcasing the best of that day's video packages. A simple button press would take the user into the selected package.

After viewing that, the user could go back to "live TV" or to the rest of the best-of-video menu. Of course the linear channel would need to collate and curate the video segments so that they were appealing to a viewer who did not have IPTV or chose not to connect through on-demand, but the ability to display the richness of content the organisation has available is surely a positive for all audiences.

Using the concept of channel-as-interface would also help to address the breaking news "Twitter has already told me that" problem. In any breaking news situation, the production team could rapidly create, via automatic hashtag or editorial curation, a stream of relevant tweets from other credible news sources. These could be displayed graphically on screen as an unverified stream. Rather than the on-air presenter in a breaking news situation appearing, as is usually the case now, to be ignoring social media, by putting that feed on screen it would force the presenter to acknowledge the contingent information and explain how the channel itself is assessing the information. That approach feels awkward and un-intuitive now. Some would even argue it breaches editorial principles. But the current approach simply drives viewers away to find news for themselves.

How could such an approach be assisted by primary or secondary screen interfaces? On an IPTV, if, for sake of argument, France 24 was reporting that the siege had ended peacefully, that should appear in a competitor's on-screen Twitter feed selection. On IPTV, the viewer would also have the ability to click through to the live stream of France 24. Traditionalists would be fearful that such an approach invites the audience to click away. Instead, I would argue that showing you have the confidence to acknowledge that others sometimes have information ahead of your channel would mean that most viewers would stick with you, being reassured by the confidence that they will be getting the best news, curated from all sources.

How would such an approach work on a "second screen"? There are two kinds of second screen possibilities. The first is "unconnected" and non-technical. It is simply acknowledging that most viewers now have their own personal device and that helping them to find relevant other news is a service. For instance during the French crisis a presenter could simply have said "We know that lots of unverified information is available on social media. We recommend the hashtag #parissiege or, if you want to see what we are attempting to verify follow #bbcchecking."

But a connected second screen offers many more possibilities. By connected second screen I mean a handheld device that is "digitally tethered" to the primary screen so that if the TV offers some extra editorial content, the press of a TV remote button would throw the relevant content to the device. In that breaking news situation, the user could select #parissiege or #bbcchecking and that feed could be automatically subscribed and presented on the connected second screen. Incidentally, we may need to revise the terminology of primary and secondary screens, at least in a news context. The fact that TV is primary and the personal device is secondary is redolent of the traditional platform hierarchies. The approach outlined in this chapter will only be achievable when news organisations think of the personal device as the primary news consumption technology.

The reform to the channel's content proposition is not just about incorporating social media sources and providing click-through pathways to other content. It requires a fundamental reworking of the linear TV mindset. Think of the phrases channel presenters habitually use: "coming up in the next half hour," "later in the programme," "later we will be bringing you Bill's special report" (Viewer: if it is so good, why can't I see it now?).

Changing those traditional formulations implies an acknowledgement of the declining importance of the TV platform, which feels profoundly destabilising for TV-trained people. I know the feeling. I was a TV person. But if TV channels aren't prepared to acknowledge their complementary function with digital platforms and instead bury their heads in the sand, they will hasten their end, rather than evolving towards an alternative role that can endure.

I have examined the challenges to channels without any reference to economics and costs. In fact, because of the promotional and soft power capabilities of channels, vanishingly few of them are required to be genuinely financially self-sufficient—i.e. that the subscription and advertising revenues attributable to the channel need to cover the full costs of the channel. However commercial revenues do play a very significant role in some of the best-known channels (CNN and BBC World News) and it is worth looking at the economics of channels now and the costs of transition to this proposed model.

Putting it simply, all revenues for TV news channels are under threat. Subscription or affiliate feed revenues have been relatively stable, but as digital news consumption increases, viewership is declining. Digital has a further undermining effect. The fact that higher proportions of viewers have digital boxes with a return path means that the cable and satellite systems providers now know how many, or few, viewers are watching and that data inevitably puts downward pressure on affiliate revenues.

On the advertising side of the house, there is a similar picture. Some years ago digital advertising was sold as a nice-to-have add-on to mainstream TV news campaigns. More recently, it has moved to digital first, with TV sometimes reluctantly bought alongside digital. Soon, in many markets it will be digital only.

Of course, digital revenues are generally less fat than TV revenues were, so commercial teams have been understandably desperate to hang on to TV revenues and some of that lack of converged thinking has held up the content and technological convergence that is required editorially. But now the writing is on the wall for TV news commercially, it is essential that all news businesses take a converged approach to the profit-and-loss account as well as the editorial proposition. The business must realise that investing in convergence and the concept of TV-as-interface makes long-term business as well as audience sense.

Nevertheless the shift to converged digital revenues is likely to mean that the economics of multiplatform video news production will get tougher. That will force a focus on costs and the wider value of the channels to their parent organisations.

Converged platform and editorial offers should generate significant cost reductions. Although there will continue to be slightly varying content requirements between platforms, the need to have real breaking news content on all platforms (with much less speculative blather from live correspondents) should simplify acquisition and ingest costs. Making the TV proposition a simpler product that is pointing viewers to digitally available content means a likely reduction in the costs of presenters and elaborate studios. In this model, all graphics would be created for multiplatform use, rather than separately for TV and online, etc. Crucially, that level of content integration can only be achieved by integrated and multi-skilled editorial teams. There are some encouraging signs that multiplatform leadership is being put in place, belatedly, in some news organisations. But there are still not enough genuinely multi-skilled editorial staff in the newsrooms of the major news organisations. If necessary, new, younger multi-skilled staff will be required.

Technology can assist in the creation of TV-like experiences. Given that so much of a channel, except in a breaking news environment, is fundamentally a repetitive curation of existing content, digital techniques can make them more cost effective.

In order to create lower-cost models, it is often not wise to start conceptually with an existing much-loved proposition and try to radically reduce its costs. Instead, try to build something new. Before I left BBC World Service, I made a speech in Delhi where I saw an increasing role for the BBC's international news services and, in particular, its language services. I envisaged the World Service making content in 50 languages (up from the current 28) and that it would have continuous video streams and channels in roughly half of those languages. Such a dramatically increased footprint could be achieved by multilingual multimedia-content production units supported by a radical approach to costs and work flow. The World Service board tasked tech and editorial teams to create a 24-hour video stream that could be staffed, at any one time, by two people—one editorial and one production. Using shared content, digital playlists, pre-recorded presenter links, an element of automatic processes in translation and cutting cloth according to resource, the team is making great progress towards the aim of a two-person channel. I look forward to seeing the video channel "BBC Burmese 24" before long.

Some of these techniques that are being piloted with new TV-like channels (for instance, also AJ+, HuffPo Live) will help existing "main" channels to cut costs. But the challenging economics I described earlier and the need for the main channels to convey visual clout to fulfil their brand function will mean that there will still be a financial gap.

For the propaganda/plaything news channels that have a political role (China, Russia, Turkey globally, Pakistan and India locally) the economics are not going to deter them. Indeed, the economics provides an opportunity for the soft power channels.

For news channels that sit within larger media conglomerates (CNN, MSNBC, Bloomberg, Fox, BBC, Sky, etc.), corporate leaders will need to decide how important a news channel is to the overall brand.

I agree with Richard Sambrook's analysis that there is declining direct news utility in the channels as they exist. But that doesn't mean the death of TV news channels. By turning them into exciting digital gateways that are set up primarily to deliver the news as it happening now, they can have a watchable function in their own right and as a navigation device to digital news. And, not only that, these "channels" (available on all devices not just TV screens) will become the entry point and point of departure for any multigenre media organisation.

Digital news, fragmented and often cold and non-human, does not have the power to convey the brand, to demonstrate warmth and to provide the front door to what a great multipurpose media organisation can offer. That's why "channels" deserve financial support from their mother ships, as classic loss leaders. It is why the leaders of the news industry need to reinvent not discard the TV experience so that it becomes the essential starting point for audiences to access news content that is richer, more varied and more relevant than ever before.

CHAPTER EIGHT

The View from the Middle East: Al Jazeera

IBRAHIM HELAL, DIRECTOR OF NEWS, AL JAZEERA

As the world is debating the relevance of 24-hour news channels, we in the Middle East are totally engrossed in creating more news platforms. Unfortunately, we are doing so for the wrong reasons.

As a TV journalist for the last 25 years, I should be happy to see more chances to work and more competition, but actually I am not optimistic at all with the current professional scene. Since the 9/11 attacks, all news organisations are living in the age of immediate reporting with all its accompanying professional problems. No one has managed to cope with or recover from the symptoms of this transition. I have many concerns about the current health of TV news around the world.

But we in the Middle East have been suffering from an even more problematic transition. Since the region became more central to the "global war against terrorism," more money and political influence have been pumped into the media scene. After Al-Qaeda and similar organisations managed to use media tools to recruit and circulate their messages, the governments and businessmen in the region decided to follow suit and use media. Al Jazeera was the only one created before 9/11 and managed to dominate the scene, which increased reasons to create more media platforms. Saudi Arabia, Saudi businessmen, UAE, and Kuwait are all wealthier than Qatar, so they felt pushed into a corner when the Doha-based channel Al Jazeera dominated the scene. Also Egypt, Iraq, Turkey, Iran and even other superpowers like the US, Russia, China, UK and France felt ashamed to sit silent watching the Qatar-owned media network growing fast.

From 2003 and until today, the Middle East has been witnessing what many called the "mushrooming media era." Tens or maybe hundreds of news and general TV channels were created, from the more professional ones like BBC Arabic, Alarabiya and Sky News Arabia, to the smaller scale ones like France 24, Iran-based Alaalam TV, TRT Arabic in Turkey, or CCTV Arabic in China.

I believe it was a trap for many of these channels as they are hardly watched and have little hope to seriously compete for the hearts and minds of the millions of desperate viewers in this volatile region. The latest studies done by IPSOS and Sigma, the most trusted survey companies in the region, showed that Al Jazeera is still leading the viewership lists far ahead of its next competitor. Al Jazeera & Alarabyia together occupy more than 60% of the Arabic TV news viewers market, the rest is fragmented into tiny percentages to tens of irrelevant "news organisations."[57]

Nevertheless, the future of our business in our region is not that bright, and actually the current status is not promising either. The entire market of TV news viewers is shrinking rapidly due to many factors and the most dangerous is that the credibility of all news organisations in the Middle East is declining even faster.

THE POLITICAL ATMOSPHERE IN THE MIDDLE EAST IMPACTS NEWS PRODUCTION

I can confidently say that most of news and current affairs content provided today in the Middle East is politically biased. After years of trial and error, objective editorial values failed to become integrated in the region's news machine.

The notion of accurate, fair and impartial reporting did not even exist in the mindset of journalists or media managers before 1994 when the BBC started its first Arabic TV, which then closed down and fuelled the launch of Al Jazeera two years later. Gradually, objective reporting became necessary for marketing and competitive reasons. But since the US invasion of Iraq in 2003, the political division in the Middle East became stronger than any goodwill; nothing could have stopped the main countries that fund the media from interfering in the news agenda.

The more tension we had in the region, the more political interference. For example, the Saudi-funded Alarabyia portrayed the 2009 uprising in Iran as if it was the end of the Islamic Revolution regime; the news and analysis provided by this channel about Iran was and still is very inaccurate and full of negative connotations. The same thing could be said about Al-Manar (the Hezbollah-funded Lebanese channel); every report about Iran, Syria, Saudi, or Israel is full of opinion rather than information.

The problem is that those who work on these channels, and also their millions of viewers, *do not mind* if they are inaccurate and politically motivated for what they believe are "the right reasons"!

There are multiple examples of how the political orientation of the country behind the media organisation impacts the editorial position. Although the Al Jazeera Network has always been the most independent one from the funder's policies, still there were some examples when Al Jazeera identified with Qatar's foreign policies—although in a smart way and by deliberate omission of facts and opinions rather than inaccurate reporting.

When the so-called Arab Spring started and the uprisings in Tunisia, Egypt, Yemen, Libya and then Syria took place, the political division in the entire Arab world became unprecedented. The coverage of the events—and civil wars—that happened in these countries became the manifest face of a cold war between regional powers that support and oppose the democratic change in regimes through revolutionary actions. All media funded by the conservative countries like Saudi Arabia, UAE and ironically Iran took a clear stance against the Arab Spring, while all media funded or supported by Qatar and Turkey kept celebrating the "Arab Revolutions."

The content of news and current affairs coverage by these channels of the events of the last four years in the Arab Spring countries was politically oriented. And I believe this was so obvious that millions in the audiences were aware and adapted by either stopping to follow all regional media or by following more than one channel to confirm information and get alternative views.

Now, since the Arab Spring apparently failed in all countries, the media coverage of the aftermath is still politically oriented and even journalists who move from one media organisation to another do so based on their personal political opinions and how much they match the media organisation's editorial position.

I have to mention another essential element, which has contributed to making news production more politically oriented in the Middle East. The cost of producing relevant and competing news content in this volatile region became higher than any possible commercial outcome. And consequently when a media organisation loses money, more political money comes to save and maintain the production, eventually interfering in, if not dominating, the editorial line.

EDITORIAL CHALLENGES WE FACE

News coverage has become more complicated, faster and more difficult to manage editorially—globally as well as in the Middle East.

Complexity of many current news stories

Let's take the war in Iraq between ISIL and the Iraqi government supported by some Western and regional countries. How can any media organisation explain such a complicated story adequately? I have been watching many Western channels and big names in TV news to see how they cover this story since June 2014, and I am generally disappointed with the lack of knowledge and shallow analysis. Yet, I cannot blame the editors of these newsrooms as it is a really complicated situation, full of lies, propaganda and editorial traps.

Another example is Libya. It is a story of global interest, but how can any TV newsroom explain in the available TV formats the division between a government in the east of the country supported by the parliament (which was later dissolved by the high court!) and another government in the capital supported by the elected national council (which extended itself after its mandate expired!)?

In such complicated stories there is a fine line between what is "fair and legitimate" and what is "true and accurate." Covering such stories based on our usual editorial values is not enough as it will, at its best, reflect the accurate facts while it may neglect the fair realties. Reporting the vicious actions of ISIL, as demonstrated in its video releases, is not enough to reflect the whole picture. Good journalism would also report the vicious sectarian actions committed by the Shiite militias supported by the Iraqi government and Iran, which indirectly feeds ISIL with more reasons and more supporters.

Another example is when hundreds of reporters were covering the war in East Ukraine and the Crimea peninsula. The complexity and the political affiliation of many Western media drove them to label the pro-Russian forces as "the bad guys" without putting more effort to get to the bottom of the story.

The outcome of this is that with more complicated stories, the news coverage is becoming shallower and misleading, but trying to fully explain the complexity risks losing viewers.

Breaking news and the need for immediate reporting

Due to the available technology and competition, most of news coverage has to be so fast and nearly live. One editor once told me: "News became faster than us… faster than our ability to think, judge and decide carefully what to do!" With live reporting, more editorial mistakes are inevitably made. Unless the media organisation has got a very strong editorial and technical team, it will fall into all kinds of traps and the output will be inaccurate and sometimes laughable.

That results in less accuracy and lower editorial standards. We rush into breaking news, rush into live coverage, and compete to grab hold of events even before they become a full story—sometimes the media is faster than the story itself! So, we all have weak editorial thinking, less depth and, consequently, less fairness.

The amount of available sources and authenticating the information and pictures

In big stories, there are more political interests and many parties want to get their views and their own narrative onto the screen. Thanks to current technology, more sources are available. If we compare the amount of information and material from the Falklands war in 1982 with the current civil war in Syria, we discover that the huge amount of detailed information and video available from Syria is more than any newsroom's ability to handle.

This situation, which I personally faced in Syria, Libya, Yemen, and Egypt, causes two problems: prioritisation and contextualisation. With all the information available, what is right to include and what should we drop? How can the newsroom draw the clear and easy-to-understand picture of events? How can we put the story into context on a daily or even hourly basis?

And we have the growing problem of verification. With new technology, it is easier than ever to fake video and still pictures. So the amount of sources and information is becoming a headache for professional editors who have had to create mechanisms to check the origin of every video or still picture, especially if the source is social media.

Impact of social media

With social media becoming more than "social" and evolving as a primary news source, we—professional newsrooms—are facing many threats that have started to shape the future of our profession: The audience—and often activists—have become reporters. Many activists who managed to accumulate hundreds of thousands of followers are able to say whatever they want in social media without having to respect any editorial values. There is no need to refer to two sources or to present the other opinion. On many occasions, the more regulated, traditional media had to react to stories which originally appeared in social media, where there are no rules or standards.

Many traditional media changed the style of presentation for specific hours of the day in order to compete with new and social media. Examples include Outside Source on BBC World and Stream on Al Jazeera English. In such shows, the traditional media organisations are putting out huge effort and investment to reach a group of their audience which is no longer there. Those who are addicted to social media are not likely to come to the TV screen; they may do when they get older, but when they do, it's probable they will be following established shows like *Newsnight* rather than Outside Source.

Many news organisations are using social media wrongly. Instead of using social media to promote the screen, they use the screen to show how good they are in social media. The ability to promote the content rather than the medium is not an easy job, especially in the Middle East.

Dictatorships are using social media in a dangerous way. For example, the Syrian regime created what it called "the Syrian Electronic Army" in 2011. Its duty, which is publicly stated, is to defame the "enemies of Syria" and spread propaganda to help the regime in its "global war." This "army" managed to spread fake pictures of many of what it considered "enemies" of the Syrian regime, spread false rumours about politicians and media figures and hack many websites of those who oppose the Assad regime. The problem here is, because the media have created a huge respect for social media and talked for the last few years about how much it will take over our jobs, we actually participated in this crime of giving this casual, non-regulated media a lot of credibility. This credibility has been misused by dictators like the Syrian, Iranian and many other regimes and groups like Hezbollah in Lebanon, and even ISIL.

For me, social media is *not* media and is *not* a positive element in the future of news. We must ensure that our efforts to use the social and new communication platforms to support our editorial values.

Repeated stories: Déjà vu

Millions of people—even those who are still loyal to the daily habit of watching the nightly news bulletins—are really fed up with what we present. For them, it is déjà vu. Although for journalists many headlines and developments seem attractive, many of the viewers are not freshly engaged by stories like "car bombs in Baghdad," "Syrian regime used barrel bombs on Aleppo," "More girls taken by Boko Haram in Nigeria," "Another black American killed by police in USA," "Clashes erupting in eastern Ukraine," etc. For millions of viewers, they can be attracted to screens to follow news stories like the earthquake in Nepal (for a few days) and the Airwings plane crash in Germany. They may follow such stories because they are unusual or unpredictable. So what can we do? The answer seems to be nothing! We just try to sell our "boring" product in more creative packages and we, sadly, use the unusual disasters to attract more viewers than those who watch us on normal news days.

Too much drama, blood, and sad detail

Even on those days when we have more unpredictable stories, most likely there is too much blood, death, drama and sad detail. Millions of people, who are already living their own dramatic lives, do not want to get more upset by watching such stories unless they really have to. The question again is what can we do to reduce these sad and dramatic elements in our bulletins? The answer is we can do very little: if we reduce the sad elements more than a certain amount, we risk compromising the factual part of the story and we may end up with shallow content.

Dull information: Meetings, conferences, routine coverage

The dry days suffer not only from the repeated déjà vu stories, but also from the dull stories, like EU meetings, the UN Human Rights Council meeting in Geneva, the 5+1 meetings in Vienna, the Japan-USA summit, etc. For millions of viewers, this news is dull, boring and a waste of time. For us, it is essential to cover and we invest a lot in explaining these events and putting them into the wider context. Again, the question is how can we make them more attractive? The answer sadly is, we can just pray to God to inspire the audiences with more patience to follow up and appreciate our efforts to contextualise these dull stories.

Local news vs. regional and/or international news

With more available sources of news (new and social media), fierce competition, less appetite to invest in serious objective reporting, and a faster speed of news, for many channels regional and international stories have become a luxury. For millions of viewers, there is no real need to follow up events that don't impact on their daily lives directly.

Local stories about elections, taxes, health services and education problems will always come before any story about Yemen, Iraq, Ukraine or even ISIL. What can we do to keep regional and international news alive? It is still possible, but difficult. It's why we adopt a human interest approach—to rely on the human feelings of our beloved audience to want to follow up what happened to their fellow humans elsewhere. Packaging the complex, dull, sad or routine stories in a simple, human way seems to be the main way of getting the audience to follow.

What the audience wants vs. what we should offer

The last point I would like to present before I come to the conclusion is the very basic dilemma we have been facing for decades: what we think we should put in the running orders compared to what the audience really wants.

For decades we managed to design our production based on our own taste, editorial lines, editors' experience, accessibility, budget and technology. Now we have to add more elements. Our relations with audiences became more interactive and we are better able to understand what they really want. The question is to what extent should we go for what the audiences want and where lies our "educational role"? We have been telling the next generations of journalists that our duty is to "inform, educate and entertain" our viewers. Now we are facing the dilemma of whether we should obey the viewers' requests and preferences and to what extent.

It is a tough call. Big organisations with a long history can always enjoy a bigger margin of freedom to "tell" the audiences what they should know. They are still able to decide the running order based on their integral editorial traditions. But we have to always be aware of the "them and we" dilemma.

- We like more details—viewers like fewer details.
- We like general news—viewers like customised news.
- We like context—viewers like simple stories.
- We like the big picture—viewers like to focus on human personal stories.
- We like loyal viewers—but the viewers are now fragmented and tend to watch more than one screen as source of news and current affairs.

WHAT WILL REALLY HAPPEN NEXT?

Taking all these issues and reservations together, what is the future of TV news? In my view, our profession will not die. Actually, it may witness a more promising future. We just need to change two things: our presentation and our expectation of the audience in ways which may be uncomfortable for mainstream news producers.

More available platforms mean less loyal viewers. More technology means more on-demand viewing. With the IPTV technology, news can be demanded like movies. We need to be ready by presenting different formats for different customers.

Viewers have become customers. We need to expect their reaction to our content like customers react to any other product. We should meet our customers' changing appetite for news by being short and to the point. We will also present news in more geographic and topical categories: more specialist news shows about Africa, Asia, Middle East, science, or health rather than a bulletin for everything. We will be more flexible and accept new formats for news production like mobile phone footage or drone cameras. More stories on TV news will be about religion-related conflicts; for many audiences in the Middle East and globally, religion is now one of the most attractive elements in news stories and it will stay like that for years to come.

More graphic images will be accepted on TV news as the same scenes are available on social media and attracting audiences. With regulatory bodies like Ofcom, it would be difficult to change the traditional rules immediately, but outside the borders of these regulations in other countries, TV news will recognise it has to show more graphic scenes to meet audience expectations.

These changing viewer expectations may mean shallower news with shorter packages and fewer background stories. We will have fewer viewers to the traditional TV news programmes and more followers to the non-regulated sources of information. We will have what we can call "the occasional viewers" when big crises happen like the earthquake in Nepal or the clashes in Baltimore in the US. All our TV news content will be available for online customers. We will see more political money in Middle East TV news production, which will mean less objectivity and more biased reporting.

In general, our profession is not regulated anymore; many amateurs are in the business and many amateurish TV news platforms are being watched for reasons like good presentation, pretty female faces, political affiliations, and religious or geographic relevance. Traditional objective news values do not draw audiences in the way they used to. More discussion is urgently needed among professionals in our business in order to maintain the minimum amount of values and standards before they get completely diluted.

NOTE

1 http://www.aljazeera.com/pressoffice/2013/05/201352291421900835.html (Accessed July 30th 2015)

CHAPTER NINE

View of the News Agencies: The Struggle for Renewal and Renaissance

DAVID SCHLESINGER, FORMER EDITOR-IN-CHIEF, THOMSON-REUTERS

The era of the news agency dinosaurs ended on the night of January 16–17, 1991; the comet that changed their world forever was the first live broadcast of the start of a war. In the flash of a cruise missile over Baghdad, in the light of a bomb blast marking the start of the first Gulf War, the now shaky foundation of the news agencies' dominance was exposed.

CNN's live coverage of the war brought it directly into the audience's living rooms. The middlemen were dis-intermediated. Worse yet, the suppliers became dependent on their customer!

The news agencies—at the time notably Worldwide Television News, Visnews, Associated Press (AP), Agence France-Press (AFP) and Reuters[1]—supplied the world's media with text, photos, and video as they had in various guises since the modern news agency era began with the founding of Agence Havas (the forerunner to AFP) in 1835.[2] That night, however, the tables were turned. Scores of breaking news alerts on the war were written not from Baghdad itself, but from New York by subeditors watching television and citing CNN's live coverage. Video clips from that live coverage beat agency video packages by substantial margins.

As I sat in the Reuters Beijing bureau watching the news come over our slow teleprinter—until that moment as cutting edge as I needed—it was clear that we had been leapfrogged: we, the news agencies, were no longer the world media's primary source for the news. Others now had the technology, the platform, the capability, and the audience to chip away at our authority.

That it was a crisis was clear: the internal message wires burned with heated debates, embarrassing to recall now, about whether it was even "right" to be using the live coverage from television to source the news instead of waiting for reports from "our own" journalists.

Of course, the agencies were not snuffed out that evening; well into the current century they still have large staffs and equally large turnovers, if not always profit.[3] Change, however, has come slowly, challenging their long-term survival. Yet, almost every single factor that brought the agencies to life in the 19th century is no longer relevant in the 21st.

The agencies were born as cost-savers that used their ubiquity and their technology to give their customers what they could not provide for themselves: an immediate presence wherever news was taking place. Where it would have been prohibitively expensive for every newspaper, or later broadcaster, to have its own correspondent in every world hot spot, the agency could deliver. Whether by sharing news costs in a cooperative arrangement or by charging competitive market rates as a commercial venture, the agencies allowed their customers to leverage tremendously their own relatively small staffs.

Technology was key to this endeavor. In the early days when text was all they did, the agencies used telegraph, pigeons, steamer ships, and trains to take dispatches in from correspondents and distribute them to customers. A single correspondent's writing went into a central newsroom and then fanned out to thousands of media outlets around the world. The agencies invested in proprietary systems of gathering and delivery, operating vast private communications networks spreading the cost among all their customers.

When the age of video dawned, the principle of being the cost-effective news wholesaler became ever more important and followed precisely the same model. The early days of video were extraordinarily expensive. The equipment cost money and required significant capital investment. The crew to operate the equipment cost money—not unusual was the crew that had cameraman, producer, soundman and possibly even a lighting man. The excess baggage fees to get the equipment from base to news hotspot cost money. Finally, the price of transmission cost significant money—even the agencies often "pouched" tapes (sending them from the field to headquarters by courier) rather than paying prohibitive fees for a few minutes of satellite time.

Broadcasters took full advantage of the agencies' reach—many subscribing to the services from both AP and Reuters to ensure that they didn't miss an exclusive. Customers used the raw footage provided by the agencies with gusto. Agency video had no logo, so broadcasters could seamlessly superimpose their own, making it look like they were the ubiquitous ones. The agencies provided the footage with natural sound and usually no narration, so customers could narrate in their own language and use whatever sound bites they desired.

It was the perfect wholesaler-retailer relationship. The agencies never thought about going direct to the consumer. The broadcasters never worried about having to provide their own coverage from everywhere. The agencies were comfortable in the knowledge that their 24/7 provision of text, photographs, and video could fill whatever space in the news budget clients' editors needed—and that without them their customers would be at a significant disadvantage. The broadcasters were comfortable knowing they could have a full supply of professionally packaged material that they could slice, dice, and process into something that looked like— and in fact served as—their own.

The agencies were the witnesses to the world—they provided news that others couldn't because they weren't there or couldn't afford to be there. The agencies were the wallpaper of the world—they provided vital background shots that others used to establish their own news credentials and credibility. The agencies were the logistics command for the world's media—they efficiently and uncomplainingly took care of the difficult details at a time when covering the news meant wrangling huge loads of heavy equipment, large and seemingly impossible distances and then maneuvering tricky and expensive technology to get the resulting shots out.

However, it is important to note that while it was the exclusives and the dangerous war shots that defined the agencies' bravado and mystique, the vast bulk of the coverage was not that—it was (and to a large part remains) the humdrum, workaday, quotidian diet of grip-and-grins at diplomatic meetings; it was the arrival ceremonies for yet another head of state or, worse yet, provincial factotum whose hometown station demanded coverage. It was the regional sports championship more often than the Olympics; it was the set piece of a diarized event more often than the spontaneous uprising. In other words, much of the coverage was and remains ripe for disruption, for substitution, and for commoditization.

Just as the agencies entered a crisis period, revolution also invaded their customers, the broadcasters. The golden age of the agencies coincided with the golden age of broadcasters, a time when they enjoyed near-monopoly status. If they were commercial broadcasters, they luxuriated in large profit margins; if they were public broadcasters, they savored budgets that were comfortable at a minimum. That easy broadcast oligopoly was destroyed, first by cable providers, which put hundreds of channels on offer (few of which were news), and then by the internet, which expanded choice to the nearly infinite.

Choice was wonderful for the consumer of news; choice was terrible for the providers of news, which found it ever more difficult to maintain the profits and the budgets they had long basked in. For some broadcasters, one choice they found easy to make was to go (or to threaten to go) from subscribing to both major agencies (AP Television and Reuters Television) to buying only one, further forcing salespeople to compete on price instead of on content and reputation.

With the profusion of choice, too, came a diminishment in the agencies' value claim—with the audience able to access myriad sources of video content, the "wall paper" of coverage was now clearly available everywhere. Pure facts—what the agencies specialized in—were commoditized. Having an agency feed no longer bolstered a local channel's claim to ubiquity in the same way it once did.

Simultaneous with this change on the customers' side, the agencies were forced to contend with their own huge—and to them both liberating and destructive—shifts in technology, ultimately affecting every key part of the news value chain: news gathering, production, and dissemination. Video cameras, which had been bulky, heavy, and clumsy, shrunk to the point where even a handheld smart phone could take credible, high-definition video. Just slightly larger than that phone were new professional-grade video cameras that gave journalists freedom to shoot broadcast-quality clips like never before. Crucially, unlike the massive gear of years past that required highly trained professionals to wrangle, the new cameras of today could be used to a high standard by anyone after only minimal training. Desktop computer cameras could beam interviews to anyone wanting to retrieve them, turning news subjects themselves into news producers.

News production, too, shifted—video news editing is no longer the preserve of the trained professional, but available to all, without the need for huge expensive dedicated equipment and extensive experience. All that's needed is a computer, a program and the will.

News dissemination has been the most revolutionized. YouTube led the charge, making everyone in the world a potential video impresario. Social media now alerts the global audience of hundreds of millions to new videos within seconds of posting. Gone are the days when a video clip had to be sent by courier, landing days later, or sent by satellite, costing many hundreds of dollars, before being transmitted on a dedicated communications network to be broadcast at a set time to an audience passively waiting for a scheduled news program.

The agencies, of course, have also benefited from this technological change. Large news services like the AP and Reuters can start to break down the walls between text and video reporters, training new generations of multimedia journalists comfortable working across boundaries that had been absolute, thus expanding their reach while reducing costs. One lone person with a camera and tripod can now cover news events—no more the team of highly paid professionals who had to move together as a pack. Live broadcasts of breaking news events are now routine parts of the agencies' offerings, making what had been CNN's Gulf War breakthrough ordinary. The sheer number of stories able to be produced and moved daily has expanded tremendously. If only the value had expanded at the same rate.

However, the benefits of the technological change have been available to all—agency, broadcaster, and public alike—chipping away massively both at the broadcasters' privileged place in the news world and at the agencies' own reason for being.

New news organizations have sprung up quickly and simply, without the huge investment in capital or people once required. Able to narrowly define key demographic slices and specialized audience groups, they attract digital advertisers eager to hone their message and reach a specific target, something broadcasters by their very name are ill suited for. There is less value today in being general and broad than in being focused and targeted. The value comes from delivering specific content to specific people—and that can be defined geographically, demographically or in terms of content or subject. A successful niche player of today can provide video news aimed at 18–24 year olds, or targeted at Brazilian women, or produced for lovers of nature programs, and have very little need for a vast, general agency product.

The agencies never understood the end consumers of the news, because the agencies had always been the middlemen. The agencies never had deep supplies of specialist niche content to deliver because they had always run on a "we cover everything in the world," broad but fundamentally shallow, model. Moreover, with technological ability and innovation as key differentiators in the new world, the old-style agencies seem at a severe disadvantage to new startups, as cutting-edge science has never been one of their core capabilities.

More importantly, however, the technological revolution has completely democratized the news process, turning every person who had been merely passively in the audience into a potentially active journalist. And what journalists—what chroniclers of the events around them—they are!

With the gatekeepers removed, every one of us can watch, with a few searches on Google or strategic follows on Twitter, dramatic or quotidian world events, both from next door and from halfway around the world from an eyewitness's perspective. Demonstrations from the point of view of a participant or a bystander, police baton charges from the point of view of a target, a human stampede filmed from inside the crowd, a bull goring a tourist, a full-length city council meeting, a neighborhood shopping mall opening—these videos and more are available and stay available.

From being the ephemeral currency that bought broadcasters legitimacy and status by its very scarcity, video footage is now the permanent, ubiquitous reminder that the old order is no more.

Of course, the agencies now monitor and distribute selected user-generated content along with their professionally produced material, adding value by their selection of clips, verification of authenticity and negotiation of rights. Selection is key because there is simply too much choice; with everyone a potential videographer, the sheer bulk of video is impossible to navigate comfortably. Verification is a necessity; the very tools that make it possible for anyone to be a witness to history also allow anyone to distort and manipulate events.[4] Rights negotiation is a fundamental task to ensure that the profit from effort is fairly dispersed.

All these worthy and vital tasks, however, are no longer the sacred preserve of the storied, historic, and entrenched agency. The democratic revolution of technology, access, and platform mean that new companies and even individuals can easily establish themselves as players in the game, and with their efforts earn status, credibility, and legitimacy. Technological algorithms can add metadata and then use it to select video for narrowly targeted user groups. Geolocation tagging and close digital examination including image and pattern recognition can provide verification. Rights management can be handled by specialized platforms, completely bypassing the traditional agencies.

From holding all the cards, the agencies are left with a few unique assets: historic brands; large professional staffs and networks of reliable stringers; accreditation to governments and sporting bodies; and codified editorial standards. All of these are vulnerable, however. To give one example, U.S. President Barack Obama in 2015 gave news-making video interviews to three new news organizations; two of them founded just the previous year, smashing both the brand and access barriers.[5]

What the agencies can still deliver is a quality mark, anointing content delivered over their systems with the mystique that it is of top production quality, journalistic value or importance. In an era of broadcasters' shrinking budgets and shrinking demand for generalist, broad, international news coverage, it is hard to assess how much that badge of honor is actually worth, however. At a time when consumers themselves prefer to make up their own minds about things after social media discussion with peers around the world, it is difficult to assess whether maintaining a hierarchical position of authority is an asset or actually a liability.

Of course, neither major video agency is simply sitting back and waiting for a slow death. Both ramped up efforts to provide ever more live shots; both are investing in high definition cameras, trying to make quality a key attribute.

The AP is making a big bet on live streams. It has an arrangement with the mobile application Bambuser that allows both professional AP journalists and any Bambuser user around the world to offer streaming live content from their mobile devices into the AP network.[6] It has also begun offering multiple live streams to its customers after tripling the number of live events it streamed between 2013 and 2014.[7]

Reuters Television is trying to shed its middleman image, offering a mobile television application directly to consumers,[8] allowing viewers to customize a stream of live and produced shots for $1.99 per month.

The switch from being business-to-business to being business-to-consumer is fraught with peril, however. Whether a traditional wholesaler of news can become a retailer in any meaningful way will be a test of the organization's people, ethos and economics. Delivering a steady diet of produced video is very different from offering raw footage to be processed by the broadcasters before use. Maintaining consistent quality to enhance your brand's reputation is very different from pushing a pipe of lightly processed news clips into a client's newsroom. Thinking of the "person on the street" as the customer is very different from selling to, catering to and editing for fellow professionals at broadcasters. Moreover, selling to your customers' customers runs the risk of being perceived as a competitor and not a partner, and losing the wholesale business faster. Where a wholesale news contract for a broadcaster could cost many hundreds of thousands of dollars per year and be locked in for years by contract and by the inertia that is fed by the complications of training and implementation, a retail service at $1.99 per month per user needs a massive uptake and dedicated retention efforts to create an impression on the bottom line.

Both agencies, moreover, remain fundamentally closed systems providing almost exclusively video produced by their own journalists or selected and curated by their own editors. This is at a time when every user of the internet is used, instead, to selecting at will from a vast potpourri of content issued by a mindboggling array of sources.

The agencies began and had their heyday when acquisition and availability of technology was limited, when access to news events was severely restricted by logistics and by tradition, when distribution of news required complex and private networks. Their customers had money to spend and were willing to spend it for content they absolutely needed. It was at that time that exclusive and sealed networks were a virtue.

Now, technology is ubiquitous and innovation tends to happen outside of the agencies. Access to newsmakers and news events is becoming swiftly democratized. Distribution is simple, effective, fast, and cheap. Customers are guarding their budgets ruthlessly, and are questioning the need for a lot of what the agencies provide. What had been a virtue looks more like a vice.

The virtue that is left, and that employees—from senior executives to sales managers to journalists—take great pride in, is the brand. These storied news names have been built and imbued with meaning for more than a century. However, the value of this virtue is open to serious question as well, not because the agencies stand for bad things but because they stand for the same things.

Here are statements from the three major traditional agencies about their essential natures:

- "One of the largest and most trusted sources of independent newsgathering…single-minded focus on newsgathering and its commitment to the highest standards of objective, accurate journalism."

- "One of the world's largest international multimedia news providers…unparalleled international and national news coverage with speed, impartiality and insight."
- "One of the world's major media organisations not only due to its network, the competence of its journalists and technical staff, but also because it has always remained true to its core values. Truth, impartiality and plurality…guarantee rigorous, verified news, free from political or commercial influence."

That the first is from the AP,[9] the second from Reuters,[10] and the third from AFP[11] is probably not obvious to any reader, or client, or potential client. If the values are fungible and the content is increasingly common to all, it becomes extremely difficult to keep prices at a premium. Moreover, the assumption that the traditional news values espoused are in fact the key values that the audience of the 21st century prizes is an arrogant one, and not necessarily borne out by examining what news reports or video clips trend on Twitter or go viral on Facebook.

The fundamental question that remains is how any broadcaster or news service or journalist can make professionalism valuable and even essential to the consumers of today. It is the testing of various potential answers that will determine if the traditional news services will find themselves confined to an ever-shrinking niche, or whether they will be able to reimagine, reinvigorate and reinterpret their mandates and revolutionize their offerings.

NOTES

1 The Associated Press (AP) founded its own television service in 1994 and bought Worldwide Television News (which had been owned by ABC in the U.S. and the UK's ITN) in 1998. See http://www.nytimes.com/1998/06/03/business/ap-buys-worldwide-television-news.html (Accessed February 16, 2015). Reuters' television arm at the time of the Gulf War was called Visnews and was a venture owned jointly by NBC, the BBC and Reuters. Reuters bought out its partners by 1992 and changed the name to Reuters Television in 1993. See http://www.independent.co.uk/news/business/reuters-set-to-buy-rest-of-visnews-1534714.html (Accessed February 16, 2015). AFP began web video production in 1991 and launched AFPTV in 2007.

2 http://www.afp.com/en/agency/afp-history/ (Accessed February 18, 2015)

3 Reuters news revenues were $319 million in 2014, down 1% on the year before http://ir.thomsonreuters.com/phoenix.zhtml?c=76540&p=irol-newsArticle&ID=2015545 (accessed February 16, 2015). The company no longer provides a separate profit or loss figure for its media ventures. The Associated Press reported a $3.3 million profit in 2013 against a loss of $25.5 million the previous year even as revenues fell 4% to $596 million. http://www.ap.org/Content/AP-In-The-News/2014/AP-posts-profit-in-2013-even-as-revenue-falls (Accessed February 16, 2015).

4 It is important to state that professionals manipulate images too, unfortunately. Fully 20% of the photographs that reached the final stages of the 2015 World Press Photo competition were

disqualified for impermissible alteration of the image. http://www.bjp-online.com/2015/02/image-manipulation-hits-world-press-photo/ (accessed February 18, 2015).

5   The three were Vox, Buzzfeed and Re/Code. Re/Code launched January 2, 2014. Vox launched April 6, 2014. Buzzfeed was founded in 2006. The interviews are available at: http://www.buzzfeed.com/buzzfeednews/full-transcript-of-buzzfeed-news-interview-with-president#.mp464oYvW (accessed February 18, 2015); http://recode.net/2015/02/13/barack-obama-recode-kara-swisher-video/ (accessed February 17, 2015); and http://www.vox.com/a/barack-obama-interview-vox-conversation/obama-interview-video (Accessed February 17, 2015).

6   http://blog.bambuser.com/2012/04/let-associated-press-contact-you-about.html (Accessed February 18, 2015)

7   https://www.journalism.co.uk/news/ap-expands-live-video-offering-as-slow-television-grows-online/s2/a564040/ (Accessed February 18, 2015)

8   http://www.reuters.tv/ (Accessed February 18, 2015)

9   http://www.ap.org/company/about-us (Accessed February 18, 2015)

10 http://thomsonreuters.com/news-services/ (Accessed February 18, 2015)

11 http://www.afp.com/en/agency/ethics/ (Accessed February 18, 2015)

PART TWO

Understanding the Past,
Present and Future of
24-Hour News: Changing
Conventions and
Journalism Practices

Revisiting the Three Phases of 24-Hour News Television in the Age of Social Media

STEPHEN CUSHION

On the 30th anniversary of the launch of CNN—the first dedicated news channel, which began broadcasting in 1980—I examined how 24-hour news television had evolved over three decades. In brief, it was suggested the rolling news genre and the way it had been studied by scholars could be interpreted within three overlapping phases (Cushion, 2010).

First, a "coming of age" phase when CNN was launched and grew in recognition, most notably in its live reporting of the first Gulf War. While other channels emerged in this period—Europe's first rolling news channel, Sky News, in 1989, for example—it took until the early 1990s before the perceived influence of CNN triggered the arrival of other channels with similar ambitions.

With the availability and penetration of cable or satellite services increasing post–Gulf War, *the second phase can be characterized as a race for transnational reach and influence.* Euronews's arrival in 1993, for example, was created in a bid to develop a European identity and present a challenge to the monopoly of American news channels with a global reach. Likewise, Al Jazeera's launch in 1996 provided coverage that addressed, for the first time, Middle Eastern audiences.

The last overlapping and ongoing phase marks a stage when news channels began to scale down aspirations, with a proliferation in national news channels competing *within* nations. The global rise of 24-hour news channels has meant stations operate in a far more crowded and competitive marketplace. Overall, the *third ongoing phase represents a regionalization of the 24-hour television news genre, encouraging news conventions that champion greater speed and immediacy,* but potentially undermining journalistic balance or accuracy.

While it was concluded not all rolling channels around the world conform to this tripartite model of 24-hour news development, it was argued that each overlapping phase helps mark out key turning points in what remains the short-lived history of dedicated news channels. But as outlined in the introduction to this book, the news landscape has dramatically changed over the last decade, with changing technologies and new content platforms creating faster, more flexible and fluid ways of not only producing journalism, but consuming and sharing it instantly. So, for example, in the first decade of the 21st century, social media use was in its infancy, with sites such as Twitter challenging the speed of 24-hour news channels without yet understanding the power behind it or their wider agenda-setting influence. For it was arguably not until the second decade of the new millennium when its worldwide connectivity was fully realized—notably during the Arab Spring—or when it became a tool used widely and regularly by broadcast journalists.

Over recent years, in many advanced Western democracies surveys have steadily shown people turning to social media for news, part of a wider trend of consuming media online and on the move with increasingly sophisticated smartphones (Reuters Institute, 2015). Although internet-enabled phones were available at the turn of the century, it was not until towards the end of the decade when more mobiles or tablets could more fluently and reliably instantly stream images or facilitate apps such as Facebook and Twitter. Whereas audiences relied on dedicated news channels or online sites, at the tap of an app today many people instantly receive breaking news alerts on their mobile devices.

With new content and social media platforms delivering news instantaneously, 24-hour news channels face existential questions about their journalistic role and wider purpose. In light of the faster delivery of news and the way it is being consumed on new platforms, this chapter revisits the three phases of 24-hour news channels (Cushion, 2010). In brief, I ask whether 24-hour news channels have changed their journalistic purpose or character in order to keep up with the wider pace of news culture. Before assessing whether there is evidence to support a fourth phase in the life cycle of 24-hour television news, the three phases are briefly summarized (see Cushion, 2010).

PHASE 1: THE "COMING OF AGE": WHEN DID 24-HOUR TELEVISION NEWS BEGIN?

When CNN launched the first dedicated news channel in 1980, its reach was relatively limited. At this point, it was available as a single US network broadcast to 1.7 million American homes. Because the development of cable television was in its infancy when CNN began broadcasting, it was not until the mid-1980s that the channel began to operate at a profit as cable penetration grew and advertising revenue followed (Flournoy & Stewart, 1997: 2). To provide a sustainable eco-nomic model of broadcasting, CNN had to quickly stamp its editorial credentials on the genre of news and justify the existence of a dedicated television news chan-nel. But it delivered an editorial distinctiveness within minutes of broadcasting. It cut away from routine items to produce live pictures of Jimmy Carter visiting a civil rights leader in a hospital in Indiana after a failed assassination attempt. The emphasis, in other words, was on communicating visual rather aural journalism. While nightly news bulletins had often gone live before the arrival of 24-hour television news, what CNN delivered was a sustained period of immediate "live-ness." Rolling news channels, in this context, could thus claim to be first with news stories (because a news agenda could be immediately interrupted), be on location when news happens (if the resources of news channels were available) and, most importantly for television audiences, provide a visual window on what was hap-pening in the world.

Since the character and nature of CNN became apparent relatively quickly, the implications of rolling conventions such as live or breaking news for journal-ists and journalism soon became a debating point within the media industry and academy. The instinct to "go live" as opposed to a more investigative approach to journalism is perhaps understandable given the limited resources under which roll-ing stations operate. As CNN's reputation grew through covering breaking stories (a convention returned to in Phase 3), including a failed assassination attempt of Ronald Reagan in 1981 and the launch of space shuttle *Challenger* in 1986, rival US networks began to voice their disquiet with the direction broadcast journal-ism was heading. With cameras broadly pointed at ongoing "newsworthy" events, many critics believed reporters behaved almost like witnesses at the scene of an accident, caught up "in the action" and forced to speculate on what they could see and relay back to audiences.

CNN's reach throughout the 1980s grew not only within America due to increasing cable penetration and its reputation for breaking news, but beyond with the 1985 launch of CNNI, an international channel available on many plat-forms globally (see Flournoy & Stewart, 1997, for history of the channel). CNN had been available in some countries before (Flournoy & Stewart, 1997: 3), but its more international reach had opened up, for the first time, the possibility of constituting a global public sphere. Almost immediately, according to Volkmer (1999: 2), CNN "reshaped the conventional agenda of international or 'foreign' news and created a platform for worldwide communication." Many scholars were eager to apply the concept of "globalization" or "global communication" (Volkmer, 1999: 2) to demonstrate 24-hour news's capacity to reach worldwide audiences.

From this perspective, CNN posed a perceived hegemonic threat with the potential to advance a global (or, worse still, American/Western) news agenda. This fast became a reality with the first US/Iraq Gulf War (1990–1991) and the role CNN played in reporting the conflict. Whereas most foreign journalists were forced to leave Iraq on the eve of the war, CNN journalists had built up diplomatic ties with the Iraqi regime, managing to capture exclusive live action of Baghdad under attack in the opening conflict. The world's media temporarily paused and looked on with jealousy at what CNNI was communicating to the world. It was, in short, a global event and, for the genre of rolling news, arguably represented the moment when 24-hour television news "came of age" and demonstrated the influence it could potentially wield. To permanently mark CNNI's contribution to journalistic folklore, the channel's achievements were even made into a Hollywood movie in 2002, *Live from Baghdad*, with an all-star cast capturing how the use of new technology and intrepid reporting could open up a window through which to see modern warfare at its most spectacular. What CNNI could achieve not only drew greater attention toward the channel itself, particularly its influence in foreign news reporting (Livingstone, 1997; Robinson, 2002, 2005), it also triggered what could be argued is the second phase of the development of 24-hour television news channels.

PHASE 2: THE RACE FOR TRANSNATIONAL REACH AND INFLUENCE

If CNNI had won the sprint for transnational reach and influence, many other channels wanted the race to turn into a marathon. In the years following the first US/Iraq Gulf War, a host of new channels burst onto the scene, with many hungry to exercise power at a time when a post-national broadcasting ecology was beginning to take shape. This quest to control the global flow of information was captured in the sinister deception of a James Bond villain in the film *Tomorrow Never Dies* (1997). Using the power of 24-hour news television, the film's plot revolves around Carver's intent to start World War III by inflaming diplomatic ties between China and the United Kingdom. Inevitably, Bond saves the day, but it is a moment in popular culture that revealed a deep sense of paranoia and suspicion about media tycoons such as CNN's Ted Turner or News International's Rupert Murdoch.

If this could be dismissed as mere Hollywood fantasy, two years later Edward Said (1999) warned, "CNN has now shot to an enviable position of almost total global hegemony." Close to 10 years on, Thussu (2002) has argued that the Murdoch-owned 24-hour news channel Star News in India, over the previous decade, has diminished the quality of Indian television journalism. What Murdoch had achieved was not so much global dominance with one news channel—the type of evil James Bond had defeated—but a more cumulative impact on the 24-hour news genre. With Star News, Murdoch had captured a key regional market, one that sat lucratively alongside other Murdoch channels around the world, such as his Sky News stations (in Australia, New Zealand, Italy, Ireland, and the UK, for example) or, of course, Fox News in America, which is almost global in its reach.

The most potent impact attributed to the power of 24-hour news television, however, is what has become widely known as the "CNN effect" or, in some quarters, the "CNN curve" or "CNN factor" (Livingstone, 1997: 1). In its original application, the term referred to the possible influence CNN was having on foreign policy making, most notably in encouraging military interventions into the Iraqi (1991) and Somali (1992) conflicts. Well before the internet age, the advent of real-time, rolling news channels had made foreign policy making almost like a televised live event, a spectacle for viewers around the world to tune into to see humanitarian crises unfolding 24 hours a day. If this was once a concern left to diplomats behind closed doors, CNN had brought international issues into sharper focus, with governments expected to act—or, at least, publicly respond—more swiftly than ever before to ongoing events. As other international news channels arrived on the scene, the "CNN effect" began to refer more generally to the influence of global forms of communications on foreign policy making. But while the perceived power of the "CNN effect" entered into political and media discourse, academic studies began to suggest the relationship between news coverage and foreign policy intervention was something of a myth (Robinson, 2002).

CNN, of course, was not the first news organization to establish a transnational news agenda. Other international news channels arrived on the scene in the early 1990s, including a European news channel. A project to develop a pan-European news service had dated back to the beginning of the 1980s (Adonnino, 1985), but impetus grew when the European Union acquired new countries and powers. The launch of the 1993 rolling news channel, Euronews, was an attempt to convey a shared political, cultural, and linguistic image of European nation-states (González Martín, 1995). While it was intended to encourage a pan-European agenda *within* Europe (Garcia-Blanco & Cushion, 2010), at the same time the channel was available beyond European airwaves and could thus offer an alternative to CNNI's American "window on the world." Other channels with transnational reach began to be launched.

If Al Jazeera crept quietly onto the international 24-hour news landscape in 1996, five years later during the second Gulf War the Qatar-based news channel aimed primarily at Middle Eastern audiences was a major media player in the American battle to win "hearts and minds." Like CNN a decade before, it gained exclusive coverage of the first few weeks of the war in Afghanistan, but unlike the American news channel, its coverage revolved around Middle Eastern concerns and issues. Opening up a new, less West-centered public sphere, Al Jazeera sought to reflect Arab public opinion even if this questioned the motives of American foreign policy making and exposed much Western hypocrisy (Miles, 2005).

Al Jazeera appeared, at least in this conflict, to have burst the bubble of Western news dominance.

In a new international multimedia, multichannel environment, it has been argued that a "CNN effect" would be difficult to isolate with audiences diversifying and news agendas diverging (Robinson, 2005). A widely shared global news agenda, in this context, would appear less likely to pressure governments to act at a time when multiple CNN syndicate channels operate from the perspective of countries all round the world. How far these news agendas diverge, of course, is open to debate. But, in my view, this led to the next phase in the development of rolling news, when channels began to scale down aspirations of global reach, with national news channels proliferating and regionalizing the genre of 24-hour television news. At the same time, the next phase is shaped by an increasingly more competitive news environment as channels compete for audiences more intensely. As has been widely claimed, this has arguably led to a more market-driven set of rolling news conventions than previous phases.

PHASE 3: THE REGIONALIZATION AND INCREASING COMPETITIVENESS OF ROLLING TELEVISION NEWS CHANNELS

In most parts of the world, 24-hour news channels operate in a very crowded and highly competitive marketplace. While several news channels have a global reach (CNN and CNNI, BBC World, CNBC, and Bloomberg TV), many more cover large swaths of the world, such as Fox News, Euronews, Russia Today, Al Jazeera and Al Jazeera International, Zee News, NDTV India, NDTV 24×7, Sun News, PTV News, ARY News, Star News, CCTV-9 (International), Channel News Asia, and ABS-CBN News Channel (Rai and Cottle, 2007). These channels, taken together, overlap in transnational reach and influence, and compete not only with each other, but with a growing army of national and state/local news channels that emerge year on year.

In some countries—India, for example—it is difficult to pin down the precise number of rolling channels because there are so many news stations offering both rolling content and current affairs programming. These are both national and local in reach with other transnational channels also available. There are many external factors at work here influencing the broadcasting ecology of rolling news stations, and the type and nature of 24-hour news coverage that emerges from different regions of the world. Indeed, as Young (2010) has argued, the "shape of 24-hour news in Australia is the outcome of broadcasting policies shaped by media ownership concentration and fierce lobbying by powerful existing players." As a result, in that part of the world the first home-grown channel did not emerge until 1996 when Sky News Australia launched. It took over a decade before two other news channels were launched (A-PAC in Australia in 2009 and TVNZ 7 in New Zealand in 2008). There is a New Zealand Sky News channel available, but this appears to be more of a branding exercise than a substantive source of local information, with very little 24/7 news content produced from the island (just afternoon or nightly bulletins as opposed to an ongoing rolling format). Even if Australia and New Zealand were relative latecomers to the genre of 24-hour television news, there appears—even in the space of the previous two years—to be a regionalization of rolling news with more choice and more local content being produced: a trend identified more globally (Rai & Cottle, 2007). This reflects a key shift from Phase 2 to Phase 3.

But as the New Zealand Sky News channel demonstrates, the regionalization of 24-hour news can be somewhat superficial, a corporate gloss to market news to the natives. It is important, in this context, that news agendas are empirically assessed. So, for example, while Jean Chalaby (2009: 172) suggests transnational news channels could be divided into rolling stations that tell stories beyond their borders, within their borders, or about their borders, at the same time this implies a typology of 24-hour news agendas that are closely tied to geographical boundaries. And yet, even a channel such as Euronews, which explicitly states it is an extended voice of the EU, is primarily concerned with news beyond Europe (even if this may be from a "European perspective" (see Garcia-Blanco & Cushion, 2010).

What this suggests is that while the global rise of news channels has weakened the notion of a so-called "CNN effect," there remain *globally binding influences* that continue to encourage a worldwide news agenda. The flow of information may be shared between a wider pool of rolling local and national news stations, but the sources drawn upon to produce routine news content remain global in scope and power. The political economy of a 24-hour news station, in other words, polices the boundaries of how local, national or international a rolling news station's news agenda is.

A regionalization of 24-hour television news on a global scale, from this perspective, appears somewhat superficial. If the "CNN effect" once represented *one* channel's potential influence of foreign policy making, today it could perhaps better capture how corporate interests—including but not exclusive to CNN spin-offs such as CNN TÜRK, CNN- IBN, CNNj, and CNN Chile—help influence the type and nature of news agendas. To recap, while Phase 3 is a period when the 24-hour news genre grew at some pace, with new national and subnational stations continuing to arrive on the scene, how far this has regionalized the content of rolling news agendas remains open to debate and ongoing scholarly enquiry. But what has been the impact of an increasingly crowded marketplace of dedicated television news channels?

In reaction to the coverage of the Mumbai terrorist attacks in November 2008, P. N. Vasanti (cited in Magnier, 2009) has suggested that the copious number of television news channels in India have resulted in an "extraordinary pressure to sensationalize, claim specious 'exclusives' and do almost anything else to attract attention." Rolling news competition, in this context, appears to have led to a more tabloid agenda of 24-hour news reporting. Fox News in America has also been criticized for encouraging other broadcasters to pursue a tabloid approach to rolling news. Because Fox News is now the most watched news channel in America, there have, in more recent years, been concerns that a "Foxification effect" is taking place. Rival US news channels, according to commentators, have begun to mimic Fox's polemical approach to news reporting and adopt some of its conventions used to sensationalize rolling coverage (Cushion & Lewis, 2009). For example, where once a "breaking news" tag was used to interrupt the news agenda because a new and important story had broken on air, Fox News raised the stakes by producing a "News Alert" *in addition* to its breaking news stories. Over time, according to the producer who created the "News Alert," it has been used far more liberally and applied to soft news, such as celebrity gossip about Hollywood movie stars (Outfoxed, 2004). It is a convention, in other words, that *connotes* rather than denotes a significant interruption to routine service.

Overall, the global rise of 24-hour news channels has meant stations have to operate in a far more crowded and competitive marketplace. This, in my view, has sparked more intense competition to be the quickest and thus most attractive to opinion formers by churning out endless "breaking news" exclusives or being live "in the action." As a result, more market-led rolling news conventions have been introduced. Of course, not all rolling news channels share the same editorial priorities and news agendas. As Khun's (2010) analysis of France 24 pointed out, the channel "accords a relatively low priority to breaking news when it comes to modifying its scheduling." Similarly, a study found Euronews reported no breaking news items and did not operate at the frantic pace of other many other rolling news stations (Garcia-Blanco & Cushion, 2010). In other words, the genre of 24-hour news has what Cottle and Rai (2008) label "communicative complexity," adopting different formats, styles and conventions in how they deliver rolling news journalism.

In very recent years, communicating news has arguably become more complex, since the social media age has accelerated the news cycle and delivered information more instantly. The next section explores how the medium of 24-hour news television has responded to the enhanced pace of news culture, asking whether this represents a break from the third phase of 24-hour news or a reinforcement of it.

A FOURTH PHASE? 24-HOUR NEWS CHANNELS IN THE SOCIAL MEDIA AGE

The arrival of social media has prompted existential questions about the purpose and continued relevance of dedicated rolling news channels. As outlined in the three phases, since the arrival of CNN many 24-hour news channels have competed to break news faster than their competitors, a race to be the first window on the world to convey events on the global stage. But, as Sambrook and McGuire (2014) have pointed out, while "24-hour TV news broke the audience away from the daily news cycle… If we want it now, we will go online and get it instantly. Twitter—and increasingly live blogs of breaking news events—consistently beat 24-hour TV channels." Although many rolling news stations continue to market themselves on breaking news first, today the competition is fought between TV channels rather than the wider culture of news.

As internet penetration has increased around the world (Reuters Institute, 2015), the global breadth of social media use has also begun to outflank the geographic reach of rolling news channels. Despite witnessing a regionalization of 24-hour news channels in recent years, even the BBC's or CNN's network of international reporters cannot possibly rival the dispersed nature of citizens with access to smartphones uploading information and images instantly in remote or restricted locations. In 2009, for instance, social media was able to break through Iran's repressive media censorship to convey the civil unrest after the election results, or during the Arab Spring in 2010–2011—from Egypt to Tunisia—when citizens posted videos and messages on YouTube and Twitter not aired by 24-hour news television. But even in the US when news broke of a terrorist attack in the 2013 Boston marathon, or in the UK when Margaret Thatcher died the same year, social media broke news well ahead of rolling news channels.

Of course, 24-hour news channels have not ignored the opportunities to interact with citizens and for them to (re)shape journalism. Since 2006, CNN has asked citizens to register for iReport, which allows people to upload breaking news pictures and images. Boasting over a million users, it draws on the eyes and ears of ordinary citizens to cover stories both domestically and internationally. In this sense, it can enhance news gathering, moving beyond established sources and accessing "ordinary" people, who could act as eyewitnesses, say, or contribute local knowledge about an unfolding event or issue (Allan, 2013).

As well as following breaking stories on the wires, social media itself has also become a more prominent source in routine rolling news coverage, such as reading out celebrity tweets or streaming YouTube footage. Moreover, journalists and editors even respond to viewers in real-time, potentially modifying how they communicate or explain news coverage. Ashley Hinson, a news anchor at the US channel, KCRG TV-9, for example, explains how they "have people who will ask us a question on Facebook and we're looking that up while there's a story on the air. You just have to learn how to function in that capacity, where you're responding to all angles, literally on air, online and on social media" (quoted in Marcoccia, 2014). The 24-hour news cycle, in this context, is less about conveying the flow of news throughout the day, but reinterpreting it according to the demands and scrutiny of viewers plugged into it a wide range of online and social media platforms. Viewed from this perspective, the social media age has accelerated the rolling news cycle, opening up a potentially more diverse universe of sources and stories.

In doing so, however, new content and social media platforms have forced 24-hour newsrooms to act at even greater speed, prompting faster editorial judgements and encouraging instant scoops and revelations. The reporting of the Malaysia Airlines Flight 370 in March 2014 is a case in point. When news broke of a missing plane carrying 227 passengers and 12 crew, it immediately received global coverage, but with no new information or updates, the vacuum of news was fuelled by widespread speculation on social media about its whereabouts.

Although this was a major international breaking story, in recent years these social media feeding frenzies can influence domestic rolling news coverage on a more regular basis. In St. Louis, when a white police officer shot a black local resident, the prosecutor investigating the case in 2014 issued a stern warning not just against the wild rumors replayed on social media, but for news channels reinforcing them and the misinformation that followed. Bob McCulloch, the St. Louis Prosecutor, pointed out that "within minutes, various accounts of the incident began appearing on social media, accounts filled with speculation and little, if any, solid or accurate information" (quoted in Feldman, 2014). Further still, he suggested the "most significant challenge encountered in this investigation has been the 24-hour news cycle and its insatiable appetite for something, for anything to talk about, following closely behind with the non-stop rumors on social media" (quoted in Feldman, 2014).

In recent years, new platforms supplied by digital media appears to have prompted traditional news outlets to enhance their journalism output. In 2015, CBS launched an online 24/7 news platform, CBSN, because, according to its chief executive Les Moonves (2014), "There's so much information that we get every day that doesn't fit into a 22 minute newscast…we can go direct to digital" (quoted in Huffington Post, 2014). Compared to the routine costs of a 24-hour news television channel, supplying news via computers, phones, tablets, smart TVs or streaming services is far cheaper and subject to less regulatory oversight or management (Sambrook and McGuire, 2014). Rather than adding editorial costs, CBSN draws on content already generated by the CBS network, such as streaming live events, or using resources from local affiliate stations. In between her anchor slot for WCBS news, for instance, Kristine Johnson was delivering live two-ways about the January 2015 snow storm in New York.

New content and social media platforms have also begun to converge with more traditional rolling news formats. In 2013, a CNN/BuzzFeed channel was launched on YouTube, a "perfect modern day collaboration" according to KC Estenson, SVP at CNN Digital. In his words, "By pairing the journalistic strength and reach of the CNN brand with BuzzFeed's unique editorial approach and young audience, our partnership will enable both organizations to engage new audiences" (quoted in CNN, 2013). Indeed, there is evidence to suggest young people in the US are far more interested in following up breaking news stories than seeking daily news compared to older age groups, a symptom perhaps of the news culture they were born into (American Press Institute, 2013).

In this and many other fast-moving breaking news stories, while social media such as Twitter can deliver greater speed and depth to a story, its reliability is highly questionable with journalists not always able to verify the source of information or meaningfully assess its veracity. Since rolling news platforms are competing to break news first, even when senior journalists working for major news organisation acknowledge the dangers of relying on social media, it can be difficult to withstand the pressures to resist following leads and bring it to viewers' attention. This was recognized by Sky News, who in 2012 issued new guidelines for journalists in their use of Twitter. The management team told editorial staff:

> don't tweet when it is not a story to which you have been assigned or a beat which you work…Where a story has been Tweeted by a Sky News journalist who is assigned to the story it is fine, desirable in fact, that it is retweeted by other Sky News staff…Do not retweet information posted by other journalists or people on Twitter. Such information could be wrong and has not been through the Sky News editorial process. (quoted in Halliday, 2012)

Aware that Sky News journalists might be compromising accuracy for speed, it is a revealing moment in rolling news history that demonstrates the limits to delivering news fast and the risk it poses to undermining an organisation's credibility.

In assessing 24-hour news television in the social media age, then, rather than reshaping the purpose of the genre, it would appear dedicated news channels remain committed to delivering news instantly regardless of whether they break the story or not. *This suggests a reinforcement of, rather than a departure from, the third phase of 24-hour news channels, defined by more intense competition and market-led conventions such as breaking news. However, the competition moves beyond out-scooping rival broadcasters, asking journalists to interpret news in the wider landscape of media.* If anything, the pace of the cycle has accelerated further, bringing viewers information from a wider universe of sources more quickly but not always with the same rigor or accuracy.

Despite predictions of the death of rolling television news, the growth and competition from new content and social media platforms appears not to have put 24-hour news channels out of business. Far from it, as Rai and Cottle demonstrate in Chapter 12, the number of 24-hour news channels has broadly remained the same around the world over recent years. As Phases Two and Three outlined, previous decades have not only witnessed an internationalization of rolling news channels, but a regionalization of them. This has been reinforced by international channels creating sister channels in different regions of the world. So, for example, Russia Today (now called RT) has launched English, German and French channels (including RT America, RT UK, RT Deutsch and RT Français). Al Jazeera, of course, is not just a Middle Eastern broadcaster; it has English, Balkans, American and Türk sister channels. Likewise, China's CCTV has American and African dedicated channels. And there are new entrants to the international and domestic news market. German broadcaster Deutsche Welle launched an English language news channel in 2015, competing alongside the BBC and France 24 in Europe. i24, an Israeli international news channel from 2013, broadcasts in English, French and Arabic. Meanwhile, ANN7—the African News Network 7—was launched in the same year and represents a third dedicated South African news channel. To counteract RT's international news influence, in 2014 Ukraine Today was launched.

Needless to say, not all these and other channels besides operate as profit-making commercial news organisations or subscribe wholesale to the ideals of public service broadcasters. Many of them aim to secure what is known as "soft power," projecting a nation's views and perspectives about international affairs, funded by states or billionaires. As a former head of the BBC's world service recently warned, the "information war" is increasingly being out-funded by powerful governments or rich oligarchs (Halliday, 2014). While some market-oriented news channels turn a profit, such as Fox, CNN and MSNBC in the US, many more (as has always been the case) remain heavily subsided. This makes it a challenging prospect for news channels from organisations with less power and resources to break into the market.

Despite the internet's global reach and ability to communicate news immediately, then, it would appear (so far) not to have outflanked the perceived power of 24-hour news television channels. Viewed from this perspective, *the life-span of television news channels will be determined not just by the availability of new technologies, but from the power and influence it can demonstrably wield in domestic and international affairs.* As Robert Picard explains in Chapter 13: "As long as their patrons are able and willing to provide funds, and as long as the broadcasters are seen as delivering the results desired by their benefactors, the channels will be able to survive."

Since both national and international 24-hour news stations continue to flourish and new channels are being launched, this suggests the end of the third phase of 24-hour news channels is not immediately imminent. While people's news consumption habits gradually move towards more online and mobile platforms, in most advanced democracies it has not replaced but supplemented television news viewing (Reuters Institute, 2015). In short, watching television remains central to people's social and cultural experiences, including tuning into fixed and, to a lesser extent, continuous news programming (Cushion, 2015).

Of course, once the online world fully morphs into most people's television sets, when online and broadcast worlds operate seamlessly on the same platform—more fluidly interacting with mobile phones and tablets than is possible today—then it is at this point the fourth (and possibly final) phase of 24-hour news channels might be triggered. But it remains to be seen whether new technology in the next phase will create continuous news platforms that can meaningfully challenge corporate or state power, or whether the format and style of the 24-hour news genre will continue to operate at the same frenetic pace. But a starting point for debating the future direction of 24-hour news and the wider challenges facing journalism in the 21st century is in the many chapters featured in this edited volume.

REFERENCES

Adonnino, P. (1985). *A People's Europe. Reports from the ad hoc committee.* In Luxembourg: Office for Official Publications of the European Communities.

Allan, S. (2013). *Citizen Witnessing: Revisioning Journalism in Times of Crisis. Key Concepts in Journalism.* Cambridge: Polity Press.

American Press Institute (2014). Social and demographic differences in news habits and attitudes. 13 August, http://www.americanpressinstitute.org/publications/reports/survey-research/social-demographic-differences-news-habits-attitudes/ Accessed 24 June 2015.

Chalaby, J. (2009). *Transnational Television in Europe: Reconfiguring Global Communications.* Networks. London: I.B. Tauris.

CNN (2013). BuzzFeed and CNN Launch "CNN BuzzFeed" News Video Channel for Millennials. CNN pressroom blog, May 28, http://cnnpressroom.blogs.cnn.com/2013/05/28/buzzfeed-to-aggressively-expand-video-operation-in-partnership-with-youtube/ Accessed 24 June 2015.

Cottle, S., and Rai, M. (2008). Global 24/7 news providers: Emissaries of global dominance or global public sphere? *Global Media and Communication*, 4(2), 157–181.

Cushion, S. (2010). The three phases of 24-hour news. In Cushion, S., and Lewis, J. (eds.), *The Rise of 24-Hour News Television: Global perspectives.* Oxford: Peter Lang.

Cushion, S. (2015). *News and Politics: The Rise of Live and Interpretive Journalism.* London: Routledge.

Cushion, S., and Lewis, J. (2009). Towards a "Foxification" of 24-hour news channels in Britain? An analysis of market driven and publicly funded news coverage. *Journalism: Theory, Practice and Criticism*, 10(2).

Feldman, J. (2014) St. Louis Prosecutor Scolds Social Media, 24-Hour News Cycle in Announcement. *Medialite*, November 24, http://www.mediaite.com/tv/st-louis-prosecutor-scolds-social-media-24-hour-news-cycle-in-announcement/ Accessed 24 June 2015.

Flournoy, D. M., and Stewart, R. K. (1997). *CNN: Making news in the global market.* Luton: John Libbey Media.

Garcia-Blanco, I., and Cushion, S. (2010). A partial Europe without citizens or EU-level political institutions: How far can Euronews constitute a European public sphere? *Journalism Studies*, 11(3), 393–411.

Gilboa, E. (2005). The CNN effect: The search for a communication theory of international relations, *Political Communication*, 22, 27–44.

González Martín, P. (1995). *Euronews. Una Television Pública para Europa*. Barcelona: Icaria.

Halliday, J. (2012). Sky News clamps down on Twitter use. *Media Guardian*, 7 February, http://www.theguardian.com/media/2012/feb/07/sky-news-twitter-clampdown Accessed 24 June 2015.

Halliday, J. (2014). BBC World Service fears losing information war as Russia Today ramps up pressure. *Media Guardian*, 21 December, http://www.theguardian.com/media/2014/dec/21/bbc-world-service-information-war-russia-today Accessed 24 June 2015.

Huffington Post (2014). CBS Eyes 24-Hour Digital Channel. 15 June, http://www.huffingtonpost.com/2014/05/15/cbs-news-24-hour_n_5334234.html Accessed 24 June 2015.

Kuhn, R. (2010). France 24: Too Little, Too Late, Too French? In Cushion, S., and Lewis, J. (eds.), *The rise of 24-hour news television: Global perspectives*. Oxford: Peter Lang.

Learnmouth, M. (2015). Not cable news: CBS' 24-hour digital news channel makes leap to TV. *International Business Times*, January 28, http://www.ibtimes.com/not-cable-news-cbs-24-hour-digital-news-channel-makes-leap-tv-1797850 Accessed 19 June 2015.

Livingston, S. (1997). *Clarifying the CNN effect: An examination of media effects according to type of military intervention*. Harvard University, John F. Kennedy School of Government Research Paper R-18. Hodder Arnold.

Magnier, M. (2009). Indian news channels criticized for Mumbai coverage. *Los Angeles Times*. 18 January, http://articles.latimes.com/2009/jan/18/world/fg-mumbaitv18 Accessed January 10, 2010.

Marcoccia, P. (2014). 24-hour news takes on a new meaning. *The Gazette Company*, November 19, http://www.thegazettecompany.com/2013/11/19/24-hour-news-cycle-takes-on-a-new-meaning/ Accessed 24 June 2015.

Miles, H. (2005). *Al Jazeera: How Arab TV News Challenged the World*. London: Abacus.

Outfoxed (2004). *Rupert Murdoch's war on journalism*. The Disinformation Company.

Rai, M., & Cottle, S. (2007). Global mediations: On the changing ecology of satellite television news. *Global Media and Communication*, 3(1), 51–78.

Reuters Institute for Study of Journalism (2015). *Digital News Report 2015*. https://reutersinstitute.politics.ox.ac.uk/publication/digital-news-report-2015 Accessed 24 June 2015.

Robinson, P. (2002). *The CNN Effect: The Myth of News, Foreign Policy and Intervention*. London: Routledge.

Robinson, P. (2005). The CNN effect revised. *Critical Studies in Media Communication*, 22(4), 344–349.

Said, E. (1999). Public spectacle, public history. *Al-Ahram Weekly*, Cairo, 18–24 February. Issue 417.

Sambrook, R., and McGuire, S. (2014). Have 24-hour news channels had their day? *Media Guardian*, 3 May, http://www.theguardian.com/media/2014/feb/03/tv-24-hour-news-channels-bbc-rolling Accessed 24 June 2015.

Thussu, D. (2002). Managing the media in an era of round-the-clock news: Notes from India's first tele-war. *Journalism Studies*, 3(2), 203–212.

Thussu, D. (2007). The "Murdochization" of news? The case of star TV in India. *Media, Culture & Society*, 29(4), 593–611.

Volkmer, I. (1999). *News in the Global Sphere: A Study of CNN and Its Impact on Global Communication*. Luton: University of Luton Press.

Young, S. (2010). Audiences and the impact of 24-hour news in Australia and beyond. In Cushion, S., and Lewis, J. (eds.), *The Rise of 24-Hour News Television: Global Perspectives*. Oxford: Peter Lang.

CHAPTER ELEVEN

Televisual Newspapers? When 24/7 Television News Channels Join Newspapers as "Old Media"

MICHAEL BROMLEY

In the third quarter of the 20th century, 24/7 television news channels could be classified as a form of "new media"; by the middle of the second decade of the 21st century they were already looking old fashioned (Miller, 2013). Technical developments dating back as far as the 1920s, which helped make possible the introduction of cable-based 24/7 news channels from 1980, subsequently contributed in part to the rapid growth of alternative online media forms with the potential to supplant TV. The uptake from about 2002–2004 of Web 2.0 capacity to host interactive, community-building social media represented an acceleration of change in media forms which threatened to eclipse existing ones. At about the same time, audio and video streaming were positioned as consumer products available on demand through devices connected to the internet. These were accompanied by a shift from wired static to wireless mobile receiving and sending tools. Globally by 2013, the number of television users (5.5bn) was only marginally greater than the number of mobile phone users (5.2bn) (Meeker, 2014: 8). There were also 3bn internet users, two-thirds of them in the developing world (Dragomir and Thompson, 2014: 11). Although 24/7 news channels were almost ubiquitous, the media environments in which they existed and to which they contributed varied considerably.

The mid- and late-1990s have been proposed as marking the end of the first phase of the development of 24/7 television news channels, chiefly in the global West (Cushion, 2010: 19–20).[1] By the end of the decade, there were such channels in at least 25 countries spread across Africa, Asia, Europe, the Indian sub-continent, Latin America, the Middle East, North America and Oceania. Many were international in scope and reach and/or originated in national public service broadcasters with external services. The privately owned Fox International Channels were formed in 1993, and Rupert Murdoch's News Corp operated news channels in Australia, China, India, New Zealand, the UK and the USA. Initially global in aspiration, channels such as Cable News Network (CNN, 1980) and its international version (CNNI [1985]), BBC World (1991), CNBC (1989), Bloomberg TV (1994) and Fox News (1996) (see Rai and Cottle, 2010: 55), were outnumbered by the end of the 1990s by regional and national channels. Local channels were also beginning to appear (Cox, 1995: 214; Rai and Cottle, 2010: 73).

Notwithstanding the near-global nature of this development, commercial, administrative and critical interest in 24/7 news channels remained skewed towards the global West. In a short introductory survey, it was not possible to pay equal attention to all regions and nations. Data for this chapter were most readily available from the West, and chiefly the US, the UK and Australia. Nevertheless, a global audit of digital media found that variations between and within regions, as well as between the developed and developing world, existed among newspapers, radio and the internet as well as television (Milojević, 2014: 129–131, 137–144). Ideological tendencies to liberalisation, deregulation and democratisation in the 1980s and 1990s had differential impact. In Africa there was an explosion of radio stations (both commercial and community); for example, in Mali about 300 came into existence and about 200 in the Democratic Republic of the Congo. Radio remained the predominant medium across much of Africa, in Latin America, and in individual countries, such as Jordan and Pakistan (Milojević, 2014: 133). Elsewhere, for example, Japan, China and much of Asia, newspapers proliferated and their circulations grew. In 2014–2015 the BBC (2015) claimed a global weekly reach of 283m people; for the first time the television audience (148m) was greater than that for radio (133m).

Where processes of liberalisation and deregulation were associated directly with democratisation, rising rates of literacy and increases in incomes, it was argued, the effect was to drive users not to novel, perhaps more attractive, media forms in commercially competitive ecologies, but to those media (old and new) which held out the greatest promise of popular emancipation (Santhanam and Rosensteil, 2011). For many around the world the main sources of news remained old media albeit often in new forms. Nor did a shift in media form use patterns necessarily signify changing content preferences. The global audit referred to above concluded that "People tend to consume the traditional media they trust, but in more convenient forms." Interestingly, the study presented newspaper websites as typical of this emergent configuration of old and new (Dragomir and Thompson, 2014: 26). On the other hand, contemporary developments may have been evening out unequal access to the internet, as the proportion of mobile phone usage dedicated to accessing the web was twice as great in Asia and Africa as it was in the US and Europe (Ewing, van der Nagel and Thomas, 2014: 9). The challenge for 24/7 news television, it seemed, came, therefore, from old media content in new forms as much as it might from new media.

In "crowded and variegated" news ecologies (Cottle and Rai, 2006: 164), scarcity began to be replaced by abundance; demand rather than supply determined use. Recipients pulled news from sources, whereas previously it had been pushed out to them. Satellite and cable circumvented an often state-regulated distribution bottleneck which limited the quantity and/or quality of terrestrial broadcasting, leading to channel proliferation. Preston (2001: 201) estimated that in 15 European Union countries the number of television channels grew from 55 in 1980 to 616 in 1995. The growth came principally from a 1990s boom in satellite transmission: there were no satellite stations in 1980, 38 in 1990 and 403 five years later. However, there were also shared continuities (Cottle and Rai, 2006: 184). Looking back over almost 35 years of 24/7 rolling news, Sambrook and McGuire (2014) argued that the channels failed ultimately to break free from the limitations of linear television. Perhaps not surprisingly, then, there was little uniformity in the formation of different media ecologies around the introduction of 24/7 news television which unfortunately can hardly be done full justice here.

NEWSPAPERS AND 24/7 NEWS CHANNELS IN THE ONLINE AGE

Newspapers appeared to find a second wind on the worldwide web and in social media. Social news leaders on Facebook were not only online services such as BuzzFeed and *Huffington Post*, but the US network ABC, and the *Guardian*, *New York Times* and *Daily Mail*. The only 24/7 television news channel in the list was Fox News (at number 10). The situation was similar on Twitter where the BBC, CNN and Fox News were joined in the top ten by the *New York Times* (2); *Time* (6); *Guardian* (7); and *Forbes* (8) (Ugander et al., 2011: 44). A list of the 15 most popular news websites named seven newspapers (the *New York Times*, [Daily] *Mail Online*, *Washington Post*, *Guardian*, *Wall Street Journal*, *USA Today* and *LA Times*); three online services; the three US television networks, but only two 24/7 news channels (CNN and Fox News) (eBizMBA, 2015).

All the same, Western orthodoxy had it that newspapers in their original printed form, already in decline from at least the 1950s, lost out most in the online digital age. Between 2007 and 2009, 20% of journalists' jobs on American newspapers were culled. By 2012, there were fewer full-time newsroom posts than there had been in 1978 (Jurkowitz, 2014). The website Newspaper Death Watch (2015) listed 12 metro papers which had ceased publication since 2007 and a further dozen titles or groups of titles which had converted to online-only or online-and-print or had reduced the number of paper-based editions. In the UK, *Press Gazette* recorded 242 closures of local newspapers between 2005 and 2012 (Ponsford, 2012). In 2012–2014, it was estimated that a fifth of mainstream journalists in Australia lost their jobs (O'Donnell, Zion and Sherwood, 2015: 1). In many European countries, including France, Portugal, Spain, Germany, Hungary, Poland, Italy and Greece, newspapers struggled to survive (Langley, 2012; Mance, 2014). However, television was not exempt. BBC News cut more than 600 jobs between 2012 and 2014 in an "efficiency drive" which included the 24/7 News Channel (BBC, 2014). Similar lay-offs on 24/7 news channels in the US began in 2007 and continued into 2015, affecting CNN, MSNBC and Al Jazeera America (Atkinson, 2014; Taibi and Hart, 2014). Free Wi-Fi, on-demand streaming content, and cheaper broadband providers were siphoning off customers from subscription-based cable TV services (Edwards, 2013; Garraham, 2015). In 2015, the Federal Communications Commission recorded its first ever year-on-year fall in the number of US pay-TV subscribers, mostly among cable users (Wheeler, 2015). One contemporaneous survey found that for news marginally more Americans turned to newspapers (66%) than to 24/7 television news channels (62%) (American Press Institute, 2014). The three leading channels—Fox News Channel, CNN and MSNBC—lost audiences between 2008–2009 and 2015 (Garraham, 2015). By 2010, US subscribers began ditching cable and bundled broadband services in favour of on-demand video, accessed on mobile and tablet devices using free Wi-Fi and specialist internet providers (Edwards, 2013). 24/7 news television, it seemed, had aged quickly (Bromley, 2010).

In 2014, only 57% of TV viewing in the US was of traditional broadcasting through a dedicated receiver (Meeker, 2014: 124). The number of broadband-only households (those without a TV set) more than doubled between 2013 and 2014 to 2.8% (Luckerson, 2014). The trend was evident not only in the US. Across six countries—Australia, China, New Zealand, Poland, Spain and Sweden—between 87% and 92% of people looked for news online; in three (China, New Zealand and Spain) it was a daily habit for a majority (Ewing, van der Nagel and Thomas, 2014: 20).[2] The number of Australian households with internet access rose rapidly from 20% to 90% in 2013. Australian internet users going online using mobile devices grew from 15% in 2009 to more than double that in 2011, and 76% in 2013 (Ewing, van der Nagel and Thomas, 2014: 1, 4). Weekly usage of news blogs also almost doubled in the period 2011–2013 (Ewing, van der Nagel and Thomas, 2014: 17). One pundit predicted that traditional television would disappear in three years (Smith, 2014). Another asked whether the "Death of the TV?" is imminent (Garraham, 2015). Tunstall (2015) concluded that at least some key television genres, including current affairs, appeared to be "in jeopardy."

A key feature of changes in media use was the speed with which new forms were taken up and threatened to supersede older ones. Television reached 35m US households in ten years; it had taken radio 25 years to do the same and the telephone 80 years (Edgerton, 2007: xi). By comparison, between 2006 and 2011, Facebook grew to reach about 10% of the world's population, including half the over-13 population of the US (Ugander et al., 2011: 2). In three years (between 2011 and 2014), visual web social networks, such as Tumblr, Pinterest and Instagram, expanded from about 10m unique US visitors to around 200m (Meeker, 2014: 38). Annual increases in use routinely reached triple figures—the messaging service WhatsApp's usage grew 100% a year for four years; the Chinese equivalent, Tencent WeChat, grew 125% for just over three years (Meeker, 2014: 36). The YouTube news channel peaked with year-on-year growth of 213% (Meeker, 2014: 109). Of particular interest to television was the take-up of streaming video through services such as Netflix, Hulu/Hulu Plus and HBO GO. At the end of 2014, Nielsen (2014) reported a 60% increase per month in accessing streaming video among Americans.

24/7 new channels faced specific challenges. Where television *per se* regained popularity, it was evident in a shift back to terrestrial viewing and away from satellite and cable services, not just in the US but also in parts of Europe (Milojević, 2014: 132). Dedicated news channels made relatively little inroad into American network news viewing habits (Baldwin, Barrett and Bates, 1992: 229). Moreover, by 2010 the 24/7 news channel viewership was decidedly ageing (Harris Polls, 2010). All the same, the idea that traditional media were being swept away by a new media tsunami was contested. Not everyone agreed that the media ecology was changing at an accelerating rate (Seidensticker, 2006). Furthermore, social practice which privileged both forms of old media and new forms of old media for news access, as noted above, seemed to run counter to the idea. However, the penetration of media forms, irrespective of their uses, seemed incontrovertible: as computing developed (from mainframes in the 1960s, through minis in the 1980s, the PC in the 1990s, desktop internet access in the 2000s to present-day mobile technologies), each new development generated about ten times the usage of its predecessor (Ewing, van der Nagel and Thomas, 2014: 11).

DISPLACEMENT OR NEWSFULNESS?

Between the 1950s and the mid-1990s, the circulation of daily newspapers in Western countries fell markedly. Among a sample of OECD countries, the downward trend was apparent in liberal (Canada, Ireland, UK and US), polarised pluralist (France) and democratic corporatist (Belgium, Denmark) media systems (Hallin and Mancini, 1984). Prior to the full emergence of web-based alternatives and the collapse initiated by the global financial crisis, in eleven out of 26 OECD countries for which data were available, the numbers of printed copies of daily newspapers circulating per 1,000 head of population had fallen between 1952 and 1996 by up to three-quarters (in Australia); in 18 there was a contraction in the numbers of daily newspaper titles (Norris, 2000: 76–78).[3] The data suggested a brief resurgence in newspaper sales in the last few years of the twentieth century, but by 2010, household penetration of daily newspapers in the US, UK and Canada had fallen again between 65% and 80.5%; none reached 40% of households where they had previously been ubiquitous (Communications Management, 2011). By 2014, 60% of readers of *New York Times* articles accessed used smartphones and tablets to do so, often finding them via social network sites, applications and search engines (Anon, 2014). More than three-quarters of connected Australians accessed news websites sometimes, but more than 60% said they would not pay to access a newspaper online (Ewing, van der Nagel and Thomas, 2014: 14, 18).

Functional displacement theory (Liu and Hsu, 2011: 262) might suggest that this extended period of newspaper decline was attributable initially to television offering what NBC first called in the 1950s its morning news magazine programme—a "television newspaper"; and then almost 30 years later CNN's view that an entire channel was a "televised newspaper" (McDowall, 2012: 10; Spigel, 1992: 81), compelling newspapers themselves to take on new roles. One such was a reversion to the 19th-century function of providing analysis and commentary on news, which was provided in greater quantity, more speedily and to more people simultaneously by television, as so-called "viewspapers"; another was the introduction of "many more non-news items" (Conboy, 2001: 90; Lewis and Cushion, 2009: 305–306; Tunstall, 1996: 59). In the 1960s, US television established its credentials as a news medium as it shifted away from entertainment (Davis, 2006: 125). News increased sevenfold as a proportion of broadcasting on the US networks between 1950 and the early 1980s (Gunter, 1987: 1–2). Individual UK television channels transmitted relatively more news from the outset and growth there came from channel proliferation (Cox, 1995: 212–215; Gunter, 1987: 6–7). The BBC, which had introduced television news only in 1954, revamped it five years later to bring it up-to-date. During the 1960s the commercial network Independent Television's (ITV) competitor nightly *News at Ten* bulletin (started in 1967) regularly featured among the ten most-watched programmes in the UK. Over time coverage of major events turned them into primarily television experiences. By the 1990s news-related programmes, such as *60 Minutes* (CBS) and *Panorama* (BBC), attracted exceptionally large audiences. For an essentially entertainment medium, there appeared to be limits, however. In the UK in particular, as 24/7 news channels began to appear, terrestrial TV withdrew from the extension of news, current affairs broadcasting and the single topic documentary (Conboy, 2011: 90; Sterling, 2009: 1385).

In the 1980s it was CNN, not network or public service broadcasters, which led the field in "breaking news," defined as "unpredictable and dramatic events unfolding onscreen, with kudos to those reporters who are first (or seen to be first) on the scene" (Lewis, 2010: 88). Covering breaking news "live," in imitation of CNN, was the key element in the development of Sky News (London) (Neill, 1996: 296). Perhaps in response, both the BBC and ITV instituted changes to their flagship terrestrial news programmes in 1999, the most drastic of which was the axing (temporarily, as it turned out) of *News at Ten*. If television had become "the primary news medium" (Gunter, 1987: 7), dedicated 24/7 channels seemed to be establishing a new kind of provision, described as more "breaking" than "news" (Cushion and Lewis, 2009: 315–317), but nonetheless trumping traditional news bulletins for immediacy. Terrestrial television in the US found a new role in providing "softer" news in the news magazine format (Sterling, 2009: 1385). Within the BBC, *Breakfast Time* (subsequently, *Breakfast News*) dropped its "serious" news approach, adopted in 1986, when it was co-transmitted as *Breakfast* by the terrestrial channel, BBC One and the 24-hour news channel BBC News.

The televisual techniques of immediacy as practised by 24/7 news channels were two-ways, on location presence, continual updates, improvisation, interpretation, studio interviews, studio discussions (Bivens, 2015: 197; Cushion and Thomas, 2013: 366–368). Tracing a form of intra-mediatisation, Cushion and Thomas (2012: 376) concluded that these channels most obviously influenced other forms of television: they found that in the UK "many of the editorial conventions and practices on fixed television news bulletins resemble the format and style of the immediacy and liveness delivered on 24/7 rolling news formats." Young (2009: 408–409) similarly detected the influence of Sky News Australia on television journalism in that country. Although Sampert et al. (2014: 290) could not correlate what they discovered to be significant changes in newspaper political journalism in Canada—fewer stories, sensationalism, a focus on celebrity, more photographs, more comment, more personalisation, more negativity—directly to the influence of all-news television, they noted the "impact" of "the 24-hour news channel" (p. 280). Patterson (2000: 247), too, asserted that 24/7 television news had led to a "steep decline in the overall size of the news audience" in the US but without explaining how this had occurred.

Indeed, research conducted in the US during the final years of the 20th century suggested that the negative impact on newspapers of cable television with its 24/7 news channels was significantly less than on news magazines and radio. In any event, access to cable news most depressed news consumption among those who were least interested in news, a category which tended to exclude newspaper readers (Cohen, 2008: 151–158). At the other end of the scale, 24/7 news television was most attractive to those who were already "news junkies" (Young, 2009: 403). All-news channels may have provided content which was more dramatic and more visually interesting, but it was also more superficial, and often "predictably repetitive" (Sjøvaag, 2013: 57). They offered a "quick-fix news diet" for short-term casual viewing (Cushion, 2012: 64), in some instances acting as a kind of appetizer for more considered newspaper reading (Vasanti, 2008). Furthermore, the 24/7 time slot was often padded out with extraneous material (Larson, 2013: 384). The impact on newspapers of 24/7 television news in this period should not be conflated with that of the internet from the mid-1990s, but which grew particularly in the first years of the 21st century following the introduction of Web 2.0 (Hanson, 2007: 117–118); nor with the economic effects of the global financial crisis, starting in 2007 (Macnamara, 2012). 24/7 television added to the supply of news at a time when the news consumption overall was declining in many places. It provided a "supplementary" source of news (Sjøvaag, 2013: 53). This may have further disaggregated the audience for news (Seib, 2001: 5–6). However, in order to truly displace newspapers, 24/7 television channels needed to outperform newspapers in satisfying the demand for news (Bucy, Gantz and Wang, 2007: 148–149). Where they had an advantage was in immediacy, in terms of both speed (breaking news) and live-ness (on-the-spot reporting) (Bivens, 2015: 197; Sjøvaag, 2013: 61), facilitated by the technological developments in the 1980s and 1990s, including videotape, microwaves, wireless and evolving satellite connectivity (Geisler, 2014; Hope, 2011). Printed newspapers could not compete on these grounds (Hanson, 2007: 117–118). Nevertheless, audiences for 24/7 television news remained selective and small (Cushion, 2012: 64; Young, 2009: 406).

A more satisfactory analysis, given that most households and individuals (at least in the West) owned multiple devices capable of accessing news, may be that of "newsfulness"—the extent to which media platforms (independent of content or even form) satisfy users' requirements for accessing news (Chyi and Chadha, 2012: 434). This may include how much pleasure users get out of accessing news through any particular media platform (Chyi and Chadha, 2012: 436). Together, these factors may indicate the news compatibility (measured in terms of use) of competing and complementary platforms. A survey of Americans found that old media (television followed by printed newspapers) were the most enjoyable to use for accessing news. Desktop and laptop computers were rated more highly than the iPhone (Chyi and Chadha, 2012: 441). Other electronic devices (iPod Touchs, e-readers, iPads) were simply not used very much for accessing news (Chyi and Chadha, 2012: 442). This suggested that, rather than one media form displacing another, each adopts a functional space in relation to the others (Friedland, 1996: 69).

CONCLUSION

The secular decline in newspaper sales was apparent for at least three decades before CNN went to air (as far back as the late 1920s in the US). As noted, that trend was particularly evident in Australia, even though pay-TV and 24/7 news channels developed there patchily and late (Young, 2010: 245–247). External social and economic factors, such as changes in labour practices and the expansion of suburbs, altered the structure of the newspaper industry independently. In both the US and Australia, evening newspapers all but ceased to exist as their mainly working-class male readership taking public transport home after a standard working day in industry disappeared (Donovan and Scherer, 1992: 300–302). Jain (2010: 165) noted the alignment of the emergence of 24/7 television news in India with "the shifting role of the middle class." By the 1980s, the emergence of the 24/7 society, driven by rising affluence and the availability of increased disposable income among higher earners, had been widely recognised. Almost all routine activities—shopping, banking, eating, travelling, working, accessing health care, leisure and, of course, communication—could be undertaken outside "normal" daytime hours (Loveday et al., 2008). In processes of communication, time became "fluid" (Castells, 2010). Although facilitated by technology and exploited economically, the emergence of the 24/7 society was a "fundamentally cultural" phenomenon (Castells, 2013: 135). Factors such as peer pressure worked to institute a process of social customisation whereby humans were re-programmed, as it were, to be in sync with the 24/7 world (Williams, Coveney and Gabe, 2013). The "news habit" changed accordingly. In the 24/7 society people no longer waited for newspapers to be printed or news bulletins to be broadcast, they began to "consume news in a steady stream of information bites" (Bird, 2009: 293). But they did not abandon old media altogether: their disappearing readers, listeners and viewers were probably not news users at all.

The modern mass media ecology, as defined by those platforms which added "reach and status to journalism's repertoire" (Conboy, 2011: 22), was constituted historically of a "family of technologies and uses" (Heinderyckx, 2013: 107). Although over the last 40 years or so of the 20th century, newspapers, magazines, television and radio constituted the core of the modern mass (news) media landscape—in the UK at least, this has often been seen as a "gilded age" of these media (Brock, 2013: 55)—the adoption of new forms and formats, including from the 1980s, 24/7 news television, could hardly be considered to be exceptional—or terminal.

NOTES

1. It may be argued that the end of the first and the beginning of the second phases were marked by the introduction of privately owned 24/7 satellite news channels in India in 1998 (Mehta, 2010: 321).
2. Switzerland was an outlier.

REFERENCES

American Press Institute (2014). 'How Americans get their news', *The Personal News Cycle: How Americans Choose to Get Their News*. Available at http://www.americanpressinstitute.org/publications/reports/survey-research/how-americans-get-news/: accessed May 2 2015.

Anon (2014). 'Read it and leap', *The Economist* (May 24). Available at http://www.economist.com/news/business/21602714-new-york-times-ponders-bold-changes-needed-digital-age-read-it-and-leap: accessed May 7 2015.

Atkinson, C. (2014). 'Al Jazeera America's ratings struggle leads to layoffs', *New York Post* (April 11). Available at http://nypost.com/2014/04/11/al-jazeera-americas-ratings-struggle-leads-to-layoffs/: accessed May 7 2015.

Baldwin, T.F., Barrett, M. and Bates, B. (1992). 'Profile: Uses and values for news on cable television', *Journal of Broadcasting & Electronic Media*, Vol. 36(2), 225–233.

BBC (2014). 'BBC News to cut a further 425 jobs' (July 17). Available at http://www.bbc.co.uk/news/entertainment-arts-28342929: accessed May 7 2015.

BBC (2015). 'BBC's combined global audience revealed at 308 million', BBC Media Centre (May 21). Available at http://www.bbc.co.uk/mediacentre/latestnews/2015/combined-global-audience: accessed May 22 2015.

Bird, S.E. (2009). 'The future of journalism in the digital age', *Journalism*, 10(3), 293–295.

Bivens, R. (2015). 'Affording immediacy in television news production: comparing adoption trajectories of social media and satellite television', *International Journal of Communication* 9, 191–209.

Brock, G. (2013). *Out of Print: Newspapers, Journalism and the Business of News in the Digital Age*. London: Kogan Page.

Bromley, M. (2010). '"All the world's a stage": 24/7 news, newspapers and, the ages of media', S. Cushion and J. Lewis (Eds.), *The Rise of 24-hour News Television: Global Perspectives* (pp. 31–49). New York: Peter Lang.

Bucy, E. P., Gantz, W. and Wang, Z. (2007). 'Media technology and the 24-hour news cycle', C. A. Lin and D. J. Atkin (Eds.), *Communication Technology and Social Change: Theory and Implication* (pp. 143–164). Mahwah, NJ: Lawrence Erlbaum.

Castells, M. (2010). *The Rise of the Network Society*, 2nd ed. Chichester, UK: John Wiley.

Castells, M. (2013). *Communication Power*. Oxford: Oxford University Press.

Chyl, H. I. and Chadha, M. (2012). 'News on new devices: is multi-platform consumption a reality?' *Journalism Practice*, 6(4), 431–449.

Cohen, J. E. (2008.) *The Presidency in the Era of 24-hour News*. Princeton, NJ: Princeton University Press.

Communications Management (2011). *Sixty Years of Daily Newspaper Circulation Trends*. Markham, Ont.: CMI Canada.

Conboy, M. (2011). *Journalism in Britain: A Historical Introduction*. London: Sage.

Cottle, S. and Rai, M. (2006). 'Between display and deliberation: analysing TV news as communicative architecture', *Media, Culture & Society*, 28(2), 163–189.

Cox, G. (1995). *Pioneering Television News*. London: John Libbey.

Cushion, S. (2010). 'Three phases of 24-hour news television', S. Cushion and J. Lewis (Eds.), *The Rise of 24-hour News Television: Global Perspectives* (pp. 15–29). New York: Peter Lang.

Cushion, S. (2012). *Television Journalism*. London: Sage.

Cushion, S. and Thomas, R. (2013). 'The mediatization of politics: interpreting the value of live versus edited journalistic interventions in UK television news bulletins', *The International Journal of Press/Politics*, 18(3), 360–380.

Davis, D. R. (2006). *The Post-War Decline of American Newspapers, 1945–1965*. Westport, CT: Praeger.

Donovan, R. J. and Scherer, R. (1992). *Unsilent Revolution: Television News and American Public Life*. Cambridge: Cambridge University Press.

Dragomir, M. and Thompson, M. (Eds.) (2014). *Digital Journalism: Making News, Breaking News*. New York: Open Society Foundations.

eBizMBA (2015). 'Top 15 most popular news websites' (May 1). Available at http://www.ebizmba.com/articles/news-websites: accessed May 22 2015.

Edgerton, G. R. (2007). *The Columbia History of American Television*. New York: Columbia University Press.

Edwards, J. (2013). 'TV is dying, and here are the stats from the US that prove it.' *Business Insider Australia* (November 25). Available at http://www.businessinsider.com.au/cord-cutters-and-the-death-of-tv-2013-11: accessed May 2 2015.

Ewing, S., van der Nagel, E. and Thomas, J. (2014). *CCi Digital Futures 2014: The Internet in Australia*. Swinburne University of Technology: ARC Centre of Excellence for Creative Industries and Innovation.

Garraham, M. (2015). 'Death of the TV?' *Financial Times* (April 29). Available at http://video.ft.com/4203968032001/Death-of-the-TV-/Companies: accessed May 2 2015.

Geisler, J. (2014). "24/7 culture: Tips from the best (and worst) of TV." Poynter (November 25). Available at http://www.poynter.org/how-tos/leadership-management/80865/247-culture-tips-from-the-best-and-worst-of-tv/: accessed May 7 2015.

Gunter, B. (1987). *Poor Reception: Misunderstanding and Forgetting Broadcast News*. Mahwah, NJ: Lawrence Erlbaum.

Hallin, D.C. and Mancini, P. (2004). *Comparing Media Systems: Three Models of Media and Politics*. Cambridge: Cambridge University Press.

Hanson, G. (2007). *24/7: How Cell Phones and the Internet Change the Way We Live, Work, and Play*. Westport, CT: Greenwood.

Heinderyckx, F. (2013). 'In praise of the passive media', I.T. Trivundža, N. Carpentier, H. Nieminen, P. Pruulmann-Venerfeldt, R. Kilborn, E. Sundin and T. Olsson (Eds.), *Past, Future and Change: Contemporary Analyses of Evolving Media Scapes* (pp. 99–108). Ljubljana: University of Ljubljana Press.

Hope, W. (2011). 'Global capitalism, temporality, and the political economy of communication', J. Wasko, G. Murdock and I Sousa (Eds.) *The Handbook of Political Economy of Communications*. London: Wiley-Blackwell.

Jain, A. (2010). "Beaming it live': 24-hour television news, the spectator and the spectacle of the 2002 Gujarat carnage', *South Asian Popular Culture*, 8(2), 163–179.

Jurkowitz, M. (2014). 'The losses in legacy' Pew Research Center (March 26). Available at http://www.journalism.org/2014/03/26/the-losses-in-legacy/: accessed May 7 2015.

Langley, A. (2012). 'Europe's newspapers are dying too.' *Columbia Journalism Review* (December 12). Available at http://www.cjr.org/behind_the_news/european_newspapers_in_dire_st.php: accessed May 7 2015.

Larson, C.U. (2013). *Persuasion: Reception and Responsibility*, 13th ed. Boston: Wadsworth.

Lewis, J. (2010). 'Democratic or disposable? 24-hour news, consumer culture, and built-in obsolescence', S. Cushion and J. Lewis (Eds.), *The Rise of 24-hour News Television: Global Perspectives* (pp. 81–95). New York: Peter Lang.

Lewis, J. M. W. and Cushion, S. (2009). 'The thirst to be first: An analysis of breaking news stories and their impact on the quality of 24-hour news coverage in the UK', *Journalism Practice*, 3(3), 304–318.

Liu, Y. and Hsu, W. (2011). 'Media platform competition: the displacement effect of IPTV on cable TV in Taiwan', Y.K. Dwivedi (Ed.), *Global Impact of Broadband Technologies: Diffusion, Practice and Policy* (pp. 258–272). Hershey, NY: Idea Group.

Loveday, D. L., Bhamra, T., Tang, T., Haines, V. J. A., Holmes, M. J. and Green, R. J. (2008). 'The energy and monetary implications of the '24/7' 'always on' society', *Energy Policy*, 36(12), 4639–4645.

Luckerson, V. (2014). 'Fewer people than ever are watching TV', *Time* (December 3). Available at http://time.com/3615387/tv-viewership-declining-nielsen/: accessed May 7 2015.

Mance, H. (2014). 'Shadows creep across the face of European newspapers', *Financial Times* (September 21). Available at http://www.ft.com/cms/s/0/49cf1598-3e56-11e4-b7fc-00144feabdc0.html#axzz3ZSYmzm48: accessed May 7 2015.

McDowall, K. (2012). *Chicken Noodle News: CNN and the Quest for Respect*. Hons thesis, Baylor University.

Macnamara, J. (2012). 'As the 'rivers of gold' dry up, what business model will save the media?' *The Conversation* (June 28). Available at http://theconversation.com/as-the-rivers-of-gold-dry-up-what-business-model-will-save-media-7956: accessed May 28 2015.

Meeker, M. (2014). *Internet Trends 2014*. Menlo Park, CA: Kleiner, Perkins, Caufield, Byers.

Mehta, N. (2010). 'India live: satellites, politics, and India's TV news revolution', S. Cushion and J. Lewis (Eds.), *The Rise of 24-hour News Television: Global Perspectives* (pp. 319–335) New York: Peter Lang.

Miller, M. (2013). 'Cutting the cable cord with streaming internet video', QUE (July 16). Available at http://www.quepublishing.com/articles/article.aspx?p=2101521: accessed May 7 2015.

Milojević, J. S. (2014). 'News choice and offer in the digital transition', M. Dragomir and M. Thompson, M. (Eds.), *Digital Journalism: Making News, Breaking News* (pp. 129–144). New York: Open Society Foundations.

Neil, A. (1996). *Full Disclosure*. London: Macmillan.

Neilsen (2014). *The Total Audience Report*. Available at http://www.nielsen.com/content/dam/corporate/us/en/reports-downloads/2014%20Reports/total-audience-report-december-2014.pdf: accessed May 22 2015.

Newspaper Death Watch (2015). Available at http://newspaperdeathwatch.com/: accessed 7 May 2015.

Norris, P. (2000). *A Virtuous Circle: Political Communication in Post-Industrial Societies*. Cambridge: Cambridge University Press.

O'Donnell, P., Zion, L. and Sherwood, M. (2015). 'Where do journalists go after newsroom job cuts?' *Journalism Practice*. DOI: 10.1080/17512786.2015.1017400.

Patterson, T. E. (2000). 'The United States: news in a free-market society', R. Gunther and A. Mughan (Eds.), *Democracy and the Media: A Comparative Perspective* (pp. 241–265) Cambridge: Cambridge University Press.

Ponsford, D. (2012). 'PG research reveals 242 local press closures in 7 years', *Press Gazette* (April 30). Available at http://www.pressgazette.co.uk/node/49215: accessed May 7 2015.

Preston, P. (2001). *Reshaping Communications: Technology, Information and Social Change*. London: Sage.

Rai, M. and Cottle, S. (2010). 'Global news revisited: mapping the contemporary landscape of satellite television news', S. Cushion and J. Lewis (Eds.), *The Rise of 24-hour News Television: Global Perspectives* (pp. 51–79) New York: Peter Lang.

Sambrook, R. and McGuire, S. (2014). 'Have 24-hour TV news channels had their day?' *The Guardian* (February 3). Available at http://www.theguardian.com/media/2014/feb/03/tv-24-hour-news-channels-bbc-rolling: accessed May 28 2015.

Sampert, S., Trimble, L., Wagner, A. and Gerrits, B. (2014). 'Jumping the shark: mediatization of Canadian party leadership contests, 1975–2012', *Journalism Practice,* 8(3), 279–294.

Santhanam, L. H. and Rosensteil, T. (2011). 'Why US newspapers suffer more than others', *The State of the News Media 2011.* Pew Research Center: Project for Excellence in Journalism. Available at http://www.stateofthemedia.org/2011/mobile-survey/international-newspaper-economics/: accessed May 22 2015.

Seib, P. (2001). *Going Live: Getting the News Right in a Real-Time, Online World.* Lanham, MD: Rowman & Littlefield.

Seidensticker, B. (2006). *Future Hype: The Myths and Technology of Change.* Oakland, CA: Berrett-Koehler.

Sjøvaag, H. (2013). 'Revenue and branding strategy in the Norwegian news market: the case of TV 2 News Channel', *Nordicom Review,* 33(1), 53–66.

Smith, A. (2014). 'Online video streaming up 60%, TV consumption down but not out', Reelseo (December 9). Available at http://www.reelseo.com/online-video-streaming-up-60-per-cent/: accessed May 7 2015.

Spigel, L. (1992). *Make Room for TV: Television and the Family Ideal in Postwar America.* Chicago: The University of Chicago Press.

Stirling, C. H. (2009). 'Television news magazines', C.H. Stirling (Ed.), *Encyclopaedia of Journalism.* Thousand Oaks, CA: Sage, pp. 1385–1388.

Taibi, C. and Hart, A. (2014). 'CNN, HLN layoffs hit 300, *Crossfire, Unguarded, CNN Monday, Sanjay Gupta MD* canceled', *Huffington Post* (14 October). Available at http://www.huffington-post.com/2014/10/14/cnn-hln-layoffs_n_5986114.html: accessed May 7 2015.

Tunstall, J. (1996). *Newspaper Power: The New National Press in Britain.* Oxford: Oxford University Press.

Tunstall, J. (2015). *BBC and Television Genres in Jeopardy.* Oxford: Peter Lang.

Ugander, J., Karrer, B., Backstrom, L. and Marlow, C. (2011). 'The anatomy of the Facebook social graph.' Available at http://arxiv.org/pdf/1111.4503.pdf: accessed May 7 2015.

Vasanti, P. N. (2008). 'The appetizer effect of Indian television news.' *Live Mint* (July 25). Available at http://www.livemint.com/Companies/d1IwYWJZ6CIYMPJt3MuhaI/The-Appetizer-Effect-of-Indian-Television-News.html: accessed May 7 2015.

Wheeler, T. (2015). Prepared remarks. NAB show, Las Vegas (April 15). Available at https://www.fcc.gov/document/prepared-remarks-chairman-tom-wheeler-2015-nab-show-las-vegas: accessed May 2 2015.

Williams, S. J., Coveney, C. M. and Gabe, J. (2013). 'Medicalisation or customisation? Sleep, enterprise and enhancement in the 24/7 society', *Social Science & Medicine,* 79, 40–47.

Young, S. (2009). 'Sky News Australia: the impact of local 24-hour news on political reporting in Australia', *Journalism Studies,* 10(3), 401–416.

Young, S. (2010). 'Audiences and the impact of 24-hour news in Australia and beyond', S. Cushion and J. Lewis (Eds.), *The Rise of 24-hour News Television: Global Perspectives* (pp. 243–262). New York: Peter Lang.

CHAPTER TWELVE

24-Hour News Channels around the Globe: Continuity or Change?

MUGDHA RAI AND SIMON COTTLE

In the last two decades, 24-hour satellite news channels have become ubiquitous around the world. We last explored the growth of this phenomenon a few years ago by comprehensively mapping the full extent and diversified complexity of satellite news ecology around the world (Rai and Cottle, 2010). More recent debates surrounding 24-hour news channels, however, suggest that the genre may be heading towards its decline (Sambrook and McGuire, 2014; Clarke, 2013; Flinn, 2015). It is increasingly argued that in an era of digital technologies with "on-demand" news needs fulfilled through social media and smartphones, the "rolling" news model of 24-hour news channels will cease to remain relevant. So far, however, there has been little attempt to measure any such decline in the status of these channels empirically. This chapter revisits and updates our findings approximately five years later. It once again systematically charts the full suite of channels around the world, highlighting major developments and departures in the last few years, thereby making a case for the extent to which 24-hour television news remains resilient in a digital media environment. The chapter also re-examines the implications of this phenomenon for traditional approaches to the formations and flows of information and culture. These findings continue, we argue, to variously challenge, endorse or qualify major theoretical positions within the field of international and global communications research. Before presenting our findings we will briefly revisit these theoretical positions.

As we have outlined elsewhere (Cottle and Rai, 2008), global news studies have invariably evaluated 24-hour satellite news channels in terms of their contribution to two opposing paradigms—"global dominance" and "global public sphere." Drawing on traditional geo-political economy approaches, global dominance theorists insist that the contemporary satellite news landscape, much like the broader international media market, continues to be dominated by the major Western players and economic processes (McChesney, 2000, 2003; Sparks, 1998; Thussu, 2003). Proliferating 24-hour satellite news channels are simply vehicles for the global expansion of corporate capitalism, engendering a new form of cultural imperialism (Boyd-Barrett, 1998; McChesney, 2000, 2003). Rather than a diversified "global public sphere," the growth of regional and non-Western news channels represents a "CNNization" of television news, with "US-style" journalism increasingly universalised in news structures and content around the world (Thussu, 2003).

Global public sphere theorists, for their part, challenge these generally pessimistic accounts (Chalaby, 2003; Hannerz, 1996; Tomlinson, 1999, 2003; Volkmer, 1999, 2003). Building on Marshall McLuhan's notions of a "global village" (1964), these theorists argue that the "deterritorializing" effect (Tomlinson, 1999, 2003) of satellite news channels coupled with the ability to simultaneously broadcast around the world and bring audiences together during key moments of "breaking news" is engendering the emergence of a genuinely "global public sphere" and laying cosmopolitan foundations of citizenship (Volkmer, 2003: 15). Furthermore, the contemporary global media scene can also no longer be characterized as a one-way flow from the West to the rest given the increasing "contraflows" emerging from satellite networks in non-Western regions (Figenschou, 2014; El-Nawawy and Iskander, 2002; Flourney and Stewart, 1997; Johnston, 1998; Volkmer, 2002, 2003).

The ongoing development of satellite news channels in the last few years, however, has given rise to a reconfigured world media ecology that continues to problematize both traditional geopolitical economy expectations and global public sphere claims. This chapter brings the story up to date, systematically mapping the full range of contemporary 24/7 channels now available around the world and to better encapsulate the regional, transnational and global complexity that defines the changing satellite news landscape. We address, in turn:

(1) the prevalence of 24/7 satellite news channels around the world today;
(2) the underlying ownership structures of these formations; and
(3) issues of 24/7 satellite news "reach" and "access."

THE CONTEMPORARY SATELLITE NEWS MAP: PROLIFERA-TION AND OWNERSHIP

Table 1: 24-hour News Channels around the World Today

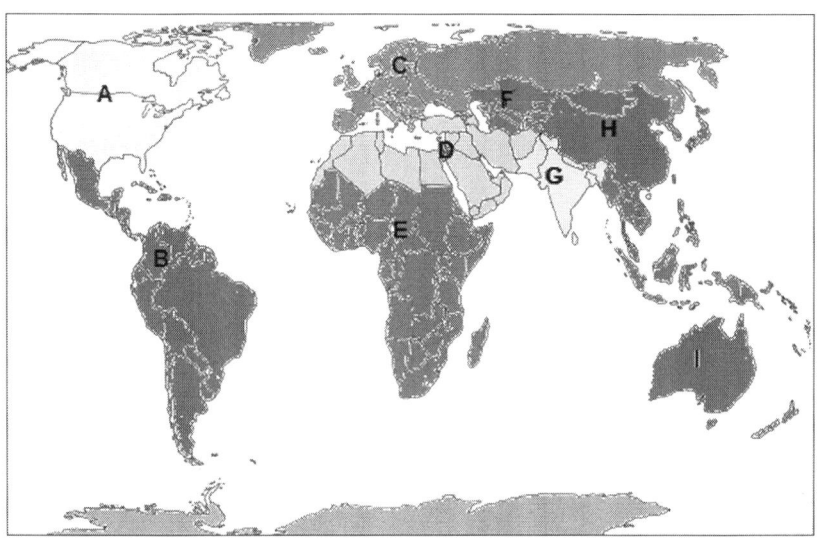

News Channel[1]	Ownership and Language	Reach[2,3]
Global[4]		
CNN & CNNI	Time Warner, various languages	All Regions
BBC World	UK Public Broadcaster – commercial, English	All Regions
Al Jazeera Network	Qatar government-financed, various languages	All Regions
CNBC Network	NBC & Dow Jones	All Regions
Bloomberg TV	Michael Bloomberg	All Regions
Fox News	News Corporation, English	A, B, C, D, H, I
North America		
MSNBC	NBC, English	A (USA)
HLN	Time Warner, English	A (USA)
C-SPAN 1, 2 & 3	US Cable Industry – non-profit, English	A (USA)
Fox Business Network	News Corporation, English	A (USA)
Fusion	ABC Group and Univision, English	A (USA)
Univision Noticias	Univision Group, Spanish	A (USA)

RT America	Russian state-owned RIA-Novosti, English	A (USA)
CTV News Channel	Canada's Bell Globemedia Company, English	A (Canada)
CPAC	Canadian Cable Industry – non-profit, English	A (Canada)
CBC News Network	Canada's Public broadcaster, English	A (Canada)
Business News Network	Canada's Bell Globemedia Company, English	**A (Canada)**
ICI RDI	Canada's Public broadcaster, French	A (Canada-Quebec)
LCN	Canada's TVA Group, French	A (Canada-Quebec)
Local/city-based news channels	Various, English, Spanish	A (US/Canada cities only)
South America		
Todo Noticias	Argentina's Grupo Clarin, Spanish	B, some A
Telesur	Latin American govts, 51% Venezuela, Spanish	B
Canal 5 Noticias	Argentina's Indalo Group, Spanish	B (Argentina)
Cronica TV	Argentine private, Spanish	B (Argentina)
Globovision	Venezuelan private, CNN affiliate, Spanish	B (Venezuela)
Canal i	Private Venezuelan network, Spanish	B (Venezuela)
Globonews	Brazil's Globo Group, Portugese	B (Brazil)
Band News	Brazil's Banderantes Group, Portugese	B (Brazil)
Record News (terrestrial only)	Brazil's Rede Record, Portugese	B (Brazil)
CNN Chile	CNN and Chile's VTR Group, Spanish	B (Chile)
Nuestra Tele Noticias	Colombia's RCN TV, Spanish	B (Colombia)
Cable Noticias	Colombian private, Spanish	B (Colombia)
Canal N	Peruvian private, Spanish	B (Peru)
RPP TV	Peruvian private, Spanish	B (Peru)
Europe		
Euronews	European public broadcasters, various langs	C, some **F**
Sky News	News Corporation majority, English	C, some **H**
France 24	France's public broadcaster, French, Eng, Arabic	C, some **D, F, E**
Russia Today – different versions	Russian state-owned RIA-Novosti, various langs	C, some **D, F**
Ukraine Today	Ukraine's 1+1 Media Group, English	C
Russia 24	Russian state-owned VGTRK, Russian	C (Russia)
BBC News Channel	UK Public Broadcaster, English	C (UK)

RTE News Now	Ireland's Public Broadcaster, English, Irish	C (Ireland)
RAI News 24	Italy's Public broadcaster, Italian	C (Italy)
n-tv	Germany's Bertelsmann, CNN (minor), German	C (Germany)
n24	Germany's SevenOne Media, German	C (Germany)
NPO Nieuws	Netherlands' Public Broadcaster, Dutch	C (Netherlands)
TVE 24 Horas	Spain's Public broadcaster, Spanish	C (Spain), some A
3/24	Catalan public television, Catalan	C (Spain-Catalonia)
RTP Informacao	Portugal's public broadcaster, Portuguese	C (Portugal)
SIC Noticias	Portugal's SIC Network, Portuguese	C (Portugal)
TVI24	Portugal's Grupo Prisa, Portuguese	C (Portugal)
CMTV	Portugal's Cotina Group, Portuguese	C (Portugal)
TV2 News	Denmark's public broadcaster, Danish	C (Denmark)
TV2 Nyhetskanalen	Norway's Egmont, A-Pressen Gps, Norwegian	C (Norway)
Hir Tv	Hungarian private, Hungarian	C (Hungary)
M1	Hungary's Public Broadcaster, Hungarian	C (Hungary)
Antena 3	Romania's Intact Group, Romanian	C (Romania)
N24	Romania's Central National Media, Romanian	C (Romania)
Realitatea TV	Romania's Realitatea-Catavencu Gp, Romanian	C (Romania)
CT24	Czech public broadcaster, Czech	C (Czech Republic)
B92 Info	Serbia's B92 Trust and local businesses, Serbian	C (Serbia)
TVP Info	Poland's public broadcaster, Polish	C (Poland)
TVN 24	Poland's ITI Media Group, Polish	C (Poland)
Polsat News	Poland's Polsat Group, Polish	C (Poland)
TA3	UK's Millennium Electronics, Slovak	C (Slovakia)
La Chaine Info	France's TF1-Bouygues Group, French	C (France)
BFM TV	France's Nextradio TV Group, French	C (France)
i-Tele	France's Canal + Group, French	C (France)
CNN Turk	CNN & Turkey's Dogan Media Group, Turkish	C (Turkey), some D,F
NTV	Turkey's NTV Media Gp, Turkish	C (Turkey), some D,F
Ada TV	Northern Cyprus' Star Media, Turkish, English	C (N.Cyprus, Turkey)
West/Central Asia, Middle East		
i24 News	Israeli private, English, French, Arabic	D, some C and E

Sky News Arabia	Sky plc and Abu Dhabi Media, Arabic	D, some F
Al-Arabiya	Saudi MBC Group, Dubai-based, Arabic	D, some F
Al-Ekhbariya	Saudi Arabian state-run, Arabic	D, some F
Al-Hurra	US-based, US-government funded, Arabic	D, some F
Al-Alam	Iran state-run, Arabic	D (mainly Iran/Iraq)
Press TV	Iran state-run, English	D, some C and E
Africa		
TVC News	Nigeria's Continental Broadcasting, English	E, some C
STV Noticias	Mozambican STV Network, Portugese	E, some C
e NCA	South Africa's etv Group, English	E, some C
SABC News	South Africa's Public Broadcaster, English	E
ANN7	South African private consortium, English	E
KTN News	Kenya's Standard Group, English	E (Kenya)
South Asia[5]		
NDTV 24x7	India's NDTV Group, English	G, D, some A,C,E,H,I
NDTV India	India's NDTV Group, Hindi	G, D, some A,C,E,H,I
CNN-IBN and News 18 India	Time Warner and India's TV18 Group, English	G, D, some A,C,E,H,I
Zee News	India's Essel Group, Hindi	G, D, some A,C,E
Sun Network – multiple channels	Indian Sun Network, various languages	G, D, some A, H
Asianet – multiple channels	India's Asianet Ltd, various languages	G, D
ETV Network – multiple channels	India's Eenadu Network, various languages	G, D
Indiavision	Private Indian network, Malayalam	G, D
TV9	Private Indian network, Telugu	G, D
PTV News	Pakistan's state broadcaster, Urdu & English	G, some F, D
ARY News	Pakistan's ARY Network, Urdu & English	G, some F, D
ABP News	India's ABP Group, Hindi	G, some C, D, H
Doordarshan News (also terrestrial)	Indian Public Broadcaster, Hindi & English	G (India)
Aaj Tak	India Today Media Group, Hindi	G (India)
Headlines Today	India Today Media Group, English	G (India)
Sahara Samay	India's Sahara Group, Hindi	G (India)
Times Now	India's Times Group and Reuters, English	G (India)
News X	India's ITV Network, English	G (India)

NDTV Profit	India's NDTV Group, English	G (India)
IBN7	India's TV18 group, Hindi	G (India)
Live India	India's SAB Network, Hindi	G (India)
India News	India's ITV Network, Hindi	G (India)
India TV	Private Indian network, Hindi	G (India)
TV24	Private Indian network, Hindi	G (India)
Star Ananda	News Corporation and ABP, India-based, Bengali	G (India)
Tara Newz	India's Tara Network, Bengali	G (India)
People TV	India's Malayalam Comm. Ltd, Malayalam	G (India)
Manorama News	India's Malayalam Manorama Gp, Malayalam	G (India)
IBN Lokmat	India's Network 18 and Lokmat Gp, Marathi	G (India)
ABP Majha	India's ABP Group, Marathi	G (India)
News Live	Private Indian network, Assamese, English	G (India)
News One	Pakistan's Interflow Group, Urdu	G (Pakistan)
Dawn News	Pakistan's Dawn Media Group, English	G (Pakistan)
Geo News	Pakistan's Jang Group, Urdu	G (Pakistan)
Express News	Pakistan's Daily Express Group, Urdu and Eng	G (Pakistan)
Indus News	Pakistan's Indus Media Group, Urdu	G (Pakistan)
Waqt News	Pakistan's Nawa-i-Waqt Group, Urdu	G (Pakistan)
Dunya News	Private Pakistani network, Urdu	G (Pakistan)
Local/city-based news channels	Various, various languages	G (Indian/Pakistani cities only)
East Asia		
CCTV News International	China's state broadcaster, English	H, some A,C,E,F,G,I
CNC World	China's Xinhua News Network, English	H, some A and C
NHK World TV	Japan's public broadcaster, English	H, some A, C, I
Channel News Asia	Singapore's MediaCorp (Govt owned), English	H, some G, D
Phoenix News	HK's Phoenix Gp, News Corp. (minor), Mandarin	H
CCTV News	China's state broadcaster, Mandarin	H (China), some F
CTI TV News	Taiwan's CTI Gp, Mandarin	H (Taiwan)
ETTV News & News S	Taiwan's Eastern Broadcasting Co., Mandarin	H (Taiwan)
Era TV News	Taiwan's Era Multimedia Gp, Mandarin	H (Taiwan)
FTV News	Taiwan's Formosa Media Gp, Mandarin	H (Taiwan)
SET News	Taiwan's Sanlih Entertainment Gp, Mandarin	H (Taiwan)

TVBS Newsnet	HK's TVB Group, Mandarin	H (Taiwan)
TVB News	HK's TVB Group, Cantonese	H (Hong Kong)
TVBN2	HK's TVB Group, Cantonese	H (Hong Kong)
TVB i news	HK's TVB Group, Cantonese	H (Hong Kong)
News 1 & 2	HK's Wharf Holdings Ltd, Cantonese	H (Hong Kong)
i-Cable News	HK's i-Cable Group, Cantonese	H (Hong Kong)
TBS News bird	affiliate of Japan's TBS & JNN Groups, Japanese	H (Japan)
Asahi Newstar	Japan's Asahi Group (Sony-minor), Japanese	H (Japan)
YTN	Korea State Electric Corp, CNN partner, Korean	H (Korea), some A
ABS-CBN News Channel	Philippines' Lopez family, Tagalog & English	H (Phil.), some **A, D**
Aksyon TV	Philippines' TV5 Network, Filipino, English,	H (Philippines)
Astro Awani	Malaysia's Astro Group, Malay	H (Malaysia)
TNN 24	Thailand's True Visions Group, Thai	H (Thailand)
Oceania		
Sky News Australia	BSkyB(News Corp), Aust's 7&9 networks, English	I
Sky News Business	BSkyB(News Corp), Aust's 7&9 networks, English	I
ABC News 24 (also terrestrial)	Australia's Public Broadcaster, English	I (Australia)
A-PAC	Australian Cable Industry – non-profit, English	I (Australia)

1. The list of 24-hour news channels was gathered by systematically going through the inventories of each satellite system around the world (www.lyngsat.com). It is accurate as of the date at which the Lyngsat sources were accessed (September 2015).

2. The reach of each channel was estimated fairly conservatively and involved an examination of both the "footprints" of carrier satellites as well as the channel's distribution networks within each region. Regions where a channel was not available through mainstream satellite and cable distribution networks were not included in its reach.

3. Country names, where in brackets, indicate the principal audience of the channel within the region.

4. Channels with a substantial reach in more than 4 different regions have been defined as "global."

5. The South Asian region has been the most dynamic, with new channels entering and exiting the market quite rapidly. This list would best be treated as indicative of the range of channels rather than exhaustive.

Table 1 comprehensively charts the ownership and reach of all global, regional, national and sub-national 24-hour satellite news channels currently broadcast in and around the world today.[1] As we previously noted (Rai and Cottle, 2010), the number of channels remains at well in excess of 100. In addition, the popularity of the genre continues to grow with over 30 new channels emerging around the world since our last update. In contrast, only a handful of channels have closed down.[2] The table also highlights the fact that these channels continue to develop in virtually every region of the globe, broadcasting in a wide array of languages, both within and outside their regions. Clearly 24/7 television news continues to proliferate as a worldwide presence. Another development that Table 1 reaffirms is that while the total number of channels continues to increase, the list of channels with a "global" reach remains limited; with the vast majority (including most of those that have emerged in the last few years) operating principally at regional, national and even sub-national levels. This confirms our earlier finding (Rai and Cottle, 2010) that a gradual "localization" of the 24-hour news genre is also taking place. To better discern the details collated in Table 1, we first provide a brief overview of its salient features, region by region, before moving on to issues of ownership.

As before, at the very top of the table we see that the small list of channels with a "global" reach remains dominated by major Western players—CNN, BBC World, Fox News and the financial channels—lending credence to the thesis of continued Western dominance in news markets. A particularly noteworthy exception to this trend, however, is the rise of the Al Jazeera network over the last few years. In our last update (Rai and Cottle, 2010), the network had limited access to audiences outside the Arab region. Since then, together with its flagship channels Al Jazeera (Arabic) and Al Jazeera International, the network has launched Al Jazeera Balkans (in 2011) and Al Jazeera America (in 2013 - since closed), gaining substantial mainstream distribution in all regions of the world (Abdelmoula, 2015). There were also plans to launch additional channels, including a new channel in Turkey—Al Jazeera Turk—in the next few years (Vivarelli, 2014). As a result of these developments, Al Jazeera has emerged as the first and only non-Western player to reach such high penetration levels around the world that it can be categorised as a "global" network.

As we move into each region, there is also discernible complexity. In North America, in addition to the global US channels, there remain several 24-hour news channels in the US and Canada which are nationally based. There is also continuing growth in an interesting phenomenon we identified previously, local, city-based 24-hour news channels that focus specifically on events within major cities, departing considerably from the "global" orientation of early 24-hour news players. In addition, there is also evidence of more diversity and "contraflow" in the North American region since our last update. New channels have emerged catering to the Hispanic communities within the region which broadcast both in Spanish and English (Fusion, Univision Noticias, local Spanish-language channels). There are also new channels in the US market that could be described as emanating from "foreign" media - RT America (Russia Today) and Al Jazeera America (until its closure) and AJ+. While both these channels base themselves and operate from within the United States, they are owned and funded by larger media networks from foreign countries—Qatar and Russia, respectively—suggesting a greater variety in flows both into and within the region. In South America, the popularity of nationally-based 24-hour news channels remains. New channels have emerged in Argentina (Canal 5 Noticias) and Peru (RPP TV) over the last few years, joining the other nationally-focused channels in Venezuela, Brazil, Chile and Colombia. The regional players in South America remain Argentina's Todo Noticias and Venezuela's Telesur.

Turning to Europe we see continued growth and diversity amongst the substantial array of 24-hour news channels. There is also ongoing evidence of the strong commitment to public broadcasting with new publicly-funded channels emerging in Ireland (RTE News Now), the Netherlands (NPO Nieuws) and Hungary (M1) alongside new private channels in Ukraine (Ukraine Today), Portugal (CMTV) and Northern Cyprus (Ada TV). The vast majority of Europe's channels remain nationally focused, broadcasting in different languages, which perhaps indicates the "limits" of otherwise "deterritorializing" satellite technology when broadcasting in regions as linguistically diverse as Europe. There are some channels emerging in the region, however, that are more "outward-looking" in their approach. In addition to the regional players—Euronews, Sky News and France 24—networks such as Russia Today, with its different language versions in the region and beyond as well as the recently launched Ukraine Today in English and Russian, have clear strategies to try and gain audiences in overseas markets across Europe and North America (TASS, 2014).

The region of West Asia or the Middle East also demonstrates a continued appetite for the 24-hour news genre. The Qatar-based Al Jazeera network remains dominant in the region alongside channels from Saudi Arabia (Al-Ekhbariya), Iran (Al-Alam and Press TV) and the US-based Al-Hurra network. In addition, the Sky News network has launched a new Arabic language channel (Sky News Arabia) in the region in 2012 in a joint venture with an Abu Dhabi media investment corporation. Another more recent and noteworthy arrival is Israel's multi-lingual i24news channel launched in 2013 and pitched to an international audience (Margalit, 2013). The channel is broadcast across the Arab world and is slowly gaining reach in Europe, Africa and some of the US; ironically, due to ownership regulations it is not yet broadcast within Israel (Perez, 2014).

Since our last study (Rai and Cottle, 2010), the region that has undergone the most substantial transformations in the development of 24-hour news channels is Africa. As noted previously, while the African region has access to a number of Western and Asian news channels, for the longest time there had not been a single indigenous 24-hour news channel available. This situation has been altered considerably in the last few years with the launch of several new channels across the continent including in South Africa (SABC News, eNCA, ANN7), Nigeria (TVC News), Kenya (KTN News) and Mozambique (STV Noticias). A number of these channels are also more internationally-focused, broadcasting across the African region and gaining audiences in Europe. These developments indicate that the 24-hour news genre is finally gaining purchase in Africa with the region gradually emerging as a more active player in global information flows.

Moving to South Asia, the news market is still indisputably dominated by the Indian media juggernaut. The appetite for 24-hour news channels continues to be insatiable in this region (Page and Crawley, 2001), with more national and regional news channels emerging across India (News X, India News, TV24, IBN Lokmat, ABP Majha, News Live) and Pakistan (Waqt News, Dunya News). The linguistic diversity of the region is illustrated not only at the regional and national levels, but also in the continuous launch of numerous new local, city-based news channels which broadcast in the different languages of the various metropolitan cities. The dominance of India's media in the South Asian region certainly provides grounds for what Sonwalker (2001) describes as "little cultural imperialisms," but in addition there is also evidence that Indian media are gaining a more global reach with channels such as NDTV and News 18 increasingly breaking into Western, African and East Asian markets.

In the region of East Asia, the 24-hour news genre has also continued to expand albeit at a slower pace. Over the last few years, new channels have emerged in China (CNC World) and the Philippines (Aksyon TV). While the number of 24-hour news channels available in the region is fairly substantial, most of these again are nationally focused. The regional players remain China's CCTV network, Hong Kong's Phoenix News and Singapore's Channel News Asia. Similar to other regions, however, the last few years have also seen the development of more "outward-looking" channels in the region. In addition to CCTV News International, China's newer CNC World channel, and Japan's NHK World are more internationally-focused and are gaining greater distribution outside the Asian region in Western markets.

While the 24-hour news genre remains pervasive in most parts of the world, the region of Oceania (including Australia, New Zealand and the Pacific islands) remains a quiet market. The long-standing solo player, Sky News Australia, has finally been joined by two new channels—ABC News 24, launched in 2010 by Australia's public broadcaster, and A-PAC in 2009, a public affairs channel along the lines of C-SPAN in America. The only 24-hour news channel from New Zealand, the publicly-funded TVNZ7 which launched in 2009, was closed down in 2012 due to funding cutbacks (Cheng, 2011). Furthermore the few channels that do remain in this region are nationally or regionally focused with little evidence of any "outward-looking" interest in audiences outside the region. These limited developments signal the continued peripheral status of the Oceania region in the global 24/7 marketplace of information flows.

The extent to which the growth of 24-hour news channels has pluralized the global news market and weakened the stranglehold of major Western players cannot really be assessed without returning to political economy considerations of ownership and control. Table 1 also charts the ownership status of all 24-hour news channels around the world today. Clearly, there is considerable evidence here to suggest the continued supremacy of major Western corporations in media markets around the globe. Time Warner, through its CNN and CNNI networks already has access to every region on the globe, and this is further reinforced by its commercial interests and affiliations with regional and national channels across South America (Globovision, CNN Chile), Europe (n-tv, CNN Turk), South Asia (CNN-IBN) and East Asia (YTN). The influence of Rupert Murdoch's News Corporation appears to be even more pervasive. The expansive reach of the Fox News network is coupled with substantial commercial interests in Europe (Sky News), West Asia (Sky News Arabia), South Asia (Star Ananda), East Asia (Phoenix News) and Oceania (Sky News Australia). Within regional and national markets, however, ownership structures appear to be more intricate. Most regions include channels owned by a range of different national and local commercial entities. The substantial presence of public and non-profit broadcasters in every region is also noteworthy, particularly in Europe where the public broadcasting tradition seems to have been carefully sustained. Furthermore, this diversity in ownership seems to be increasing—very few of the new 24-hour news channels that have emerged since our last update are owned by or associated with major global media corporations. The vast majority are either managed by public broadcasters or owned by nationally-based private interests. While the dominance of major Western corporations at the global level certainly cannot be overlooked, at regional and national levels ownership configurations reveal an increasing complexity and heterogeneity that offers a less Western-dominated reading of news flows and formations than has been proffered by traditional geo-political economy approaches. This is not to suggest that regional and national formations of capital as well as political elites are not also involved in the regionalized "colonization of communications space" (Boyd-Barrett, 1998), but it does challenge relatively fixed ideas about Western media dominance and opens up a more complex and dynamic field of regional and transnational media organization than traditionally conceived.

Table 1 reveals, then, that the contemporary satellite news landscape remains a dynamic and rapidly expanding one with information flows increasingly overlapping and intersecting both within and across regions. While the major Western players retain a foothold in media markets around the world, traditional configurations of "centre" and "periphery" seem to be increasingly shifting. On the one hand, a handful of non-Western players are increasingly reaching audiences outside their regions generating media flows from the "rest" to the "West." At the same time, there is also evidence exemplifying Sinclair et al.'s suggestion that "'each geo-linguistic region is itself dominated by one or two centres of audio-visual production" (2002: 8). Furthermore, in respect of 24-hour news these configurations are not always aligned with established patterns of media flows within each region. While traditional media centres such as India may be strengthening their media capacity to wage "mini-imperialisms" (Sonwalker, 2001), in other regions, such as South America, the Middle East and East Asia, new and unpredictable centres of media activity have also emerged.[3] This expanding range of channels, however, should not automatically be construed as indicating an increasingly diversified global news market with a "level playing field" available to all. As mentioned earlier, the number of channels with a "global" reach remains small with the vast majority restricted to audiences at regional and national levels.

REACH VS. ACCESS

For many theorists, the arrival of satellite television has revolutionized notions of territory and space, and the ever-increasing range and plurality of channels broadcast over the air space of every region around the world demonstrates the emergence of a genuinely diversified "global public sphere" (Hannerz, 1996; Volkmer, 2003). While satellite technology's ability to transcend borders cannot be disputed, such claims need to be tempered by the enduring structural and economic barriers to access to satellite television that remain in most markets. As outlined in our earlier work (Rai and Cottle, 2010), terrestrial television is "free-to-air" with a well-developed distribution infrastructure in most nations around the world; satellite television, on the other hand, is generally accessible only at a subscription cost (even if individual satellite channels may be "free-to-air") and faces considerably more structural hurdles when it comes to distribution (Parker, 1995). The US and Canadian markets, together with parts of Western Europe, are amongst the most developed with cable and satellite television penetration rates at over 70%, but in most other countries, including the UK and Australia, penetration rates only range between 20 and 50% (Business Wire, 2015; Ofcom, 2015; Burgess, 2015). It is within this subset of homes, then, that the myriad of satellite channels available worldwide compete for access.

Furthermore, gaining access to a satellite television market is not simply a matter of entering its air space, but is contingent on a number of additional factors including partnerships with local pay television providers, subscription rates and the packages of channels on offer (Cozens, 2005). It is at this level that the major global players with entrenched distribution networks in most markets hold considerable advantages over their regional counterparts. These structures of distribution and access arguably reinforce traditional political economy arguments highlighting the continuing supremacy of the major Western players in satellite news markets around the world. While an increasing number of non-Western channels are beamed over the air space of Western nations, the extent to which they are distributed and available within Western households is often fairly limited. To provide a simple analogy, CNNI is available in virtually every Indian household with satellite television, while in the US it is considerably more difficult and expensive to gain access to ZEE News broadcasts. These distribution structures are creating an interesting paradox in which the news markets of the non-Western world, in many cases, are more pluralized, offering a mix of regional and national channels alongside the major Western players, while Western news markets remain dominated by their own channels with few non-Western choices.

In this context it is worth noting, however, that while in many instances the reach of news channels is restricted by the structural limitations described above, the vast majority of satellite news channels around the world today operate principally at national and even local levels, with little interest in accessing global markets. As elaborated in our earlier work (Rai and Cottle, 2010), while satellite technology's ability to broadcast across borders remains undiminished, these channels are choosing to narrow their reach and focus on audiences within national and local markets.[4] Within a number of regional satellite news markets this micro-level approach is more successful with audiences for national and local news channels far outstripping those of global players, whose audiences remain confined to wealthy elites and upper-class professionals across different regions (Chalaby, 2002; Hope, 2004; Parker, 1995; Sonwalker, 2001).

Nonetheless, over the last few years, as outlined in the previous section, more and more news channels from non-Western regions are developing an "outward-looking" approach and trying to make headway in gaining access to international audiences. The prime example of this successful "contraflow" is the Al Jazeera network, which over the last few years has moved beyond the Arab region to successfully negotiate access to media markets around the world including Western regions (Figenschou, 2014). Similarly other regional players such as Russia Today, China's CCTV and CNC World, Japan's NHK World and India's NDTV and News 18 channels are now routinely carried by mainstream satellite and cable distribution networks in Western markets as well. Even relatively new channels that have emerged in Africa and the Middle East (such as i24 News) have quickly gained distribution in parts of Europe.

In this context, the role of digital media technologies and the synergies they offer for satellite television cannot be overlooked. While on the one hand digital media in the form of online news and social media platforms may be seen as in competition with 24-hour television news channels, on the other hand, the improved capabilities of these same digital technologies also offer new opportunities for the distribution of these channels and their content. For example, the majority of 24-hour news channels today are not only available through satellite networks, but also via online streaming from their own websites or YouTube channels. This online distribution has allowed them to enter media markets and gain audiences even in the absence of partnerships with satellite distribution networks. Similarly, as broadband infrastructure has rapidly improved, particularly in advanced media markets, satellite television channels including news channels are increasingly available through IPTV platforms which have much lower set-up costs than traditional satellite TV infrastructure. For smaller, regional and non-Western news players, therefore, these digital distribution technologies have helped reduce barriers to entry into Western markets.

These steadily increasing streams of "contraflow" to the Western world, however, continue to be driven largely by growing populations of non-Western diasporas within most Western states. Mobilizing linguistic and cultural ties, a number of satellite news channels from Asia, Africa and the Arab-speaking world are expanding beyond their regions to reach their diasporic communities within the West (Chan, 2002; Ray and Jacka, 2002; Sinclair et al., 2002; Sonwalker, 2001; Volkmer, 2003). News channels from India and China for example, as Table 1 illustrates, are increasingly available in the markets of the US and Europe where the populations of these ethnic groups are particularly concentrated. For the most part, therefore, the "contraflows" generated by these non-Western channels are limited to diasporic groups within Western markets.

At the same time, some regional players are seeking more active participation in global news flows and agendas. Channels such as CCTV News International, NHK World, Russia Today, Ukraine Today and i24 News, are targeted beyond their diaspora groups to try and reach mainstream Western audiences. They are considered essential strategic tools of cultural diplomacy to extend a nation's "soft power" on the global stage (Jirik, 2010; Kramer, 2010; Hayden, 2012). These 24-hour news channels, then, act as symbolic representatives of disparate national and regional identities within the global news market, advancing their distinct perspectives and interests. By gaining access to Western satellite distribution networks, some of these channels such as Al Jazeera, Russia Today and CCTV News, are starting to break through to mainstream Western audiences, albeit in relatively small numbers, and establishing themselves as alternative sources of news and information (Figenschou, 2014; Abdelmoula, 2015; Pompeo, 2013; Kramer, 2010).

Our findings and the discussion above relating to the structures and ecology of the satellite news landscape indicate that there is some evidence to support both "global dominance" and "global public sphere" positions and interpretations. While the dominance of major Western corporations through their ownership and distribution networks remains pervasive, the rise and expanding reach of today's

24-hour news channels, both from Western and non-Western regions, with their distinct regional, national and local approaches is providing multiple and overlapping tiers of programming. This rapidly expanding satellite news landscape, then, is characterized by an increasing heterogeneity and plurality that encompasses a range of alternative perspectives from around the world, thereby lending support to assertions of the emergence of a genuinely diversified "global public sphere."

CONCLUSIONS

As documented above, 24-hour satellite news today continues to reveal a dynamic, rapidly expanding and increasingly differentiated ecology. In this chapter, we have mapped and updated, comprehensively, the full extent of its proliferation and pervasiveness around the world. With well over 100 channels available today, comprising commercial, public and non-profit broadcasters, at global, regional, national and even sub-national levels, and the number of channels still growing in all regions of the world, the continued popularity and resilience of the 24/7 news genre is beyond dispute. Our research has also underlined once again the increasing complexities of the contemporary satellite news landscape, with overlapping and intersecting flows and formations that problematize conventional theoretical approaches to international communications and news flows in particular. At a transnational level, our evaluations of ownership and reach lend credence to traditional political economy arguments underlining the continued supremacy of major Western players (Boyd-Barrett, 1998; McChesney, 2000, 2003; Thussu, 2003). The continuing growth and expansion of channels from disparate regions around the world, however, could be interpreted as "globalizing," with a more varied assortment of sounds and images creating the beginnings of an increasingly diversified "global public sphere" (Volkmer, 1999, 2003).

The case for "global Western dominance" and cultural imperialism, we suggest, is further weakened by a tendency to overlook the dynamics *within* non-Western news cultures. New channels have continued to emerge in Asia, the Middle East and belatedly, Africa, with no ownership ties to major Western media corporations. These developments have contributed to an increasing plurality, with non-Western and Western players competing side-by-side in 24-hour news markets, thereby challenging discourses on blanket cultural "invasion" and Westernization (Sinclair et al., 2002; Sonwalker, 2001). As our research bears out, in a number of regions around the world today, the news landscape is characterized not by "one-way" flows but rather "multi-directional" flows (Sinclair et al., 2002), which, in some instances, can be interpreted as reinforcing regional "mini-imperialisms" (Sonwalker, 2001), while in other areas, as elevating and empowering erstwhile minor media players (e.g. Singapore, Middle East).

As our discussions on reach and access suggest, however, this increasing complexity should not automatically be construed as indicating "contraflows" back to the West (Volkmer, 1999, 2002, 2003). In spite of satellite technology's transnational broadcasting capabilities, non-Western news players continue to face considerable structural inequalities in their access to Western markets. Having said that, our research suggests that the volume of "contraflows" from non-Western to Western regions is on the rise. Over the years, concerted efforts and negotiations have elicited an opening up of distribution networks with major news channels from Asia, the Middle East (most notably the Al Jazeera network) and Africa increasingly gaining access to Western news markets. As discussed, in this context satellite television news has also benefited greatly from the advanced distribution capabilities of digital media technologies. These developments have stimulated increasing plurality within Western news markets, too, with a greater range of non-Western channels available in North America and Europe alongside their local counterparts.

Our research also underlines increasing complexity in the outlooks and approaches to 24-hour news channels around the world. For example, a number of channels have been conceived from the outset as "outward-looking," often with the express mandate of representing national perspectives and identity to international audiences (e.g. CCTV News, Russia Today, NHK World). At the same time, our findings reaffirm a continued trend towards localization of the satellite news genre with more and more channels around the world catering specifically to audiences within national and even sub-national (US, Canada, India) spheres. Thus, in such ways, the contemporary ecology of satellite news reveals complex stratifications with multifarious flows and formations that can both span across regions stimulating greater global integration and convergence, and concentrate and intensify within regions engendering localization, regionalization and even fragmentation. These intricacies and complexities problematize theoretical orientations towards 24-hour news as essentially emissaries of "global dominance" or a "global public sphere." The ongoing growth and development of the contemporary satellite news landscape revealed in this chapter suggests that the genre of 24-hour news not only remains resilient, but continues to evolve and diversify.

NOTES

1. See note 1 of Table 1 for an explanation of how this list was collated.
2. These are CNN Plus in Spain, TV1 in Malaysia and TVNZ7 in New Zealand.
3. In South America, Venezuela and Argentina as opposed to Mexico and Brazil; in the Middle East, Qatar and Saudi Arabia as opposed to Egypt; in East Asia, Singapore as opposed to Taiwan.

4. Their decision to select a satellite platform could, ostensibly, be explained by the lower barriers to entry that exist in most parts of the world with respect to the satellite television industry as opposed to terrestrial television (licensing, regulation, etc.).

REFERENCES

Abdelmoula, E. (2015) *Al Jazeera and Democratization: The Rise of the Arab Public Sphere*. New York: Routledge.

Boyd-Barrett, O. (1998) 'Media Imperialism Reformulated', in D.K. Thussu (ed.) *Electronic Empires: Global Media and Local Resistance*, pp. 157–76. London: Arnold.

Burgess, J. (2015) 'Will media streaming kill Pay TV?', APCMag.com 18 February 2015, URL (consulted September 2015): http://apcmag.com/opinion-will-media-streaming-kill-pay-tv.htm/

Business Wire (2015) 'Worldwide Pay-TV Penetration to Exceed 50% in Next 2 Years, Says ABI Research', *Business Wire* 18 June 2015, URL (consulted September 2015): http://www.businesswire.com/news/home/20150618005037/en/Worldwide-Pay-TV-Penetration-Exceed-50-2-Years#.Vg1Wkv2hciR

Chalaby, J. (2002) 'Transnational Television in Europe: The Role of Pan-European Channels', *European Journal of Communication* 17(2): 183–203.

Chalaby, J. (2003) 'Television for a New Global Order: Transnational Television Networks and the Formation of Global Systems', *Gazette: The International Journal for Communication Studies* 65(6): 457–72.

Chan, J.M. (2002) 'Television in Greater China: Structure, Exports, and Market Formation', in J. Sinclair, E. Jacka and S. Cunningham (eds.) *New Patterns in Global Television: Peripheral Vision*, pp. 126–60. Oxford: Oxford University Press.

Cheng, D. (2011) 'Axe falls on last public service channel', *New Zealand Herald* 7 April 2011, URL (consulted September 2015): http://www.nzherald.co.nz/nz/news/article.cfm?c_id=1&objectid=10717622

Clarke, S. (2013) 'Whither 24-Hour News Channels in Digital Age?', *Variety* 3 October 2013, URL (consulted September 2015): http://variety.com/2013/tv/news/whither-24-hour-news-channels-in-digital-age-1200685925/

Cottle, S. and Rai, M. (2008) 'Global 24/7 News Providers: Emissaries of Global Dominance or Global Public Sphere?', *Global Media and Communication* 4(2): 157–181.

Cozens, C. (2005) 'Al-Jazeera Finds US Tough Market to Crack', 11 November 2005, *The Guardian*, URL (consulted September 2015): http://www.guardian.co.uk/media/2005/nov/11/broadcasting.iraqandthemedia

El-Nawawy, M. and Iskander, A. (2002) *Al-Jazeera: How the Free Arab News Network Scooped the World and Changed the Middle East*. Cambridge, MA: Westview.

Figenschou, T. U. (2014) *Al Jazeera and the Global Media Landscape: The South is Talking Back*. New York: Routledge.

Flinn, G. (2015) 'Are 24-hour news channels on their way out?' HowStuffWorks.com. 20 February 2015. URL (consulted September 2015): http://entertainment.howstuffworks.com/are-24-hour-news-channels-on-their-way-out-.htm

Flourney, D. and Stewart, R. (1997) *CNN: Making News in the Global Market*. Luton: University of Luton Press.

Hannerz, U. (1996) *Transnational Connections: Culture, People, Places*. London: Routledge.

Hayden, C. (2012) *The Rhetoric of Soft Power: Public Diplomacy in Global Contexts*. Lanham: Lexington.

Hope, W. (2004) 'Global Television News and the Ideology of Real Time', conference paper at *The International Association of Media Communication Research*, Porto Alegre, Brazil, 23–27 July.

Jirik, J. (2010) '24-Hour Television News in the People's Republic of China', in S. Cushion and J. Lewis (eds.) *The Rise of 24-Hour News Television: Global Perspectives*. New York: Peter Lang.

Johnston, C.B. (1998) *Global News Access: The Impact of New Communications Technologies*. London: Praeger.

Kramer, A. (2010) 'Russian Cable Station Plays to U.S.', *New York Times* 22 August 2010, URL (consulted September 2015): http://www.nytimes.com/2010/08/23/business/media/23russiatoday.html?_r=1

LyngSat (2015) (consulted September 2015): http://www.lyngsat.com/

Margalit, R. (2013) 'The Israeli Answer to Al Jazeera', *The New Yorker* 29 July 2013, URL (consulted September 2015): http://www.newyorker.com/news/news-desk/the-israeli-answer-to-al-jazeera

McChesney, R. (2000) *Rich Media, Poor Democracy: Communication Politics in Dubious Times*. Urbana-Champaign: University of Illinois Press.

McChesney, R. (2003) 'Corporate Media, Global Capitalism', in S. Cottle (ed.) *Media Organization and Production*, pp. 27–39. London: Sage.

McLuhan, M. (1964) *Understanding Media: The Extensions of Man*. New York: McGraw Hill.

Ofcom (2015) *The Communications Market: Television and Audio-Visual*. Ofcom Report 2015.

Page, D. and Crawley, W. (2001*) Satellites over South Asia: Broadcasting Culture and the Public Interest*. New Delhi: Sage.

Parker, R. (1995) *Mixed Signals: The Prospects for Global Television News*. New York: Twentieth Century Fund Press.

Perez, G. (2014) 'Drahi asks Netanyahu to air i24News in Israel', *Globes* 20 February 2014, URL (consulted September 2015): http://www.globes.co.il/en/article-drahi-asks-netanyahu-to-air-i24news-in-israel-1000918748

Pompeo, J. (2013) 'Russia Goes Viral: How the RT Network Built a U.S. Audience', NYmag.com 20 September 2013, URL (consulted September 2015): http://nymag.com/daily/intelligencer/2013/09/how-the-rt-network-built-a-us-audience.html#

Rai, M. and Cottle, S. (2010) 'Global News Revisited: Mapping the Contemporary Landscape of Satellite Television News', in S. Cushion and J. Lewis (eds.) *The Rise of 24-Hour News Television: Global Perspectives*. New York: Peter Lang.

Ray, M. and Jacka, E. (2002) 'Indian Television: An Emerging Regional Force', in J. Sinclair, E. Jacka and S. Cunningham (eds.) *New Patterns in Global Television: Peripheral Vision*, pp. 83–100. Oxford: Oxford University Press.

Sambrook, R. and McGuire, S. (2014) 'Have 24-hour TV news channels had their day?' *The Guardian*, 4 February 2014. (consulted September 2015): http://www.theguardian.com/media/2014/feb/03/tv-24-hour-news-channels-bbc-rolling

Sinclair, J., Jacka, E. and Cunningham, S. (2002) 'Peripheral Vision', in J. Sinclair, E. Jacka and S. Cunningham (eds.) *New Patterns in Global Television: Peripheral Vision*, pp. 1–32. Oxford: Oxford University Press.

Sonwalker, P. (2001) 'India: Makings of Little Cultural/Media Imperialism?', *Gazette* 63(6): 505–19.

Sparks, C. (1998) 'Is There a Global Public Sphere?', in D.K. Thussu (ed.) *Electronic Empires*, pp. 108–24. London: Arnold.

TASS (2014) 'International News Channel Ukraine Today Can Start Broadcasting August 24', *TASS Russian News Agency* 14 August 2014, (consulted September 2015): http://tass.ru/en/world/745034

Thussu, D.K. (2003) 'Live TV and Bloodless Deaths: War, Infotainment and 24/7 News', in D.K. Thussu and D. Freedman (eds.) *War and the Media: Reporting Conflict 24/7*. London: Sage.

Tomlinson, J. (1999) *Globalization and Culture*. Chicago: University of Chicago Press.

Tomlinson, J. (2003) 'Globalization and Cultural Identity', in D. Held and A. McGrew (eds.) *The Global Transformations Reader*, pp. 269–77. Cambridge: Polity.

Vivarelli, N. (2014) 'Al Jazeera Expands Global Footprint With Turkish Digital Operation', *Variety* 21 January 2014, URL (consulted September 2015): http://variety.com/2014/digital/news/al-jazeera-expands-global-footprint-with-turkish-digital-operation-1201065912/

Volkmer, I. (1999) *News in the Global Sphere: A Study of CNN and its Impact on Global Communication*. Luton: University of Luton Press.

Volkmer, I. (2002) 'Journalism and Political Crisis in the Global Network Society', in B. Zelizer and S. Allan (eds.) *Journalism After September 11*, pp. 235–47. London: Routledge.

Volkmer, I. (2003) 'The Global Network Society and the Global Public Sphere', *Development* 46(1): 9–16.

The Political Economy and Journalisms of 24-Hour News Culture

CHAPTER THIRTEEN

Financial Challenges of 24-Hour News Channels

ROBERT G. PICARD

News channels, like all broadcast channels, face fundamental cost and revenue challenges that require them to produce and make available content at costs that can be covered by their sources of revenue. News channels typically operate at lower cost per hour for programming than other channels, however, because they do not have to invest in risky—but highly demanded—original drama and comedy or to engage in heavy rights competition for sports rights and desirable contemporary films.

The amount of money required to operate a 24-hour news channel varies significantly among news channels because it is dependent upon the scale and scope of operations, where the broadcasts will be distributed, and the number of languages in which programming is produced. Costs are thus influenced by whether it is a domestic or international operation, who pays for the distribution of the channel, how much of the programming involves live broadcasts, how much live remote broadcasting is undertaken that requires use of satellite links, the number of news packages reused during the broadcast, repetitious use of broadcasts from earlier in the day, the number of bureaus and correspondents, the extent of multiplatform operation, and other operational factors.

Nonetheless, the basic financial requirements for operating a 24-hour news channel are rather straightforward. An international broadcaster, or one in a large nation, will require £20–35 million ($30–50 million) for start-up costs including facilities, equipment, personnel, and service contracts. In addition, a large news operation will require £65–250 million ($100–400 million) annually for operations. Channels operating in small nations, provincially or locally, require significantly less money, usually requiring £3–15 million ($5–25 million) in start-up costs and £10–65 million ($15–100 million) in annual operating costs.

How 24-hour news broadcasters obtain the financing required also varies significantly and is complicated by the increasing number of channels and online news providers, declining average audiences and advertising revenue, and the extent to which operating costs must be covered by the channel or another party.

Almost all 24-hour news channels face the common challenge that viewers and advertisers alone generally do not provide sufficient funding and that other arrangements to support their operations are necessary. The importance of non-market funding to 24-hour news has been a characteristic since its inception and owners of channels have been induced to operate because of policy requirements or because they are willing to endure operating losses on news channels because they are deemed important for company brand purposes. In other cases, subsidies are often provided by governments and politically active owners in order to achieve some desired foreign or domestic influence.

Particular challenges faced by broadcasters are that 24-hour news channels are a highly niche offering that tend to attract a more limited audience than entertainment channels, and they tend to be viewed for fewer minutes a day than other variety and thematic channels. The size of audiences is also influenced by domestic populations in their countries of origin and by their abilities to aggregate audiences internationally. Even in the United States, with a population of 320 million people, total average daily viewership of Fox News Channel, CNN, and MSNBC combined is only 1.8 million (Cable News Audience, 2015).

Nevertheless, the availability of satellite transmission capability, improved cable capacity, and the development of digital television have induced broadcasters in many countries to consider and launch news channels. The switch to digital terrestrial broadcasting led many public service broadcasters in Europe to enter 24-hour news provision on one of the new multiplex channels they gained in the transition. Doing so was seductive because it fit the criteria of their remits and journalists desired more broadcast time for their journalism. Many saw 24-hour channels as a tactical choice supporting strategic moves to provide news across platforms, including internet and mobile services.

Because the channel was essentially free to European public service broadcasters and the news channel could be built on the infrastructure of existing news operations, many believed they could effectively operate the news channels, but most have done so with mixed success. The primary challenges have been attracting audiences and paying the costs for operations at a time when most public service broadcasting budgets have been contracting. An example of the challenge is seen in the experience of YLE in Finland, which launched its 24-hour news channel (YLE24) in 2007, but shut it down within a year because the costs of operation could not be justified by the audiences it generated and because major news events would go live on YLE 1, the broadcaster's primary channel, where most people would turn to watch.

State-supported, privately supported and fully commercial 24-hour news channels typically operate under different financial arrangements that influence their start-up and operating capital, revenue, operating costs, audience performance, and how financial losses are handled. Most broadcasters are not very transparent about their funding and expenses, either not making budgets available or obscuring them by reporting them as consolidated expenditures with other enterprises. Nevertheless, industry sources regarding subscriptions and advertising and other information provides insight into their financial structures and performance.

COMMERCIALLY-OPERATED CHANNELS

Commercially-operated channels are typically operated in hopes of profit or for reputational gains by a media company with other profitable operations. Start-up capital is typically provided by an existing company or a joint venture of two or more companies in hopes of the future financial or brand benefits. The enterprises perceive opportunity and are willing to invest in pursuit of that opportunity.

The development of cable television in the United States created the first real opportunity to offer 24-hour news because cable bypassed limited radio spectrum and for competitive reasons needed to provide channels that were not available in terrestrial television. Although 24-hours news could easily have been provided as an early cable service, along with movies, music, and sports, no enterprise began doing so because the potential financial rewards were low. In addition, no one offering cable channels had existing news operations and experience and having to operate throughout the day was even beyond the capacity of the existing terrestrial broadcast networks' news departments. That situation was changed, however, when regulators began requiring that cable providers include public affairs/new channels in their services. Rather than start their own operations, the thousands of independent cable systems spread across the US began looking for a supplier of news and public affairs programming that would meet the requirements.

That demand was met with the establishment of CNN in 1980. The initial business model for CNN in the US was to provide the channel to cable systems at about a penny per subscriber, allowing cable systems to offer 24-hour news as part of the basic tier services as an inducement for more households to subscribe to their cable services. This created a large, stable aggregate revenue stream for CNN. Existing broadcast networks with news operations remained reticent to establish news channels because they saw cable as a competitor, so an entrepreneurial regional broadcaster—Ted Turner, who operated a mediocre television station in Atlanta—saw the possibility and created CNN, which became financially successful within 5 years. The cable subscription model, however, was not available when CNN began the CNN Airport Network (primarily funded by travel-related advertisers) and began expanding internationally, so start-up funding was provided by shareholders and joint ventures with domestic partners in other countries.

When Fox News Channel was launched in 1996 it understood that access to cable systems was crucial and it immediately gained access to 17 million cable subscribers by agreeing to pay the cable providers for carrying the channel. It has since grown to reach 87 million households and cable systems that now pay for it to be included in their offering, but the channel nevertheless attracts only about 1.1 million average daily viewers—still enough to attract significant advertising revenue.

Getting the audience and financial arrangements coordinated to support operations is a challenge and in the past decade numerous commercially operated national news channels have failed to do so and left the market. These include national channels such as ITV News in the UK, TV4 News in Sweden, and Sun News Network in Canada, as well as local/provincial channels such as News 24 in Houston and CityNews in Toronto. The common factors across them have been the inability to attract sufficient audiences to cover costs through subscription or advertising revenue and that they have all operated in crowded news marketplaces with multiple national and international television news providers (Britain's ITV News, 2005; ITV News Channel, 2005; De Vynck, 2014; Sun News Network, 2015; Quebecor's Sun News Network, 2015; Sweden TV 4, 2013; Parks, 2004; Marotte, 2013).

Potential rewards can be high for successful commercial news providers. Larger commercial 24-hour operations in the US have large cash flows and returns as high as 60%, according to analysts. The three main 24-hour channels were projected to produce total revenues of more than $3.5 billion (£2.25 billion) in 2014–15 (Figure 1).

	Subscriber revenue	Advertising Revenue	Profit/Profit margin
CNN	$710 million	$339 million	$327 million/29%
Fox	$1.2 billion	$794 million	$1.2 billion/61%
MSN	$274 million	$221 million	$206 million/41%

Figure 1. Revenues and profits of leading US news channels.
Source: State of the Media, 2015

CNN was projected to experience the strongest rate of growth in advertising, up 6% to $339 million (compared with Fox's 2% growth to $794 million and MSNBC's projected decline of 5% to $221 million). Of the three main news channels, only Fox News was projected to grow its profits in 2014 (up 10% to $1.2 billion). CNN was projected to experience a 5% decline in profit to $327 million, and MSNBC was projected to decline by 8% to $206 million. Still, all three channels continued to enjoy double-digit profit margins: 61% at Fox, 41% at MSNBC and 29% at CNN (State of the Media, 2015).

Internationally, CNN has 31 foreign bureaus, as well as the 9 in the US, and it has national partnerships with local broadcasters and news organizations around the world that create a network of news and video sharing affiliates. Overall, about half the total revenue for CNN is produced by subscriptions and the second half from advertising and ancillary services ("Cutbacks at CNN", 2014)

STATE-SUPPORTED CHANNELS

State-supported channels receive funding from governments or through government-controlled or-supported entities that provide funds for their operations. The bulk of internationally operating 24-hour news channels fall into this category. The degree of independence from the state in these relationships varies considerably, but state funding influences structural and operational choices and finances even in broadcasters that enjoy a high degree of editorial independence. This occurs because the interests of state funders vary over time with changing priorities and geopolitical focuses of their attention.

This issue is clearly illustrated by the BBC World Service and its relationship with the Foreign and Commonwealth Office, which help fund its operations. Although generally independent in its daily news decisions and provision, the World Service clearly conveys a British perspective on the world and plainly articulates British government positions. The strongest influence from the state support has historically been evident in the audiences that it pursued internationally and how its resources were spent. It was influenced to provide services that matched the Foreign Office's interests, which varied over time to include services to colonies in the Empire, then to then-Commonwealth nations, and ultimately to regions of interest because of geopolitical developments. These policies at various times emphasized or de-emphasized regions such as Asia, Africa, Central/Eastern Europe, and the Middle East. The influences altered the BBC's positioning and use of terrestrial and satellite transmitters and distributions systems, locations of foreign bureau, the amount of staffing, and the languages used of news output.

In recent years, the BBC's £250–300 million annual costs for operations of its 24-hour global channels have been progressively shifted from grants from the Foreign and Commonwealth Office and the Cabinet Office to general license fee funds and commercial revenue that it is able to generate (BBC Full Financial Statements 2012/13).

France 24, which operates French, English, and Arabic news channels, has struggled with financing for the two decades of its existence since President Jacque Chirac determined that France needed a 24-hour news channel to improve presentation of France and French perspectives on global and European developments (Hare, 1995). After being restructured and relaunched several times, France 24 became a wholly-owned subsidary of the French Government, operated though a holding company. In 2012, its budget was approximately €100 million (£72 million) per year ("Faits et chiffres," 2012). Capital for the venture came from French taxpayers, as does annual subsidies to cover operating losses. Its annual budget under its new incarnation in 2006 was set at €80 million and the French government expected that it would lose money annually for many years, perhaps as much as €100 million annually (Carlin, 2006). Today the channel broadcasts in French, English and Arabic, reaching 250 million TV households and an average weekly audience of 41.7 million viewers in 177 countries (France 24, 2015).

RT (Russia Today) was established as a purportedly independent Russian international broadcasters with about $30 million capital provided by the Russian news agency RIA Novosti in 2005 (Knobel, 2005; Painter, 2006). Nearly half the network's budget came from the state and the other half from banks and companies aligned with the government, with its budget rising to $380 million in 2011, but being reduced to $300 million in 2012 due to national economic challenges. Russian President Vladimir Putin prohibited funding to be reduced further in autumn 2012 because of its importance in representing Russia abroad (Maczka, 2012; Burrough, 2013; Putin forbids, 2012).

One of the best-known 24-hour news broadcasters from outside Europe is Al Jazeera, which was founded in 1996 and began 24-hour operations in 1999. It is owned by the state of Qatar and private investors. It is highly influential in the Arabic-speaking world and attempted to expand its influence further by launching an English-language channel in 2006 that is now available in 130 countries. Start-up financing of $137 million was provided by the Emir of Qatar (Miles, 2004). It spent $500 million in 2005 to buy an existing channel in the US to convert it into an Al Jazeera America Channel—an unsuccessful initiative that cost about $2 billion (Atkinson, 2015; Koblin, 2015).

Despite the challenges facing international operations, no 24-hour broadcaster today is expanding more rapidly than China Central Television, a state-owned broadcaster that is internationalizing as a "soft power" influence (Nelson, 2013). It now broadcasts 7 channels (including a 24-hour news channel launched in 2012) in English, French, Spanish, Arabic and Russian, as well as Chinese, to more than 140 countries. Financing for the news channel is not transparent, but in

2012, CCTV as a whole received $2.6 billion in advertising revenue—about 1/3 of all advertising revenue in China—and it is funding its growth with that income and hundreds of millions of pounds in special government grants to help its global expansions (Si Si, 2012).

PRIVATELY-SUPPORTED CHANNELS

Privately-supported channels are operated by private individuals and enterprises, sometimes associated with ruling elites, and differ from fully commercial news channels in that they typically are not expected to make a profit because they are operated for other economic or political purposes.

The country with the most extensive array of such news channels is India, where broadcasters offer about 400 news channels and about 125 operate 24-hours a day providing service in at least 14 major regional languages. Most are operated by owners of real estate firms, industrial conglomerates, and politicians to promote regional political interests and parties and they tend to be highly politicized and commercialized. Despite tendencies to pander to advertiser interests, most still lose money annually, with losses covered by the owners to support their other ambitions (Mehta, 2008, 2009).

A similar situation is found in Arabic-language broadcasting today, where many individuals are willing to underwrite the costs of channels for domestic and regional political and social influence and personal aggrandizement. A Saudi Arabian prince started the Al-Arab news channel in Bahrain 2015 to support his view of the region and challenge other Arab news channels, but immediately ran into difficulties with Bahrain authorities (Black, 2015).

OPERATING EXPENSES OF CHANNELS

Some understanding of the operating expenses of 24-hour channels can be obtained by observing expenditures of financially transparent broadcasters such as the BBC and CBC. They show that the costs are primarily content, administrative overhead costs, and distribution costs.

The domestic 24-hour BBC News Channel's costs for operations were £61.5 million ($95.9 million) in 2013, for example, about £168,500 ($262,735) per hour. Three-quarters of the costs were attributable to content, 14% for overhead costs, and 12% for distribution (Figure 2).

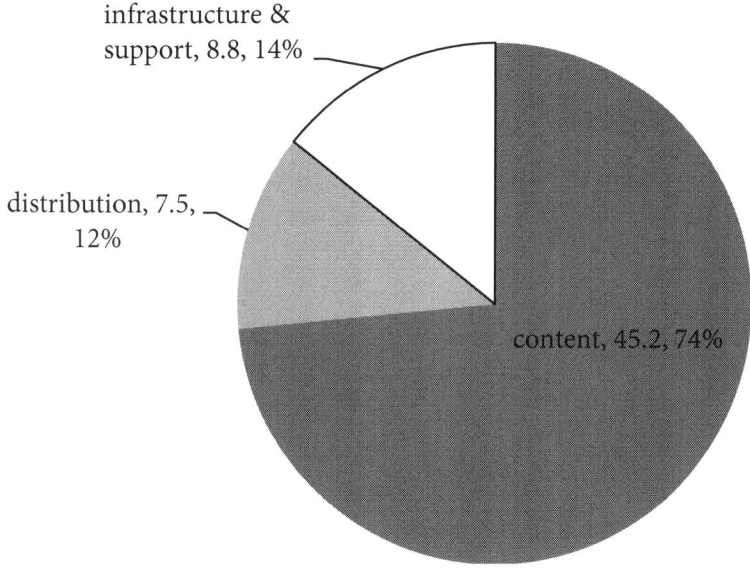

Figure 2. Expenses for operating BBC News Channel, 2013 (£m and % of total).
Source: BBC Full Financial Statements 2012/13

Operating costs for the BBC World Service, the international 24-hour channels, were £284 million. Distribution costs accounted for £33.8 million, 12%, of its expenses in 2013.

Across the Atlantic, the Canadian CBC News Channel operates as a commercial-funded entity and in 2013 earned a pre-tax profit of CAN$20.4 million, a 23.5% return on its revenue of CAN$86.5 million, while serving 11.3 million subscribers. The cost per hour for operations was thus about CAN$176,712 (£92,025). The channel receives the bulk of its revenue from cable and satellite subscription fees and about a quarter from national advertising. The news channel obtains some benefit and synergies from its relationship to the license-fee funded CBC news operations, but it is also highly supported by state policy that requires the national news channels to be carried by cable and satellite service providers and thus provide payments to the broadcasters. Because its costs of distribution are borne by those cable and satellite operators, its expenditures are primarily for content and about 90% of its expenditures are for new content (see Figure 3; Canadian Radio-television and Telecommunications Commission, 2014; Wingrove, 2013).

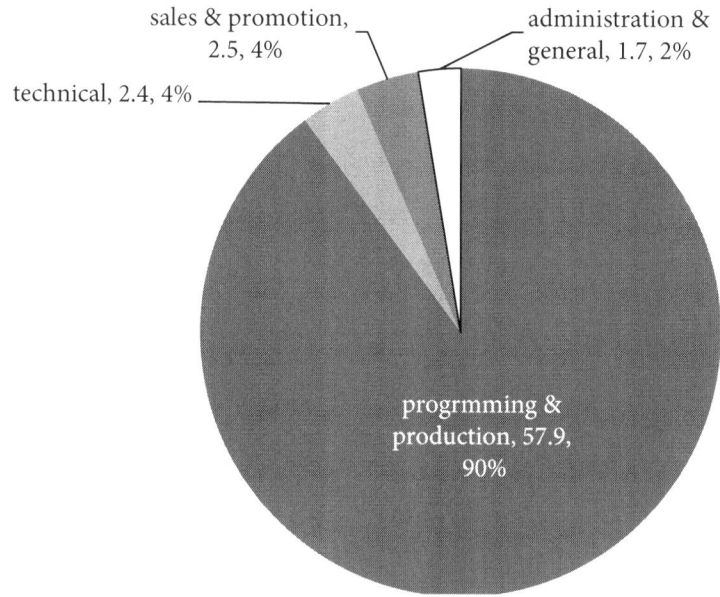

Figure 3: Expenses for operating CNC News Channel, 2013 (CAN$ and % of total).

Internal subsidies from other company operations can be significant for 24-hour channels. In the years after its 1989 establishment, Sky News operated on a budget of £40 million with a £10 million share of overhead contributed by the company—about 20% of its total costs (Chippendale and Franks, 1991). Such cross-subsidization helped it develop into an international 24-hour news channel reaching 100 million homes in 118 countries.

DISCUSSION

Given the financial realities of 24-hour news, it is apparent that the market can support only a small number of international and domestic channels. Nevertheless, the number of channels is proliferating globally, regionally, and in a number of national settings. Consequently, financial consideration of 24-hour channels requires contemplation of what the broadcasters are trying to do, who they are trying to reach with their content, and how their content is distributed. Only with that context can one fully judge the financial performance or efficacy of their operations.

A common performance feature across all types of 24-hour news channels is that they generate relatively low audiences and in most cases incur financial losses. The challenge of traditional financial analyses, however, is that the small audience might be the best audience if the intent of the broadcaster is to gain prestige or influence rather than financial gain. Non-pecuniary rewards are especially high for operators of 24-hour news channels, so those must be considered in understanding why channels remain in operation and why new entrants are appearing.

Even when owners are willing to bear financial losses, many news operations appear to moderate those losses by practices that reduce costs. Most operate only a limited number of domestic and foreign new bureaus and attempt to reduce the costs of production by repetition of previously broadcasted news segments and packages. Although these practices reduce costs, they also reduce the quality of the broadcasts in ways that can reduce audiences or the amount of time they spend watching a broadcaster.

Cost of distribution increases when global audiences are sought because continual uses of satellites incur significant costs and reliance on broadband distribution systems often requires payments to their operators, unless there is clear benefit for them to carry the channel or they are required to do so by domestic policy.

Because distribution can account for 10–15% of operating costs, opportunities to reduce those costs by alternative means are attractive to managers. Broadband internet distribution has distinct cost advantages over broadcasting, cablecasting and satellite distribution, and has the potential for global reach. The availability of broadband is highly skewed toward developed nations, however, making it less effective in reaching audiences where the majority of the world's population lives and broadband can be subjected to domestic control that blocks channels a government does not wish distributed in the country by that mechanism.

Because so many of the world's 24-hour news channels are dependent upon governments and individuals willing to bear operating costs and losses for their operations, they face the twin challenges of funding stability and organisational sustainability. As long as their patrons are able and willing to provide funds, and as long as the broadcasters are seen as delivering the results desired by their benefactors, the channels will be able to survive. Given their challenges in attracting audiences and concurrent difficulties in attracting advertising and market funding, however, the stability of that support is questionable and the ability of channels to survive without it is negligible.

REFERENCES

Atkinson, Claire (2015). Al Jazeera America angers staffers with shift in direction, *New York Post*, April 21. http://nypost.com/2015/04/21/al-jazeera-america-angers-staffers-with-shift-in-direction/

BBC Full Financial Statements 2012/13. London: British Broadcasting Corporation, 2013.

Black, Ian (2015). Saudi prince's Al-Arab news channel goes off air hours after launching, *The Guardian*, Feb. 2. http://www.theguardian.com/world/2015/feb/02/saudi-prince-alarab-news-channel

Britain's ITV News channel to shut down, UPI, Dec. 14, 2005, http://www.upi.com/Entertainment_News/2005/12/14/Britains-ITV-News-channel-to-shut-down/47301134590550/

Bullough, Oliver (2013). Inside Russia Today: counterweight to the mainstream media, or Putin's mouthpiece? *The New Statesman*, 10 May, 2013 http://www.newstatesman.com/world-affairs/world-affairs/2013/05/inside-russia-today-counterweight-mainstream-media-or-putins-mou

Cable News Audience Shrinks (2015). State of the News Media 2015. http://www.journalism.org/2015/04/29/cable-news-fact-sheet/

Canadian Radio-Television and Telecommunications Commission, CBC News Network, 2009–2013 Financial Summaries: Pay & Specialty Services. 2014. http://www.crtc.gc.ca/eng/publications/reports/BrAnalysis/psp2013/individual/62.htm

Carlin, Dan (2006). CNN, BBC, Al Jazeera…and France 24? Dec. 4, Bloomberg. http://www.bloomberg.com/bw/stories/2006-12-04/cnn-bbc-al-jazeera-dot-dot-dot-and-france-24-businessweek-business-news-stock-market-and-financial-advice

Chippindale, Peter and Franks, Suzanne (1991). "Tie Me Kangaroo Down Sport". *Dished! The Rise and Fall of British Satellite Broadcasting*. New York: Simon & Schuster. p. 262.

Cutbacks at CNN highlight the cable news paradox. Pew Research Center. Oct. 9, 2014. http://www.pewresearch.org/fact-tank/2014/10/09/cutbacks-at-cnn-highlight-the-cable-news-paradox/

De Vynck, Gerit (2015). Canada's Sun News Shutting Down After Failing to Find Buyer, Feb. 13, 2015 http://www.bloomberg.com/news/articles/2015-02-13/canada-s-sun-news-channel-shut-down-after-failing-to-find-buyer

"Faits et chiffres". Audiovisuel Exterieur de France. Retrieved March 1, 2012.

France 24 (2015). The Company, http://www.france24.com/en/company

Hare, Geoffrey (1995). All-news television channels in France: LCI (La Chaîne d'information) and Euronews, *Modern & Contemporary France*, 3(2):188–190.

ITV News Channel signs off, *Broadcast*, Dec. 23, 2005 http://www.broadcastnow.co.uk/itv-news-channel-signs-off/1033842.article

Knobel, Beth (2005). "Russian News, English Accent: New Kremlin Show Spins Russia Westward", CBS News, Dec. 12.

Koblin, John (2015). Al Jazeera America, Its Newsroom in Turmoil, Is Now the News, *New York Times*, May 5. http://www.nytimes.com/2015/05/06/business/media/al-jazeera-network-in-turmoil-is-now-the-news.html

Painter, James (2006). *The boom in counter-hegemonic news channels: a case study of TeleSUR*, Reuters Institute for the Study of Journalism, University of Oxford.

Maczka, Marcin (2013). The Propaganda Machine, New Eastern Europe website, July 9, 2012, originally published in *New Eastern Europe: New Europe, Old Problems* 3(IV), 2012.

Marotte, Bertrand (2013). Rogers dumps CityNews channel after 20 months, *The Globe and Mail*, May 30, http://www.theglobeandmail.com/report-on-business/rogers-dumps-citynews-channel-after-20-months/article12259496

Mehta, Nalin (2008). *India on Television: How Satellite News Channels Have Changed the Way We Think and Act*. New Delhi: HarperCollins.

Mehta, Nalin, ed. (2009). *Television in India*. London: Routledge.

Miles, Hugh (2004). *Al Jazeera: The Inside Story of the Arab News Channel That Is Challenging the West*. London: Grove Press.

Nelson, Ann (2013). *CCTV's International Expansion: China's Grand Strategy for Media?* Report to the Center for International Media Assistance, Oct. 22.

Parks, Louis B. (2004). News station Channel 24 shuts down, *The Chronicle*, July 23, http://www.chron.com/news/houston-texas/article/News-station-Channel-24-shuts-down-1523878.php

Putin forbids funding cuts to state-run media, gazeta.ru, http://en.gazeta.ru/news/2012/10/29/a_4828917.shtml

Quebecor's Sun News Network channel shuts down, Reuters, Feb. 23, 2015, https://www.yahoo.com/tv/s/canadas-sun-news-network-shut-down-permanently-cbc-025215351-sector.html

Si Si (2014). Expansion of International Broadcasting: The Growing Global Reach of China Central Television, RISJ Working Papers, Reuters Institute for the Study of Journalism, University of Oxford.

State of the Media 2015, Cable News: Fact Sheet, Pew Research Center. http://www.journalism.org/2015/04/29/cable-news-fact-sheet/

Sun News Network to shut down, CBC, Feb. 13, 2015, http://www.cbc.ca/news/canada/toronto/sun-news-network-to-shut-down-sources-1.2955664

Sweden TV 4 shut down its 24-hour digital channel in 2013, *Digital TV Europe*, http://www.digitalt-veurope.net/88342/tv4-to-shut-down-news-channel-expand-online-coverage/

Wingrove, Josh (2013). CRTC ruling on news channels boost for Sun News, The Globe and Mail, Dec. 19, http://www.theglobeandmail.com/report-on-business/television-providers-must-carry-sun-news-crtc -rules/article16062408/

CHAPTER FOURTEEN

Quick Quick Slow: From Fast News to Slow News

JUSTIN LEWIS

INTRODUCTION

In a previous chapter (Lewis, 2010), I explored the relentless drift towards "breaking" and "live" news, a trend pioneered by news channels but increasingly a part of all forms of broadcast news (Cushion, 2015). My point of departure was a body of research that suggested the growth of what we might call "fast news," prompted by the growth of 24-hour news channels, has impoverished the quality of information we receive (Lewis, 2010). At around the same time, Dan Gillmor posted an article on the Mediactive site, describing the inevitable unreliability that flows from what we might call the currency of currency. He ended by proposing a philosophical alternative, a move towards the idea of "slow news" (Gillmor, 2009).

This idea has its antecedents: perhaps most notably Susan Greenberg's article in *Prospect* magazine, which drew upon a range of historical literary traditions to outline the case for "slow journalism" (Greenberg, 2007). While it is tempting to unravel a clear, linear genealogy of these overlapping terms, the almost simultaneous emergence of the terms "slow news" or "slow journalism" from different sources—often without reference to one another—suggests a collective response to the technological rush that has shaped early 21st-century journalism. We are seeing the early stirrings of a counter to the 24-hour news culture of "speed it up and spread it thin" (Fenton, 2010: 561).

In *Beyond Consumer Capitalism: Media and the Limits to Imagination*, I examined the political economy of a "fast news" culture, suggesting that if we are to sustain the democratic value of journalism, we need to reverse the trend towards the treatment of news as an instantaneous and disposable commodity (Lewis, 2013a). We can—as Greenberg, Gillmor and others have suggested—learn from a "slow food" movement which stresses, with metaphorical aptness, nourishment over speed, the quality of ingredients rather than cheaply assembled efficiency and the integrity of sources.

In this chapter, I want to develop the idea of slow news in the context of our fast news culture. I will consider what it might look like and what it could offer, assess the obstacles in its path and consider where, in the news firmament, it might begin to thrive.

FAST AND LOOSE

Although "fast news" is associated with the rise of 24-hour news channels, it has its origins in forms of 19th-century newspaper production (Lewis, 2013a). The slow news zeitgeist may also have its influences (Greenberg, 2007), but as a self-conscious movement it is very much a 21st-century phenomenon, a *revolt against* today's fast news culture. Megan Le Mesurier, in her review of the green shoots of a slow journalism movement (2015: 138), describes it as a "counter discourse and practice [that] has been emerging in recent years from journalists, editors, publishers and commentators interested in slowing journalism down." For some, the idea of slow news evokes a romanticism for a recent but bygone era before 24-hour news channels, Twitter and lean, mean business models employing a dwindling body of journalists to churn out a conveyer belt of stories. Although there is ample evidence that the time available for journalists to produce stories *has* declined in recent decades (Davies, 2007; Lewis et al., 2007), the economic and cultural roots of the fast news culture can be traced back to the industrialisation of news.

In the second half of the 19th century, newspaper production went from being a relatively open cottage industry with low start-up costs to become a more technologically sophisticated enterprise that required considerable capital outlay. In the 1830s a raft of titles flourished, often produced by people of modest means. A century later the manufacture of news had become a plutocratic oligopoly dominated by wealthy businessmen (Curran and Seaton, 1997; Williams, 1998).

In the first half of the 19th century, newspaper editors like William Cobbett (*Weekly Political Register*), Henry Hetherington (*The Poor Man's Guardian*) and Thomas Wooler (*The Black Dwarf*) saw journalism as, first and foremost, a form of political and civic education. They were part of a new social movement, a thriving radical press that fuelled the democratic demands of an increasingly literate working class. They embraced a form of journalism that questioned conventional wisdom and authority with such vigour that governments of the day tried—and failed—to repress them with punitive tax and legislation.

Once news became an industrial commodity, the spread of ideas was swiftly underpinned by an overriding commercial logic. It quickly became apparent that the profitability of news production depended on high levels of built-in obsolescence (Lewis 2010, 2013a). If people were to be persuaded to replace a newspaper on a daily basis, news needed to be defined not by its usefulness or its importance, but by its temporality. The worth of a news story had to lie less in any enduring quality of information than in its being up to date. Regardless of its content, yesterday's news was, by this definition, of little value.

This business model soon became inculcated into the culture of news, and journalism was increasingly defined by this onward rush towards the latest and the newest (Rantanen, 2009). News, in this sense, became about living in the present moment, a place where little endures, and attention is always diverted to the here and now. The development of a range of instantaneous technologies, formats and platforms—notably telephony, live broadcasting, 24-hour news channels, online news sites and Twitter—intensified this momentum.

It is, perhaps, symptomatic of the state of our fast news culture that parts of this history—such as the various 19th-century government attempts to stamp out the radical press, which saw many newspapers banned and a succession of editors jailed—has been entirely forgotten. In the wake of the 2013 Leveson inquiry into the British press, many newspapers (moulded in the more recent 20th-century tradition of the Press Barons), suggested that the post-Leveson arrangements would be the *first* attempt by government to control the press for *three centuries* (predating the crackdown on the radical press by over a hundred years). "This week we are facing the end of more than 300 years of Press freedom," claimed *the Sun; The Daily Mail*, wrote in sombre tones of "curbs that end three centuries of Press freedom" and the *Telegraph* opined that "Britain's new media regulator will 'stifle' journalism and end 300 years of press freedom."

Leaving aside the hyperbolic descriptions of the post-Leveson proposals, this failure to understand (or even know) the history of British journalism—one in which press freedom has been *repeatedly* infringed by government—went entirely unchallenged by other journalists (Lewis, 2013b). The idea of "300 years of press freedom" was repeated *ad infinitum*, each repetition adding to the weight of the apparent truth. Indeed, the *Sun* newspaper—bizarrely, for media historians—invoked (again, without correction) the spirit of Winston Churchill as a champion of the free press. As many media students will know, this is a display of wilful ignorance: Winston Churchill's wartime government banned the *Daily Worker* (which opposed the war) and attempted to close down the *Daily Mirror* (which backed the war effort but not the class divisions they saw entrenched within it). More recently still, the newspapers who led the Leveson backlash tended to be the loudest voices in support of the Thatcher Government's broadcast ban on the IRA in the 1980s.

The tendency of certain newspapers to play fast and loose with the historical record is unremarkable. But the failure of British journalism, *en masse*, to correct a series of gaffes about its own history displays the way in which, in the age of fast news, repetition passes for verity—regardless of context or the historical record.

WHAT FAST NEWS TELLS US ABOUT THE WORLD

The idea that we can become witnesses *as* news is happening has an obvious appeal. Rolling news channels thereby found themselves caught in a race to be first with "breaking news"—an ersatz genre that proliferated to such an extent that it became routine (Lewis and Cushion, 2009). Breaking news went from being unusual to ubiquitous: a banal series of updates rather than history unravelling before our very eyes. Traditional news bulletins have also found ways to increase the *impression* of immediacy—most conspicuously with the live "two-way" between presenter and reporter, the latter invariably standing in front of a scene or location (such as the White House or 10 Downing Street) that symbolically (and often *only* symbolically) places them close to the action (Cushion, 2015).

News is thus increasingly crafted by conventions constructed around—or signifying—the present moment. The *Now Show*—the name of a BBC Radio 4 satirical news review—captures this sense living in the present, and most news bulletins could now be so described. Both qualitative and quantitative research suggests that what we might call the "nowness" of news obscures more than it reveals. So, for example, by a range of measures, "breaking news" stories tend to thin rather than thicken the quality of information on display. When compared with a traditional, more retrospective TV news item, they contain less context, less background, and less analysis of all kinds (Lewis and Cushion, 2009). They are also less accurate and reliable: independent verification, which takes time, takes a back seat to speculation, which is instantaneous (Gillmor, 2009). Rosenberg and Feldman put it more bluntly: "media and inaccuracy, after a flirting through the ages, are now in a steamy lip lock" (Rosenberg and Feldman, 2008, p. 4).

The privileging of the present moment accentuates journalism's long-standing tendency towards what has been called "event-orientation" (e.g., Cottle, 2006), favouring those stories that can be packaged or—better still—*witnessed* as a series of episodic events. An emphasis on the present tends to side-line history, context and causality—news becomes a narrative in which beginnings quickly disappear. In this truncated narrative, it becomes difficult to understand why things happen or what their broader significance might be (Iyengar, 1991; Lewis, 2001).

Vicky Dando's comparison of the reporting of terrorism and climate change over a decade exemplifies this structural bias (Dando, 2015). Terrorism as a topic is given a generous volume of coverage: the terrorist event is dramatic, deadly and deeply adversarial, it ticks every box for conventional news value (Lewis, 2004). While such events in the UK are rare, they are surrounded by a string of other smaller events (such as political speeches or stages in the criminal justice process) as well as being punctuated by a series of set-pieces in the arenas of politics, war and criminal justice. But there is a chronological as well as a stylistic imperative driving terrorism's popularity as a news topic: the risk of a terrorist attack in most countries may be small but it is ever-*present*. The reporting of risk is less a matter of scale than of timing.

By contrast, the risks of climate change may be alarming but they are constantly deferred. The time-frame of global warming is extraordinarily brief on a climatological calendar, but it is so far in excess of the 24-hour news cycle as to be almost beyond the imagination of a fast news culture. The narrative of climate change, played out against the intertwined upward lines of graphs measuring global temperatures and carbon emissions, works only in longer time-spans. The significance of temperature change is measured not in days, weeks, months, or even years, but over decades.

While climate change is linked to extreme weather events, these links are in the realm of probability rather than certainty. We can say that increasing our carbon emissions makes these events more likely, but we cannot link a defined volume of greenhouse gases (let alone the emitters of those gases) to a particular flood or a hurricane. Responsibility is thus indirect and unclear. Without simple connections and a clear linear cause and effect, news falls back on tried and tested frameworks for covering such events, which, like much present-day reporting, focus on more immediate concerns. Where blame is attributed, it is within a much smaller, more time-limited frame: on local councils, for example, for failure to dredge rivers or build flood defences, or on governments for an inadequate response in the moment of crisis. The bigger picture—the persistent failure of governments to act to limit (or even limit the growth of) greenhouse gas emissions—tends to exist outside the news frame.

Dando's research shows that over the last decade, terrorism not only receives consistently more coverage than climate change, but is treated with more urgency, more certainty and with much less debate or nuance. While the terrorism narratives peaks after dramatic events (notably in 2001 and 2005), as a story it remains resilient and enduring. Climate change coverage, by contrast, peaked in 2007 and has declined steadily since. Richard Thomas's study of the topics covered by BBC and ITV found that environmental stories in general had, by 2014, almost disappeared from the news agenda (Lewis, 2015).

Fast news as a credo thereby acts as a kind filter favouring a particular set of episodic narratives. In terms of our understanding of the world, it is not only unsatisfying and unfulfilling, but it only permits certain kinds of stories to be told. It has delivered a public that is only superficially engaged, and that, despite the wealth of information available, remains, on a range of issues, deeply uninformed (Lewis, 2001).

ENTER SLOW NEWS

In the journalist's lexicon, a "slow news day" means that nothing much is going on—nothing, at least, to raise the reporter's pulse. Slow news, in this conventional sense, means a lack of urgency, a lowering of the news threshold, a licence to admit quirky, off-beat and less newsworthy fare into the daily routine of news production. In the thirst to be first, slow news is seen as second-hand and second rate. This is slowness in its most pejorative sense: slow-witted, ponderous and dull.

The slow news movement begins from a very different premise. A slow news day is an opportunity to enrich the daily news fare. It gives journalists the space to cover stories that do not push us towards a short-lived present, or to consider issues that are not structured around a series of news events. Slow news is a more relaxed space amidst the relentless rush of updates that populate our 24-hour news cycle.

Like fast food, fast news appears to be the product of a successful market logic. Television news, in particular, is a form of cultural production that is especially nervous about losing the viewer's attention (many viewers dip in and out of rolling news channels, for example). News editors, uniquely amongst television programme makers, are aware that many viewers find some of the topics they cover difficult or dull. They are under pressure to create interest in subjects—notably politics and current affairs—in which large sections of the population declare disinterest or disenchantment.

Fast news, with its relentless pace and focus on the present, is a way of negotiating this tension. Some, like Luke Good, have questioned the idea "that market realism dictates an inevitable drive towards faster, softer, more bite-size news" arguing that this "constitutes fatalism rather than realism. It is simplistic at best and condescending at worst to fall back on the assumption that few outside the chattering classes want serious long-form news and current affairs any longer" (Good, 2012: 37). Good's critique suggests a familiar dichotomy, which pitches commercial, fast, "bite-sized" news against worthier, serious, slower forms of journalism.

Good also prompts us to consider how and on what basis audiences engage with news. Many of the assumptions about the more compelling nature of "breaking" or "live" news are often based on hunches rather than serious audience research. So, for example, it is not clear that audiences share the journalistic obsession with "live" or "breaking news," or would not prefer news stories that more clearly revealed how the world worked. But it is also reasonable to question whether a straightforward return to traditional serious news and current affairs *would* engage audiences "outside the chattering classes."

In quieter online spaces, away from the remorseless drum-beat of fast, disposable news, we see the beginnings of a fledgling slow news movement. Publications like *Mission and State* in Santa Barbara, *Long Play* in Finland, *De Correspondent* in the Netherlands and the distillation of academic expertise through op-ed style journalism that defines *The Conversation* offer a striking contrast to the dominant fast news culture. While not all these outlets embrace the language of slow news— *The Conversation*, for example, might best be described as cultivating a considered topicality—they all suggest a turn away from a fast news norm. Perhaps the most overt expression of slow news—set up by the self-defined 'Slow Journalism Company' in London—is *Delayed Gratification:* "the world's first Slow Journalism magazine…which revisits the events of the previous three months to see what happened after the dust settled and the news agenda moved on." *Delayed Gratification* is the product of a rounded critique of the pitfalls of fast news, boasting on the front page of its site that it is "proud to be 'Last to Breaking News'" (http://www. slow-journalism.com).

Megan Le Marusier considers many of these examples to identify various features of "slow journalism," which, in sharp contrast to the prevailing culture:

- Is not "scoop driven," but involves time for reflection and analysis and/or investigation;
- May be longer in form;
- Tends to avoid sensationalism;
- Is never a product of "herd reporting";
- Stresses the importance of accuracy and transparency in the use of sources— like slow food, its origins should be traceable.

Le Marusier identifies two other characteristics of slow news: its association with "alternative" or more communal spaces, and the preference for "telling stories using narrative techniques, not just the mechanistic expository style of hard news stories" (Le Marusier, 2015: 143). This observation undoubtedly describes where slow news presently resides and how it unfolds, but both features merit further consideration.

THE SPACE FOR SLOW NEWS

Despite the growth of online spaces for slow news, it remains, in its more self-conscious forms, marginal to our dominant systems for delivering information. The opportunities created by online and social media present problems as well as possibilities. If the multichannel television age made us impatient—there is often something more entertaining available at the click of a remote control—this was merely a preview to the ever-present distractions that define much of the online world. The cluttered screen with its multiple click-aways and the rapid-fire forms of social media do not encourage us to linger, and the rising tide of online news forms—such as Buzzfeed, Twitter, or Google News—are, in many ways, the epitome of fast news. The newspaper and the traditional news bulletin seem positively languid by comparison.

On the surface, the growth of slow news in this distracting environment seems an unlikely proposition. We should, however, be wary of the McLuhanesque notion that technology determines form. Neil Postman's well-known critique of television, for example, tends to assume that the US television system—largely constructed around the needs of advertisers—was symptomatic of a set of *technological* rather than *commercial* imperatives (Postman, 1985). Postman's critique was written at a time when television systems built around an ethos of public service broadcasting (in, for example, the UK) were qualitatively different and distinct—not least because they served the needs of audiences rather than advertisers. Television, in other words, *could be* absorbing as well as diverting, long as well as short form, compelling as well as interruptible.

It is, in the same way, *possible* to imagine online forms which encourage discovery rather than distraction. The computer game industry may be instructive here. Many of its commercial successes are based on activities that, even if they use imaginative landscapes, are both repetitive and addictive—in the same way that fast news may be for the archetypal "news junkie." Aficionados will also know games that have been designed to encourage exploration, imagination and creativity. The critique of the computer gaming industry—that, for example, its pleasures are limited but all consuming, and its content is all too often violent, sexist and derivative—is less about the technology than the genres favoured by a commercial industry.

So while the commercial development of the internet and social media has tended to favour fast, disposable forms of news, it has also allowed other forms of information delivery which create other possibilities. So, for example, the growth of mediated spaces of expertise, such as *The Conversation*, Ted, TedX and the Do Lectures can be seen as forms of journalism: they involve editorial judgment and carefully structured storytelling designed to tell us about the world. They are topical in the broader sense.

There are also new spaces emerging that have the potential to redefine news, both in terms of style and purpose, as the industry withdraws from some of its traditional territory. So, for example, the drift towards regional consolidation in news provision and the growth of cheap, online advertising in many developed countries has precipitated an abrupt decline in the staffing of local newspapers (Franklin and

Williams, 2012; Williams, 2013). Most cities now do well to support one local newspaper, while many smaller locales have no systematic news provision at all.

This has created space for a new generation of community, local and hyper-local news services (Radcliffe, 2013; Williams et al., 2014). While this new generation of news outlets takes many forms, surveys of their content suggests that many hyper-local or community news producers are not attempting to replicate the fast news model. Some are redefining the purpose and style of local news—echoing the traditions of the radical press (in style if not in substance), produced for the community by the community. There is no well-defined or sustainable business model for this new generation of community news services, but smaller revenue streams *may* be enough to support a mix of professional journalists supported by community volunteers with the smart use of mobile technologies (such as *Newstori*, developed by the Centre for Community Journalism at Cardiff).

This may not be slow news in the self-conscious sense, but it is news at a different pace. Its purposes and imperatives clearly distinguish it from the prevailing fast news culture. Journalism history is, nonetheless, replete with the failure of well-intentioned new ventures, and we cannot assume the market will support these new forms of news. To prosper, the sector will require support from agencies (like the Centre for Community Journalism) and indirect forms of subsidy. The danger, without such provision, is that the more commercial aspects of community journalism will be swept up by sophisticated, large-scale social media platforms. Their ability to deliver local markets to national and global advertisers may enable them to absorb local sources of advertising revenue leaving only the scraps for local, independent news provision.

There may also be space for slow news in more traditional settings. It has been clear for some time that if the newspaper industry is to continue in its printed form it will be increasingly difficult for it to compete in fast news terms. Newspapers will still seek to set news agendas, but there is at least the potential to do so in ways that embrace slower forms of news.

So, for example, on March 7, 2015, the front page of the *Guardian* newspaper in the UK was not a conventional collage of the main headlines, but a long-form news article by Naomi Klein about climate change. Inside this four page pull-out (on the more traditional front page) was an article by the newspaper's editor, Alan Rusbridger, explaining this departure from convention. With only a few months left before stepping down as editor, he wrote, he decided to use his position to rectify what he saw as an endemic flaw in contemporary British journalism, with its event-driven routines which meant ignoring some of the key concerns of 21st-century life: "we had not done justice to this huge, overshadowing, overwhelming issue of how climate change will probably, within the lifetime of our children, cause untold havoc and stress to our species."

Journalism, acknowledged Rusbridger, had been culpable of ignoring the issue not because of its importance but because of its incompatibility with a fast news culture. He then committed the paper to a different kind of journalism: "The climate change threat features very prominently on the front page of the *Guardian* even though nothing exceptional happened today. It will be there again next week and the week after." This approach is, as Jen Birks has shown, reminiscent of many examples of campaigning journalism (Birks, 2014)—and, indeed, a campaign by the *Guardian* to promote action on climate change followed soon thereafter.

Campaigning journalism often takes more overtly populist forms than the *Guardian's* intervention (a local newspaper campaigning to save a local school or hospital, for example), which, unlike many such campaigns, involves a self-conscious inversion of traditional news values. It is easy to see why the newspaper felt that such an inversion was too radical a gesture on its own, and the adoption of a more tried and tested model of campaign journalism gave it context and credibility. This, of course, takes us back to the era of the radical press, who were seeking not only the spread of certain new ideas, but were part of a movement to change society. This may suggest that for slow journalism to prosper, it needs to engage with a politics of social change.

SLOW NEWS AND AUDIENCE ENGAGEMENT

The phrase "slow news," I have suggested, carries with it a series of negative associations: it risks being seen as long-winded and esoteric, worthy but dull. If slow news is simply a form of serious long-form news and current affairs that retains existing news conventions, it is unlikely to be engaging. These forms of "serious" journalism rarely appeal to audiences outside the virtuous circles of those already engaged by news and current affairs. For this reason, Le Marusier's stress on the use of narrative techniques in slow news storytelling is, I would argue, fundamental to its viability as a popular genre.

Over 20 years ago, after conducting a series of in-depth interviews with viewers following a half-hour news bulletin, I concluded that the failure of many conventional broadcast news stories to either spark or sustain anything more than a superficial interest in their content was partly due to, in Le Marusier's words, the "mechanistic expository style of hard news stories" (Lewis, 1991). Most narrative forms use an enigma/pause/resolution structure to hook audiences in and sustain their interest. Most news bulletins, by contrast, neither entice viewers to become curious about a subject nor use narrative techniques to develop a storyline that shapes and then satisfies that curiosity. They simply lay out a series of facts, often in declining order of importance. On the few occasions news reporters did make use of narrative techniques, it appeared to increase audience interest and people's ability to correctly recall details of a story.

In recent years, the evidence in favour of the use of narrative in news storytelling has hardened. So, for example Machill, Köhler and Waldhauser conducted an experiment in which audiences were shown two versions of the same news story, one using a traditional format and the other using narrative devices. They found that: "the application of the narrative news concept…can significantly and clearly increase the retention and comprehension of news content. In particular, our study gives initial indications that a narrative presentation has particularly strong effects in the segment of viewers up to 30 years old who have less prior information at their disposal, are less interested in the topic and overall rarely take notice of the news on the television or in newspapers" (Machill, Köhler and Waldhauser, 2007). The use of narrative, in other words, enabled news to reach precisely those audiences who have become famously disengaged with traditional journalism.

The failure of the traditional fast news format to inform public understanding has been apparent for some time—not least in surveys of public knowledge that show consistently low levels of knowledge of politics and public affairs (Lewis, 2001; Hargreaves et al., 2003; Poucher, 2015, forthcoming). This is seen as of particular concern for young people: annual Ofcom news surveys repeatedly show evidence of falling news consumption amongst younger age groups, while other indicators of democratic engagement—such as falling voter turnout—show an increasing age divide. The traditional fast news response to this is to speed things up still further, in the hope that up-beat, bite-sized chunks of news might keep their attention (Lewis and Cushion, 2009). The positive response of younger people to the use of narrative in news storytelling suggests that a different, slow news approach—in which stories develop understanding of an issue rather than assume it—may be a more intelligent response.

If slow news is to develop as a genre, it will need to be crafted in ways that embrace popular forms of storytelling. In this sense, slow news is not simply a space for reflection, it is a chance to engage and excite people who have given up on—or learn little from—the fast news culture.

REFERENCES

Birks, J. (2014) *News and Civil Society: The Contested Space of Civil Society in UK Media,* Ashgate.

Cottle, S. (2006) *Mediatized Conflict,* Open University Press.

Curran, J. and Seaton, J. (1997) *Power without Responsibility,* London: Routledge.

Cushion, S. (2015) *News and Politics: The Rise of Live and Interpretive Journalism,* London: Routledge.

Dando, V. (2015) *Mapping the media contours of global risks: A comparison of the reporting of climate change and terrorism in the British press*, PhD thesis, Cardiff University.

Fenton, N. (2010) 'News in the Digital Age'. In Allan, S. (ed.), *The Routledge Companion to News and Journalism,* 557–567. New York: Routledge.

Franklin, R. and Williams, A. (2012) 'Local Newspapers Mean Business: A Political Economic Account of the Local and Regional Newspaper Industry'. In Petley, J. and Williams, G. (eds.), *Media in Contemporary Britain: A Critical Approach*. New York: Palgrave Macmillan.

Gillmor, Dan (2009) Towards a slow news movement, Mediactive, November 8. Available online at http://mediactive.com/2009/11/08/toward-a-slow-news-movement

Good, L. (2012) News as conversation, citizens as gatekeepers: Where is digital news taking us? *Ethical Space: The International Journal of Communication Ethics*, 9(1), 31–40.

Greenberg, S. (2007) 'Slow Journalism', *Prospect* at http://www.prospectmagazine.co.uk/opinions/slowjournalism

Hargreaves, I., Lewis, J. and Speers, T. (2003) *Towards a Better Map: Science, the Public and the Media* (co-authored), Economic and Social Research Council Report.

Le Mesurier, M. (2015) 'What is Slow Journalism?' *Journalism Practice*, 9(2), 138–152.

Lewis, J. (1991) *The Ideological Octopus: An Exploration of Television and its Audience*, New York: Routledge.

Lewis, J. (2004) 'September 11th'. In G. Creeber (ed.), *50 Television Texts*, London: Arnold.

Lewis, J. (2010) 'Democratic or Disposable: 24-hour News, Consumer Culture and Built-in Obsolescence'. In Cushion, S. and Lewis, J. (eds.), *The World of 24-hour News*. New York: Peter Lang.

Lewis, J. (2013a) *Beyond Consumer Capitalism: Media and the Limits to Imagination*, London: Polity.

Lewis, J. (2013b) Leveson and the use and abuse of history, http://www.jomec.co.uk/blog/leveson-and-the-use-and-abuse-of-history/

Lewis, J. (2015) For those interested in a sustainable future, http://www.jomec.co.uk/blog/for-those-interested-in-a-sustainable-future/

Lewis, J., Williams, A., and Franklin, B. (2008) 'Four Rumours and an Explanation: A Political Economic Account of Journalists' Changing Newsgathering and Reporting Practices', *Journalism Practice*, 2(1), 27–45.

Lewis, J. and Cushion, S. (2009) 'The Thirst to be First: An Analysis of Breaking News Stories and Their Impact on the Quality of 24-hour News Coverage in the UK'. In *Journalism Practice*, 3(3), 304–318.

Machill, M., Köhler, S. and Waldhauser, M. (2007) 'The Use of Narrative Structures in Television News: An Experiment in Innovative Forms of Journalistic Presentation', *European Journal of Communication*, 22, 185–205.

Postman, N. (1985) *Amusing Ourselves to Death: Public Discourse in the Age of Show Business*. New York: Penguin.

Poucher, R. (2015) *Hyper-local journalism and the democratic deficit*, PhD Thesis, Cardiff University.

Radcliffe, D. (2013) 'Hyper-local Media: A Small but Growing Part of the Local Media Ecosystem'. In Mair, J., Fowler, N. and Keeble, R. L. (eds.), *What Do We Mean by Local? The Rise, Fall and Possible Rise Again of Local Journalism*. Bury St Edmonds, UK: Abramis Academic Publishing.

Rantanen, T. (2009) *When News Was New*. London: Wiley-Blackwell.

Rosenberg, H. and Feldman, C. S. (2008) *No Time to Think. The Menace of Media Speed and the 24-hour News Cycle*. New York: Continuum.

Williams, A. (2013) 'Stop press? The crisis in the Welsh media and what to do about it'. *Cyfrwng: Media Wales Journal*, 10, 71–80.

Williams, A. J., Harte, D. and Turner, J. (2014) 'The Value of UK Hyperlocal Community News'. *Digital Journalism* DOI: 10.1080/21670811.2014.965932

Williams, K. (1998) *'Get Me a Murder a Day!' A History of Mass Communications in Britain*. London: Arnold.

CHAPTER FIFTEEN

Journalism in the Age of the "Interface"

INGRID VOLKMER

Over the last decade our understanding of "24-hour news" has undergone a significant transformation. The term itself seems to have become obsolete at a time when all types of "news" continuously stream across multiple layers of increasingly dense—and personalized—digital networks. The tremendous transformation of journalism within such a digital scope requires new conceptual frameworks to not only critically identify new content formats, "social media" journalism and user practices, but also to assess roles of journalism within dense transnational multidirectional news topography.

Such a news topography is not only digital, but shapes in today's advanced digital sphere an enlarged scope of diverse spatially globalized ecologies. Even the smallest local news outlets and individual journalistic comments reach—via websites, newsfeeds, blogs and social networks—users across world regions. To assess these enlarged horizontal ecologies, conceptual frameworks are needed which position journalism within these transnationally no longer just "connected," but rather fine-lined "interdependent" spheres of political communication and public discourse. Yet, despite a current debate of digitalization of journalism in mainly Western world regions which, as Steenssen & Ahva (2015) argue, has now reached a "fourth" phase where debates go beyond traditional "institutions," a focus on transnationalization of journalism remains a "blind spot." Steenssen & Ahva argue that larger philosophical issues, such as ethics, ontology and epistemology (Steenssen & Ahva, 2015:15) need to be incorporated into conceptual debates of digital journalism. Such a broader sphere is specifically required as journalism is no longer national or international, but is situated within an enlarged transnationalized sphere of communication across different society types. Journalism research within these enlarged transnational ecologies requires a solid grounding in debates of public discourse. However, until recently, it seems that most approaches rely on facets of Jurgen Habermas's conception which is strongly related to the tradition of public legitimacy of Western European nation-states. The referential methodologies in today's advanced digital sphere are no longer "networks" and "connectivity," but rather the emerging of layers of globalized public spheres, new fine communicative contours which shape discourse among like-minded users and create ontological horizons of "reflective interdependence" across a globalized peer-to-peer densities. The advanced digital density has shaped new communicative dimensions of fine-lined communication between individuals anywhere in the world with internet access. It is this fine-lined dimension which constitutes a new type of epistemic discourse, relevant to the understanding of the transformation of public spheres and conceptual frameworks of journalism.

Assessing journalism research in digital spheres reveals that conceptual frameworks are mainly based on assumptions of national public spheres of (Western) nation-states and global journalism is understood in the methodology of national comparison, often in contexts of traditional institutions, such as mainstream news outlets (e.g., Wahl-Jorgensen & Hanitzsch, 2009). It is quite interesting that, *despite* the fact that communication drives the formation of globalized digital landscapes and is restructuring societal communication in developed and developing regions of the world, the consequences of these globalized formations for inclusive paradigmatic and conceptual debates are mainly addressed in the larger social sciences. It seems that there is a disparity between disciplines regarding paradigmatic approaches to globalization. For example, sociological debates which have addressed globalized transformations for about three decades suggest a strict move away from "methodological nationalism" to assess the "nuances" of the "breaking" of the nation-state as a crucial component of social theory (Chernilo, 2007; Beck, 2009). Saskia Sassen's analysis of denationalizing phenomena reveals that the "national" and the "global" are not dichotomous categories, but that the nation is the site of multi-layered globalization processes. In consequence, multiple "denationalizing dynamics" cut across "institutional hierarchies" through "rescaling" processes (Sassen, 2007a: 91). "Rescaling processes" are set in motion by "multiple specialized cross-border circuits on which different types of places are located" (Sassen, 2007a: 91). In Sassen's view, these processes "take place deep inside territories and institutional domains that have largely been constructed in national terms in much of the world." She argues "what makes these processes part of globalization even though localized in national, indeed subnational, settings is that they involve transboundary networks and formations connecting or articulating multiple local or 'national' processes and actors" (Sassen, 2007a: 82). Communication and, specifically, journalism is not only deeply situated within these new formations, but rather one of the drivers of these rescaling processes. The notion of "trans-border circuits" could serve here as a working term, as a first framework to assess specific dimensions of the emerging transnationalization of journalism. Taking these conceptually advanced debates into our field allows us to address communicative spheres beyond territorial "boundedness" in "national" or "inter-national" frames and conceptualize emerging communicative dimensions in transnational densities, "de-bracketing the state society nexus," a process which is incorporating all society types (Volkmer, 2014: 25). Not only communication theory, but specifically conceptions of journalism need to be understood within these larger structures of social and political theory which are rapidly overcoming the demarcations of the nation-state and, through this process, are remapping the traditional parameter of journalism. Journalism is no longer conceptually embedded between national and globalized institutions, but within new scopes of a globalized civil society, which—through digital reach—includes new dimension of "blended" news "flows" *across* societies.

However, when reviewing debates of "digital" journalism, it could be argued that transnationally "blended" scales of interdependence appear in our field mainly in the lens of "connections." Early debates of digital journalism, which emerged about ten years ago, have addressed the "dissolution" of traditional boundaries of news structures and platforms. McNair describes these structures as accelerated types of news cycles through the lens of a "chaos" paradigm (McNair, 2006). Subsequent debates have focused transnational connections in the contexts of "user-generated" content, the digital newsroom (e.g., Deuze, 2008; Erdal, 2009) and 'produsage,' enabled by the amalgamation between formerly segregated public domain sectors within societies (e.g., Bruns & Highfield, 2012)—processes which are now more specifically addressed in contexts of conflict and the mobilization capacities of social media. A second debate identifies transnational "connections" mainly across "vertical" dimensions of "citizen journalism" vis-a-vis mainstream media and the way in which social media produce the framing of conflict and crisis journalism (Carr, 2014; Moon & Hadley, 2014). In addition, the conception of "network journalism" (Heinrich, 2011) constitutes a framework of transnational connections for the understanding of "dense" formations of nuanced "decentralization" of news production. An approach which has allowed us to assess the specific implications on news organizations, understood as "nodes," within an unlimited potential of "connectors." A third debate identifies transnational connections in contexts of peer-to-peer interaction enabled by "social" media. Overall, it seems that these debates reveal a focus of transnational connections mainly in the lens of Western world regions. Only recently is a debate in non-Western world regions emerging, for example, in African and South Asian perspectives from Zimbabwe (e.g., Mabweazara, 2014) and Ethiopia (e.g., Skjerdal, 2014). Mabweazara argues for a focus on social constructivist methodologies in a new debate of digital journalism which would allow us to connect "our indigenous knowledge" with "international systems of communication" and "sensitivity to context" in order to define African journalism in the digital age (Mabweazara, 2014:108, 114).

It is also interesting to assess the conception of transnational journalism as it has emerged in the debate of "global" journalism from the early 1990s onwards. For example, the study of the "global newsroom" at the European news agency Eurovision (Cohen et al., 1996) has provided a first pattern of trans-border news. Other approaches began to assess "glocal" journalism (Volkmer, 1999) and the notion of a "global journalist" (Reese, 2001). Despite the debate of the (globalized) network society, which emerged in sociology with Castells' work, networked formations seem to have informed research approaches to the understanding of transnational "connectivity" in journalism studies (see above), but have not led to larger paradigmatic frameworks which would allow contextualization of journalism beyond the nation-state. The conceptual notion of "global" journalism is either understood vis-a-vis the nation-state, in contexts of new interactive formats such as blogs, in the focus of globalized crises (Cottle, 2006; Cottle & Nolan, 2007; Cottle, 2009; Pantti et al., 2012) or in the angle of the dissolving of national "frames" as "globalizing media and journalism" means that "creators, objects and consumers of news are less likely to share the same nation-state frame of reference" (Reese, 2010). As Berglez argues, the concept of global journalism is "still being undertheorised as a news style which makes it difficult to analyse empirically" (Berglez, 2008: 845–846) and he argues that a "global news style arises when transnational and global powers are included in news reporting" (Berglez, 2008:851).

Others begin to assess the dimensions of "globalized" journalism within larger conceptions of transnational civic communication. For example, Berglez & Olausson argue that the national and transnational are not exclusive and note that "news media studies have primarily contributed to the theoretical understanding of the relationship between national and transnational political identities," however, only a "few empirical studies demonstrate their conflicting, dialectical relationship in news discourse" (Berglez & Olausson, 2011: 36). Only very recent debates begin to assess the inclusion of the "news user" in journalism studies. For example, Picone et al. argue that "a radical user perspective should go beyond merely acknowledging the importance of the user within journalism studies, but should consider it to be an intrinsic part of the epistemology of journalism studies"(2015: 36).

However, in today's advanced phase of "digital" journalism, conceptual frameworks are needed to no longer assess "global" journalism as a "comparative" dimension, but rather situated within the densities of horizontal scales of "cross-border circuits" which allow us to identify new spheres of the "fourth estate"—spheres of deliberation, communicative power, modes of participation and civic discourse within a globalized civil society.

TRANSNATIONAL JOURNALISM AS CROSS-BORDER CIRCUITS

Despite the role of news agencies in the formation of "live" news (e.g., Boyd-Barrett & Rantanen, 1998), the first type of dense cross-border circuits emerged in the era of transnational continuous 24-hour journalism, covering all dimensions of the unfolding story in the context of satellite news channels in the early 1990s. Advanced satellite technology—which provided dense low-cost platforms for direct-to-home delivery, accompanied by neoliberal deregulation of national media markets enabled by the formation of the World Trade Organization (WTO)—set the stage for new forms of live transnational news spheres. Giddens's term "time-space distanciation" (1990) as a feature of the globalization of the "second modernity" and Robertson's (1992) term "glocalization" seemed to conceptually reflect the emerging dialectic of global and local cultural, social and political structures and, in retrospect, also highlighted cross-border circuits in broad dimensions of "proximity" and "breaking news"—"magnification."

However, a few years later, satellite news channels began to focus on specific globalized audiences through "regionalization" and thematic orientation. In this period, cross-border circuits appeared no longer as broad globalized spheres, but rather as globalized fractured dimensions. In Europe, transnational satellite 24-hour news channels, such as France 24, Deutsche Welle or Russia Today (RT) were focusing on a national agenda—despite being delivered across the European continent—and even beyond; in North Africa and the Middle East, cross-border circuits of news were delivered via satellite news channels in order to implement political interests within the larger region. Despite the fact that the aim for "objectivity" was constantly exposed to "conflicting pressures from all sides" (Sakr, 2007: 61) transnational news spheres have influenced journalism within the region and have, for example, shaped models of "contextual objectivity" as a regional specific model (Sakr, 2007: 52; El-Nawawy & Iskandar, 2002); in Asia, satellite news channels have emerged in India and Pakistan as bilateral cross-border circuit of political influence within this geopolitically relevant world region. Diverse cross-border circuits are appearing across the European continent where two hundred highly diverse Arabic channels can be received live in exactly the same version as in their country of origin, Morocco, Syria and Lebanon parallel to national European channels and in addition to Russian, Slovenian, Croatian, Turkish, Farsi, Hindi channels—many of which are aired live as in the country of origin. Despite these dense dimensions of "transborder circuits," it is surprising that these processes are addressed in journalism research. Instead, these configurations of "rescaling" processes of national news spheres are only addressed in a few debates of satellite communication. These are addressed through a focus on a specific satellite news channel (e.g., Wu, 2013). Another approach is suggested by Rai & Cottle (2010) who have argued for the broadening of theoretical frameworks for positioning journalism within such enlarged "communicative architecture" (Rai & Cottle, 2007), other approaches discuss satellite news in contexts of either "footprint cultures" (Parks, 2012), regional contexts (Volkmer & Chouliaraki, 2009) or in diaspora research. For example, Georgiou suggests that we should understand this sphere of parallelism as "mediated spatialities" (Georgiou, 2006); others suggest the term of "local cosmopolitanism" as an ontological frame for subjective news spheres among migrant citizens (Chin, 2015).

Today, dense transborder circuits are fully unfolding in digital spheres constituting new sets of public, political and societal implications on a global scale. Today, transborder circuits appear in a transnational "all at once-ness," in a news cartography, accessible from anywhere in the world with a digital connection. It is a transborder circuit where, returning to Sassen's argumentation, the global and national no longer intersect, but rather the global and the lifeworld (Schuetz & Luckmann, 1979) and where the lifeworld becomes the site of globalization and transborder circuits constitute subjectively chosen micronetworks. This is a radical shift which requires not only a "user-centric" perspective in journalism research (see Picone, 2015), but rather the refining of methodologies for conceptualizing transnationally accessible communicative formations which are the embedding landscape of digital journalism.

FROM CIRCUITS TO THE "INTERFACE"

Only recently are we beginning to conceptually assess the parameters of such a new communicative landscape in broader terms. For example, Mirca Madianou, building on her work of diasporic communication, suggests the term "polymedia" for assessing communicative trajectories across such an "all at once-ness." She understands "polymedia" as a larger conceptual approach to address all sorts of "integrated environments of affordances," wherever they appear. Madianou suggests that the approach of "polymedia" allows, for example, understanding of smartphones as an "integrated environment of affordances" (Madianou, 2014: 667). A second—similarly important—recent debate focusses on the new types of data management, production and, as some argue, "mediation" in the dimension of the "interface," understood as a key component of this new communicative ecology. For example, Galloway (2012) defines the interface in broad terms less as a "surface" but as a "gateway that opens up and allows a passage to some place beyond" (Galloway, 2012: 31). Hookway understands the interface as a type of mediation between production and consumption. In his view, the main new dimension of the interface is not the "lineages of devices or technologies," but rather "the qualities of relation between entities" (Hookway, 2014: 4). In addition, Halpern proposes to understand the interface as a way to "communicative objectivity" to manage and analyze data (Halpern, 2015). In times of big data and the datafication of content, these organizing structures are equally important for new conceptual frameworks of news production as well as for subjective news consumption practices. Furthermore, and returning to our overall debate, the interface provides a new framework, indeed a new paradigm, for inclusive cross-border circuits which are no longer existing side by side, but are merging into a subjective news screen, with the potential of being shared not only by like-minded peers but in civic discourse across societies.

The paradigm of the interface might be of use here as it allows assessment of journalism in inclusive densities of digital circuits on macro- as well as microstructures—from the arrangement of news by digital news outlets to the production of subjective news screens on smartphones and pads. The dimension of the interface could serve as a macrostructural framework for the assessment of new transnationally accessible news "content systems" such as apps. Whereas in a first phase, apps resembled content of a news organization, such as the app by BBC World that reflected all features of the BBC World news site, today apps are no longer just resembling the websites of news outlets, but are used as an interface for the delivery of specific news. For example, the Australian Broadcasting Corporation (ABC) has just launched an app to specifically target expatriate Australians worldwide. The increase of news apps in Apple's App Store with a dedicated "newsstand" section reveals not only about 20,000 news apps, most of which provide live updates, including increasingly audiovisual live content streams, but also highly specific app interfaces of news organizations from all world regions, such as the BBC, ABC7 Los Angeles, News India Lite, Kenyan news, Pakistani TV, Palestine newspapers, Ukraine news to Madagascar radio live, of NGO's in addition to even local political organizations of many world regions, such as Cambodian Voter Voice. It is such a digital reach of diverse forms of political journalism across world regions and the parallelism of these landscapes as experienced by the users which requires conceptual frameworks for assessing no longer transnational but rather "scalar" news journalism, reaching individual users across continents. It is a news ecology which has implications not only on journalism, but constructs new forms of public spheres and perceptions of civic identity—a new "logic" of political communication in a globalized scope, and shaping new forms of public spheres and civic identities. A similar interface, however less diverse, is available for Android infrastructures on Google Play.

A closer look reveals a second type of interface, specifically used by news aggregators, such as Google News and YouTube. Aggregators bundle thematic news streams, providing up-to-the-minute thematic news items from a great diversity of digital sources. The news interface of the Saudi air strikes in Yemen[1] includes, for example, streaming news from ODN (On Demand News, formerly ITN), Al Jazeera English, Press TV, an English-language streaming website based in Iran, and RT (Russia Today). We need to pay attention to these developments where journalism is not only "rebranded," away from source brands and their websites, towards a thematic interface, provided by aggregation and social media sites, such as Facebook, downloaded on tablets and smartphones in any world region with digital access. Such a new parallelism of news sources shapes a new dimension of "relational" spheres within a personalized interface where news coverage can be directly compared.

A third type of interface is constituted by personalized "smart" news aggregators, such as News360, which creates custom feeds, which generates interest graphs from Facebook and other social media sites, which select news from about 100,000 sources. A third type of interface emerges in African regions, the Middle East and Asian regions and provides insights into regional conflicts, but also perceptions of globalized conflicts. For example, the site Standard Digital, a news site based in Kenya, is a combined news platform, providing general news as well as The Nairobian website, launched in 2013. News is delivered via text, streaming radio and video, and social media and shapes a new type of journalism for a region where the smartphone constitutes the main device for assessing digital content. SABC, the South African public service broadcaster and national television from Uganda, streams news live on YouTube, along with CCTV Africa, the Chinese transnational broadcaster. About 80 specialized news channels from Pakistan are streamed 'live,' such as Duny News live, Waqt News as well as Samaa News TT.

A fourth type of interface is constituted on the subjective level, where not only different news sources appear, but they are embedded within a relational scope of subjective news production. For example, the BBC's app up-to-the-minute coverage of the Middle East conflict is related, within a subjective micronetwork, to a Facebook newsfeed of a Palestinian blogger and tweets from Human Rights Watch.

Taking these debates further, I argue that we specifically require a paradigmatic debate which would allow contextualization of journalism in a new dialectic between a transnational terrain and the interface, the digital micronetwork produced by users in various world regions.

It seems that these scalar contexts are specifically reaching today's youth generation in developed and developing countries. An international comparative study, conducted in nine developing and developed countries[2] among 14–17-year-old youth has assessed not only their perception of the digital sphere of news in national contexts, but rather the way in which these spheres produce a specific notion of "cosmopolitanized" orientation. The study revealed that news is consumed in distinct combinations of digital sources and, for example, national television sites are considered a "corrector" across the diverse range of countries. The majority of the respondents (n=6400) prefer digital sources, such as Google News, due to the "kind of information" which is not limited to "country boundaries" and provides the opportunity to "look" for further information (Volkmer, 2014).

It is such an inclusive perspective which is needed to identify new dimensions of journalism no longer only in digital spheres, but within new formations of inclusive public discourse in the 21ˢᵗ century.

REFERENCES

Beck, U. (2009). *World at Risk*. Cambridge, UK: Polity.

Berglez, P. (2008). "What is Global Journalism?" *Journalism Studies*, 9(6), 845–858.

Berglez, P. (2013). *Global Journalism: Theory and Practice*. New York: Peter Lang.

Berglez, P. & Olausson, U. (2011). "Intentional and Unintentional Transnationalism: Two Political Identities Repressed by National Identity in the News Media." *National Identities*, 1, 35–49.

Boyd-Barrett, O. & Rantanen, T. (1998). *The Globalization of News*. London: Sage.

Bruns, A. & Highfield, T. (2012). "Blogs, Twitter, and Breaking News: The Produsage of Citizen Journalism." In R. A. Lind (Ed.), *Produsing Theory in a Digital World: The Intersection of Audiences and Production* (15–32). New York: Peter Lang.

Carr, D. et al. (2014). "Cynics and Skeptics: Evaluating the Credibility of Mainstream and Citizen Journalism." *Journalism & Mass Communication Quarterly*, 91(3), 452–470.

Chernilo, D. (2007). *Social Theory of the Nation-State*. London: Routledge.

Chin, E. (2015). *Migration, Media, and Global-Local Spaces*. New York: Palgrave Macmillan.

Cohen, A. et al. (1996). *Global Newsroom, Local Audiences: A Study of Eurovision News Exchange*. London: John Libbey.

Cottle, S. (2006). *Mediatized Conflict: Developments in Media and Conflict Studies*. Maidenhead, England: Open University Press.

Cottle, S. & Nolan, D. (2007). "Global Humanitarianism and the Changing Aid-Media Field: Everyone was Dying for Footage." *Journalism Studies*, 8(6), 862–878.

Cottle, S. (2009). *Global Crisis Reporting*. Maidenhead, UK: Open University Press.

Deuze, M. (2008). "The Changing Context of News Work: Liquid Journalism and Monitorial Citizenship." *International Journal of Communication*, 2, 848–865.

El-Nawawy, M. & Iskandar, A. (2002). *Al Jazeera: How the Free Arab News Network Scooped the World and Changed the Middle East*. Cambridge, MA: Westview.

Erdal, I. J. (2009). "Crossmedia (Re)Production Cultures." *Convergence*. 15(2), 215–231.

Galloway, A. (2012). *The Interface Effect*. Cambridge, UK: Polity.

Gearing, A. (2014). "Investigative Journalism in a Socially Networked World." *Pacific Journalism Review*, 20(1), 61–75.

Georgiou, M. (2006). *Diaspora, Identity and the Media: Diasporic Transnationalism and Mediated Spatialities*. Cresskill, NJ: Hampton Press.

Giddens, A. (1990). *The Consequences of Modernity*. Cambridge, UK: Polity.

Halpern, O. (2015). *Beautiful Data: A History of Vision and Reason since 1945*. Durham, NC: Duke University Press.

Hanitzsch, T. & Berganza, R. (2012). "Explaining Journalists' Trust in Public Institutions Across 20 Countries: Media Freedom, Corruption, and Ownership Matter Most." *Journal of Communication*, 62(5), 794–814.

Heinrich, A. (2011). *Network Journalism: Journalistic Practice in Interactive Spheres*. New York: Routledge.

Hookway, B. (2014). *Interface*. Cambridge, MA: MIT Press.

Mabweazara, H. M. (2014). "Zimbabwe's Mainstream Press in the 'Social Media Age'" (pp. 65–88). In H.M. Mabweazara et al. (Eds.), *Online Journalism in Africa* (pp. 65–88). New York: Routledge.

Mabwearzara, H. M. (2014). "Charting Theoretical Directions for Examining African Journalism in the 'Digital Era.'" *Journalism Practice*, 9(1), 106–122.

Madianou, M. (2014). "Smartphones as Polymedia." *Journal of Computer-Mediated Communication*, 19(3), 667–680.

McNair, B. (2006). *Cultural Chaos. Journalism and Power in a Globalised World*. New York: Routledge.

Moon, S. & Hadley, P. (2014). "Routinizing a New Technology in the Newsroom: Twitter as a News Source in Mainstream Media." *Journal of Broadcasting & Electronic Media*, 58(2), 289–305.

Pantti, M. et al. (2012). *Disasters and the Media*. New York: Peter Lang.

Parks, L. (2012) "Footprints in the Global South." In I. Volkmer (Ed.), *The Handbook of Global Media Research*. Malden (pp. 123–142). Malden, MA: Wiley Blackwell.

Picone, I. et al. (2015). "When News Is Everywhere." *Journalism Practice*, 9(1), 35–49.

Rai, M. & Cottle, S. (2010). "Global News Revisited: Mapping the Contemporary Landscape of Satellite News." In S. Cushion & Lewis, J. (Eds.) *The Rise of 24-hour News Television* (pp. 51–80). New York: Peter Lang.

Reese, S. D. (2001). "Understanding the Global Journalist: A Hierarchy-of-Influences Approach." *Journalism Studies*, 2(2), 173–187.

Reese, S. D. (2010). "Journalism and Globalization." *Sociology Compass*, 4(6), 344–353.

Sakr, N. (2007). *Arab Television Today*. London: I. B. Tauris.

Sassen, S. (2007a). "Places and Spaces of the Global." In D. Held, David & A. McGrew (Eds.), *Globalization Theory* (pp. 79–105). Cambridge, UK: Polity.

Sassen, S. (2007b). "Introduction." In S. Sassen (Ed.), *Deciphering the Global* (pp. 1–20). New York: Routledge.

Schmitz-Weiss, A. (2015). "The Digital and Social Media Journalist: A Comparative Analysis of Journalists in Argentina, Brazil, Colombia, Mexico and Peru." *The International Communication Gazette*, 77(1), 74–101.

Schuetz, A. & Luckmann, T. (1979). *Strukturen der Lebenswelt*. Frankfurt: Suhrkamp.

Skjerdal, T. S. (2014). "Online Journalism under Pressure. An Ethiopian Account." In H. M. Mabweazara et al. (Eds.), *Online Journalism in Africa* (pp. 89–103). New York: Routledge.

Steensen, S. & Ahva, L. (2015). "Theories of Journalism in the Digital Age." *Journalism Practice*, 9(1), 1–18.

Uldam, J. & Askanius, T. (2013). "Online Civic Cultures: Debating Climate Change Activism on YouTube." *International Journal of Communication*, 7, 1185–1204.

Volkmer, I. (1999). *News in the Global Sphere*. Luton, UK: University of Luton Press.

Volkmer, I. & Chouliaraki, L. (Eds.) (2009). "European Satellite Cultures." (Special issue) *Global Media and Communication*, 9(3).

Volkmer, I. (2014). *The Global Public Sphere. Public Communication in the Age of Reflective Interdependence*. Cambridge, UK: Polity.

Wahl-Jorgensen, K. & Hanitzsch, T. (2009). "Introduction: On Why and How We Should Do Journalism Studies." In K. Wahl-Jorgensen & T. Hantizsch (Eds.), *Handbook of Journalism Studies* (pp. 3–16). New York: Routledge.

Willnat, L. & Weaver, D. (2012). *The Global Journalist in the 21.Century*. New York: Routledge.

Willnat, L., Weaver, D. & Choi, J. (2013). "The Global Journalist in the Twenty-First Century." *Journalism Practice*, 7(2), 163–183.

Wu, S. (2013). "Assessing the Potential of Channel NewsAsia as the Next 'Al Jazeera': A Comparative Discourse Analysis of Channels NewsAsia and the BBC." *Global Media and Communication*, 9(2), 83–99.

NOTES

1   The air strikes took place on April 6, 2015.
2   Countries include Australia, Germany, Japan, Kenya, Mexico, New Zealand, South Africa, Trinidad & Tobago.

CHAPTER SIXTEEN

Networked Reporting on Al Jazeera English: Context, Challenges and Comparative Advantages

TINE USTAD FIGENSCHOU

What does a broadcast news organization look and feel like when it's no longer trying to just be broadcast? (promo clip for AJ+, 2015)

Since its launch in November 1996, Al Jazeera has expanded rapidly from being an obscure outsider and alternative Arabic news channel into a multi-language, multi-channel, multi-platform global media network. In addition to the news channel in Arabic and Al Jazeera English (2006), Al Jazeera Network has also launched a variety of specialized sports, documentary, children's and live channels. In more recent years it has attempted to target regional markets outside the Arab region, such as Al Jazeera Balkans (2011), Al Jazeera Turk (2014) and Al Jazeera America (2014). Throughout this expansion the network has been surrounded by controversy, and its image problem has been a defining aspect of the network for some time (King & Zayani, 2008). Overall, Al Jazeera can be characterized as a fascinatingly unpredictable and unconventional media network in the contemporary global news landscape.

It is not only the Al Jazeera Network that has undergone a dramatic development over the last 20 years. The broader Arab media landscape has changed remarkably, with the development of a new Arab public sphere. In the competitive contemporary environment, Al Jazeera has expanded internationally, entering new markets in direct competition with major Western news media. In addition, legacy media are continuously challenged and increasingly intertwined with newer networked media, both in the Arab region and globally. The contemporary media landscape is characterized by hybridity and increased interactions between traditional, mainstream, elitist media logics and newer online, social and interactive media logics (Chadwick, 2013). The development towards a networked journalism, where the newsgathering process is more participatory and collaborative, represents a significant redefinition of journalistic professionalism (Singer et al., 2011) and potential redistribution of political power (Chadwick, 2013).

There have been numerous initiatives towards a networked journalism by Al Jazeera English since it was launched. Operationalizing networked journalism on AJE's news broadcasts, Youmans (2014: 66) foregrounded as key characteristics the involvement of web journalists or producers in broadcast news, citizen-consumers as sources or co-creators, social media as news source or actor in the ongoing stories, and the broadcast of user-generated content (interface for uploading video clips, etc.). Selected AJE current affairs shows have included questions and analysis by viewers and, since mid-2011, the channel has had a dedicated daily interactive debate show, *The Stream*. It is particularly the online news desk of AJE that has been innovative, experimenting with new interactive tools. During major news events, from the 2008–9 Gaza war onwards, the channel has highlighted user-generated information and interactivity in online live blogs, event timelines and maps to provide updated reports from those affected by crisis (Powers & Youmans, 2012), war (Bridges, 2013) and elections (Youmans, 2014). From 2010–11, AJE online started to release footage under a Creative Commons licence as one of many strategies to push AJE content on networked media platforms (Bridges, 2013). Recently, AJE has redesigned its website to highlight real-time news feeds and live blogs, and the Al Jazeera Network has launched the news app AJ+ targeting younger, global audiences on social media platforms.

Like most news channels discussed in this volume, AJE is in the process of designing and implementing a digital strategy. To illuminate the particular potential and challenges facing Al Jazeera English, the present chapter first provides a brief outline of the Arab satellite media revolution and the development of the Arab public sphere as a hybrid media space. It then exemplifies the power of the integrated Arab media space through the unique case of the 2011 Arab uprisings before the potential and limitations for networked journalism on AJE is discussed in the final part of the chapter.

A STATE-BACKED MEDIA REVOLUTION

To understand the contemporary hybrid Arab media landscape, it is vital to recognize the complex and contradictory characteristics of the wider Arab public sphere. From the mid-1990s, the mushrooming of Arab satellite networks, with geographical reach over national borders, the ability to speak to the wider Arab community and to reunite regional communities scattered by war, exile and migration, nurtured optimistic claims of *satellite democracy* (Sakr, 2001). The call to drive regional political reform and social change has also been documented in survey studies among Arab journalists and editors (Pintak, 2011). For years, the direct political effects of satellite television seemed to be limited and the academic debate rather emphasized the satellite channels' contribution to the emergence of a new Arab public sphere.[1]

A defining aspect of the Arab public sphere is the continued political ownership and influence in the regional satellite media market. An extensive body of literature demonstrates that Arab satellite television was launched, developed and financed by regional governments or entrepreneurs with close ties to governments, with an overall ambition of regional political influence (see, among others, Kraidy & Khalil, 2009; Rugh, 2004; Sakr, 2001). Moreover, researchers underline the remarkable concentration of satellite channels in Saudi hands (Kraidy, 2007). The Al Jazeera Channel (Al Jazeera Arabic) was developed in opposition to Saudi media dominance and could therefore be understood as Qatar's voice in a regional power play. Like most Saudi channels, the channel grew out of the combination of local (Qatari) capital and pan-Arab (largely Palestinian, Lebanese and Iraqi) expertise. Qatar, a small peninsular state bordering Saudi Arabia in the Persian Gulf, is an absolute monarchy governed by the Al Thani family. The Emir wanted an Arab station with Arab talent and expertise, and the bulk of the editorial staff consisted of Arab journalists trained in Western newsrooms and Arab newsrooms abroad, and the diversity within the staff helped create the channel's pan-Arab identity (Miles, 2005). Al Jazeera received its initial funding ($140 million) from the Qatari Emir to help launch and subsidize the channel over a five-year period to November 2001, and has since then been supported with an annual grant of about $100 million, plus a $1 billion investment to launch AJE (estimates from Powers, 2012).

Both government and Al Jazeera officials maintain that the network is independent and that its only connection to Qatari authorities is through its funding. There is a broad consensus that the network has revolutionized Arab television news as the first Arab channel based on Arab soil that is explicitly critical of Arab regimes and governments. Furthermore, the level of individual editorial freedom within the Al Jazeera Channel has been emphasized in studies of the channel's organizational culture (Zayani & Sahraoui, 2007). Having said that, in the authoritarian Qatari political context—where the ruling family and elite monopolize executive political power, regulate the contradictory Qatari media system and operate as the de facto "owner" of the network—it is obvious that this independence is relative and conditional. In recent years, the limits of this independence have been challenged by Qatar's new position as an unconventional regional power through its involvement and military intervention and its complicated relations to its Gulf neighbors. At the same time, as a public diplomacy tool, the Al Jazeera

Network only serves Qatar as long as it can maintain its editorial independence, or at least is perceived to be editorially independent by the outside world (see Figenschou, 2013, for analysis).

THE NEW ARAB PUBLIC SPHERE: A HYBRID MEDIA SPACE

Although it is still largely owned and sponsored by political and business elites, the new Arab media landscape is characterized by an unprecedented plurality of media outlets. The growth and intertwinement of social and satellite media in the Middle East and North Africa have gradually opened up the Arab public sphere in three interrelated ways.

First, researchers have called attention to a feeling of common identity among Arabs congregating around the new media, which has contributed to build a common, core Arab narrative or a new sort of pan-Arabism. Among the core "Arab" issues were Palestine, Iraq and the question of political reform, and all elections in the Arab and Muslim world received considerable attention in numerous debate programmes (Lynch, 2006; Pintak, 2011).

Second, the new Arab media are broadening the range of topics that people in the Arab world can talk about publicly. Over time, Lynch (2005) emphasized, the legitimacy of disagreement demonstrated in Arab satellite media would strengthen the long-term foundations for a more pluralistic political culture in the autocratic Arab region.

Third, there has been a fundamental change in the dynamics of the traditional sender–receiver relationship. Since the mid-1990s, the Arab public have been invited to call in their questions and views to studio debates, participate in phone or online polling and comment directly on screen by text messaging in an increasing number of media outlets (see, among others, Kraidy & Khalil, 2009). Moreover, the relatively rapid diffusion of the internet in the Arab world and the falling costs of mobile phones with video, photo and internet capability (Tufecki & Wilson, 2012) facilitated the audience's ability to document and share content (Kraidy, 2007). In terms of format, programmes became increasingly interactive and participatory, qualities generally lacking in regional politics (Kraidy 2007, 2010; Kraidy & Khalil, 2009). According to Kraidy (2007, 2010), the pluralistic, non-hierarchical *Arab hypermedia space* has served as an incubator of social change as Arab protesters have employed, tested and refined their media literacy tools in popular protests. Most of these street mobilizations have been over foreign-policy issues, such as the Israeli occupation and the US-led Iraq war (Lynch, 2006), but there has also been a growing number of internal political protests in Lebanon, Egypt and Iran (Lim, 2012; Pintak, 2011; Sakr, 2013).

The literature on the Arab public sphere synthesized here demonstrates that although the pace, scale and profound political changes caused by the 2011 Arab uprisings stood apart from previous protests, the ideas, the pan-Arab "media-fuelled narrative of change" had been circulating in the Arab public sphere for almost a decade (Lynch, 2011). In the now-voluminous academic literature about the Arab uprisings, researchers call for a more integrated examination of the media's role during the dramatic events and underline the need to go beyond the legacy media—networked media divide (Aday et al., 2013; Alterman, 2011; Aouragh & Alexander, 2011; Figenschou, 2013; Lynch, 2014; Robertson, 2012; Sakr, 2013; Sarnelli, 2013). Although employing different analytical concepts, these studies find that the Arab public sphere, in spite of its state-sponsored media ownership, systematic censorship practices and authoritarian political regimes, had developed into a particular, yet compelling, version of a hybrid media system.[2]

THE 2011 ARAB UPRISINGS

In hybrid media systems, political power is exercised by those who can create, tap, or steer information flows to suit their own agenda and modify, enable, or disable others', across and between a range of older and newer media settings (Chadwick, 2013: 207). In general terms, Lynch (2014) finds that the new media environment is challenging Arab authorities in six interrelated ways: (1) it promotes connective collective action by lowering the barriers of communication and organization and increasing the visibility of protest; (2) it causes diffusion and demonstrates effects by linking together protests across the region into a single protest narrative through using spilt screens and similar language (satellite media) and hashtags, linking and retweets (social media). The new media landscape (3) facilitates and enhances state repression, for Arab authorities working systematically to limit internet access; monitor online activity, repress and intervene in online activities; but at the same time it (4) increases the possibility to capture, document and expose violent crackdowns of the demonstrations and thus affect the calculus of violence; (5) it affects international attention and relations; and (6) undermines the Arab authorities' control over the public sphere as increasing numbers of people strive to access information, communicate, organize, engage and circumvent state surveillance.

Analysis of this new media environment has emphasized different aspects of the interaction, but there is broad agreement that the Al Jazeera Network played a vital role. Internationally, the AJE top news management pushed their editorial staff to their limits, deployed massive human and technical resources and aggressively pushed Al Jazeera reports on Twitter, Facebook and other platforms—to become *the channel of reference* during the dramatic events (Bridges, 2013). Regionally,

Al Jazeera Arabic and other regional satellite channels drove the protestors, framed them, legitimized them, and broadcast them to a larger audience (Abdelmoula, 2015; Alterman, 2011). Particularly during the early phases of the 2011 uprisings, AJA served as the main source of information for a vast majority of the population (Kalander, 2013; Wilson & Dunn, 2011) and consequently played a decisive role in shaping public opinion (Aouragh & Alexander, 2011). Al Jazeera presented the various local protests as part of the same revolution (Alterman, 2011: 111), and employed a common frame on the protests (Lynch, 2014), to the extent that the channel's emotional coverage and validation of the protest movements sparked criticism of campaign journalism.[3]

Both the network's news channels had numerous staff on the ground, an extensive network of sources and contacts, and regional experience and expertise that proved valuable during the dramatic uprisings (Bridges, 2013). Moreover, the channels had broad experience navigating and working in the often dangerous, difficult and contradictory regional political context and extensive practice in how to get around government censors (Lynch, 2014). During the Egyptian uprising, the network's correspondents on the ground—reporting undercover, embedded in the crowds, hidden among the protestors, took great personal risks and became part of the story.[4]

A key element in Al Jazeera's coverage of the uprisings was the network's raw, direct and constant visual documentation of the events on the ground. The live streaming (monitoring the events 24/7) and the live blogs (where the very latest developments on the ground were updated frequently on a 24/7 basis) became particularly popular (Mair, 2011; Mir, 2011). The increased visibility of such events, documented by mobile phones, amateur cameras and the network's live cameras, contrasted the massive popular, peaceful mobilizations with police brutality and violence (Alterman, 2011; Lynch, 2014).

Another way in which the Network amplified the protestors' message was through the authority accorded to social media sources and the inclusion of networked media in its coverage. AJE broadcasts included more such sources and incorporated more information from blogs and tweets than its international competitors and explicitly focused on social media's importance and role in the uprisings (Robertson, 2013; Sarnelli, 2013). Insider accounts from the Doha headquarters state that social media was used to source and corroborate information (Fisher, 2011), a practice confirmed by prominent activists on the ground (Aouragh & Alexander, 2011). AJE set up a desk dedicated to monitoring networked media for updates, conversations and trends in the material, summarized in hourly "web desk" updates (Aouragh & Alexander, 2011; Bridges, 2013). During the uprisings, Al Jazeera set up a website to facilitate the uploading of clips and images directly (Sarnelli, 2013). Networked media was most important in the coverage of those areas where the channel was banned or restricted, such as Tunisia, where AJE had to cover the conflict through a network of bloggers. According to AJE sources, the network of bloggers had been checked as credible sources in the weeks before the uprising—channel staff made direct contact with them to keep communication lines open in case they were blocked or shut down (Fisher, 2011; Mir, 2011).

Overall, the activists largely cooperated and interacted more extensively with mainstream journalists than is recognized in the literature (Bossio, 2014). During national communication blackouts, when regional governments shut down internet and mobile phone networks and/or censored national media, activists set up an operation to physically collect images and information and bring it to Al Jazeera and other international news media for redissemination (Aouragh & Alexander, 2011). Reporting through social media and cell phones became vital in those phases and areas when AJE's regular broadcast operations were closed down (Fisher, 2011). While the Al Jazeera Network validated and integrated social media in their newscasts, social media posts on the Arab uprisings largely reflected, referred to and linked to established mainstream media sites (Aday et al., 2013). Al Jazeera's live cameras and live blogs from the protests were among the most-read links during the uprisings in Tunisia, Egypt, Libya and Bahrain (Aday et al.: 910–911).

The intense regional satellite media and social media coverage of the uprisings drew massive international attention to the early phases of the Arab uprisings (Alterman, 2011). Analyzing location-specific linking patterns, Aday et al. (2013) find that Twitter mainly functioned as a megaphone, broadcasting information to a wider international audience. Similarly, translating the protesters' voices, AJE played a major role in bringing the drama playing out in the Arab streets to audiences, politicians and journalists across the world (Sarnelli, 2013). Traditionally, protest and radical movements in the Arab and Muslim world have received minimal coverage in mainstream Western media but, here, AJE broadcasts and online news played a key role in representing the demonstrations as legitimate, peaceful movements for democracy (Alterman, 2011; Robertson, 2012).

ORGANIZATIONAL INCENTIVES AND OBSTACLES

Within AJE, the lessons from the Arab uprisings demonstrated the popularity and potential of online and networked journalism (Bridges, 2013). A number of studies have illuminated how networked journalism is instigated by a combination of technological innovations, economic interests, democratic ideals and a cultural shift in editorial newsrooms (see, among others, Singer et al., 2011). In the final section of this chapter, I will outline some of the particular challenges and competitive advantage AJE faces in fully integrating social media and participatory practices into their news production in the wake of the Arab uprisings.

Starting with economic interests, it is imperative to stress that the Al Jazeera Network's unique model of funding largely shields it from commercial pressures. Except for top management, editorial staff have limited insight into the financial state of the channel as all budgetary information has been scarce and opaque. At the same time, interviews with staff document that they share a perception of financial stability (Figenschou, 2013; Usher, 2013). This funding model has arguably had both positive and negative impacts on innovation processes within AJE. First and foremost, it is a privileged position, in which the newsroom (including the online news desk) has had comparatively more human resources than most competitors (Bridges, 2013) and the freedom to experiment with cost-intensive organizational models, such as its decentralized news production and presentation (Figenschou, 2013). Moreover, from the beginning, the financial situation has given it the opportunity to invest in state-of-the-art technology, including video walls and HD technology (Powers & Youmans, 2012), although it remains challenging and time-consuming to recruit technological knowhow to the Doha headquarter (author interview, Doha, May 2015).

On the other hand, the financial stability may also have delayed interactivity as AJE has not had a defined target audience throughout its first years on air, and consequently the audience may remain a vague, undefined group about which AJE staff have little knowledge of or interaction with (see Figenschou, 2013, for discussion). This is also a major finding in Usher's (2013) study of the AJE website, which was characterized by a lack of incentives to use metrics to systematize and improve the understanding of their audience (Usher, 2013: 343). The systematic analysis of audience data is only now gradually being introduced into the AJE newsroom, primarily with the help of key members of the team behind the networks news app AJ+, an initiative in the Network that had audience engagement as a primary aim and with team members able to use sophisticated analysis of the network's online audience data (author interview, Doha, May 2015).

Due to a combination of political and market sensitivities, distribution to the North American market has been a major obstacle for AJE. For about seven years, from the launch of Al Jazeera English in 2006 to the launch of its US sister-channel Al Jazeera America in 2013, the American public primarily accessed AJE online through its YouTube channel or its website. Indirectly, this struggle to access the North American television news market expedited the channel's online streaming and distribution (author interview, Doha, May 2015). In addition to its attempt to launch AL Jazeera America, the network has also stepped up its efforts to reach American Millennials aged 18–34 through the launch of the news app AJ+. The AJ+ teams in San Francisco and Doha have a defined target group, continually monitoring, analyzing and systematically integrating audience data in their production (author interview, Doha, May 2015).

Where the channel's high-end technology and relatively stable and roomy news budget could be incentives for networked journalism, some characteristics of its newsroom culture can be said to have obstructed the process. The first English-language online initiative, launched in 2003, was harmed and paralyzed by a series of internal conflicts until it was re-launched as part of Al Jazeera English three years later (Powers, 2012). Similarly, frequent changes in the management of AJE online have disrupted its overall digital strategy (author interview, Doha, May 2015). When the management initiated an institutionalization and renewal process in 2008–9, the channel's "digital leap" was one of the key issues (AJE Renewal Project Report, March 2009). As part of the Report's process, internal staff surveys identified two major concerns with the channel's online service at the time—lack of alignment to the broadcast channel and lack of innovation, explained by poor management, negative working environment and lack of communication. AJE online journalists were largely isolated from the main newsroom, rarely conducting any original reporting, participating in editorial planning or receiving little professional feedback. The report called for a more consistent and ambitious digital strategy that integrated web and broadcast divisions more thoroughly. Recommendations emphasized systematic integration of channel output and social media, more depth and context through infographics, maps, timelines and pictorials, and, more participation and interactivity by inviting in bloggers and individual experts (AJE Renewal Project Report, March 2009). From 2013, AJE has stepped up its initiatives to bring digital services to the core of the channel by founding an innovation division aimed at becoming a hub for open innovation. Many of the recommendations from the renewal report are reflected in the editorial priorities of AJ+, although it remains challenging to extend interactive production routines and audience engagement into the main newsrooms, and the process of implementing a digital strategy on the network level is still forthcoming (author interviews, Doha, May 2015).

Perhaps *the* strongest incentive for AJE's networked journalism, is found in its progressive editorial agenda: Although the editorial vision of Al Jazeera English has been negotiated and reformulated throughout almost ten years on air, its emphasis on being a *voice of the voiceless* has remained strong. From the start, it was an expressed editorial aim within AJE to strive to find ways to let people on the ground report on their own lives (Figenschou, 2013: 56). This fundamental idea of voicing the subaltern remains key in the channel's new brand campaign, *Hear the human story* (2014), where stories of regular disenfranchised people represent the channel instead of its professional correspondents. The strong democratic ideals incorporated in the discourse of giving voice strongly encourage interactive initiatives. Within AJE, the use of new social media has been emphasized as a key editorial strategy to give voice to ordinary people. Voicing the voiceless has been a strong editorial imperative, although the channel has struggled to reduce established elite domination in its newscasts (Figenschou, 2013). Secondly, the anti-establishment spirit which has characterized Al Jazeera English's editorial priorities, emphasize alternative perspectives to distance the channel from international power elites and mainstream media. AJE aims to be systematically critical towards authority and elites, to invite all sides of the story into the studio, including *the other opinion* denied access to mainstream news (Figenschou, 2013: 99). Over the years the channel has accorded great authority to alternative or oppositional elites such as representatives from international organizations and non-governmental organizations (NGOs), media or cultural personalities, analysts and academics. Many of these voices with which AJE has established solid ties over the years have become distinct voices in the transnational hybrid media landscape.

NOTES

1 For in-depth analyses of this, see, among others, Abdelmoula (2015), Hafez (2008), Lynch (2006), Sakr (2007).
2 What Kraidy (2007, 2010) labels *hypermedia space*, and Aouragh & Alexander (2011) label *media synchronization*.
3 See, among others, Mair (2011) and Mir (2011) for further discussion.
4 See Bridges (2013) for a riveting inside account of Al Jazeera English's coverage.

REFERENCES

Abdelmoula, E. (2015) *Al Jazeera and democratization: the rise of the Arab public sphere.* Routledge: London & New York.

Aday, S., Farrell, H., Freelon, D. et al. (2013) 'Watching from afar: media consumption patterns around the Arab Spring', *American Behavioral Scientist*, 57(7): 899–919.

Alterman, J. B. (2011) 'The revolution will not be tweeted', *The Washington Quarterly*, 34(4): 103–116.

Aouragh, M. & Alexander, A. (2011) 'The Egyptian experience: sense and nonsense of the Internet revolution', *International Journal of Communication*, 5: 1344–58.

Bossio, D. (2014) 'Journalism during the Arab Spring: interactions and challenges', pp: 11–32, in S. Bebawi, & D. Bossio, (Eds). *Social media and the politics of reportage.* New York: Palgrave Macmillan.

Bridges, S. (2013) *18 days: Al Jazeera English and the Egyptian revolution.* Braddon: Editia.

Chadwick, A. (2013) *The hybrid media system: politics and power.* Oxford & New York: Oxford University Press.

Figenschou, T. U. (2013) *Al Jazeera and the global media landscape: the South is talking back.* New York: Routledge.

Fisher, A. (2011) 'The "Arab spring", social media and Al Jazeera', pp: 149–159, in J. Mair & R. L. Keeble, (Eds). *Mirage in the desert? Reporting the 'Arab spring.'* Suffolk, UK: Abramis Academic Publishing.

Hafez, K. (Ed.) (2008) *Arab media: power and weakness.* New York & London: Continuum.

Kalander, A. A. (2013) 'From TUNeZINE to Nhar 3la 3mmar: a reconstruction of the role of bloggers in Tunisia's revolution'. *Arab Media and Society* 17.

King, J. M. & M. Zayani (2008) 'Media, branding and controversy: perceptions of Al Jazeera in newspapers around the world', *Journal of Middle East Media*, 1(4): 27–43.

Kraidy, M. K. (2010) *Reality television and Arab politics: contention in public life.* Cambridge University Press.

Kraidy, M. K. (2007) 'Saudi Arabia, Lebanon, and the changing Arab information order', *International Journal of Communication*, 1(1): 139–156.

Kraidy, M. K. & Khalil, J. F. (2009). *Arab television industries.* London: Palgrave Macmillan.

Lim, M. (2012). Clicks, cabs, and coffee houses: social media and oppositional movements in Egypt 2004–2011. *Journal of Communication*, 62: 231–248.

Lynch, M. (2014) 'Media, old and new', pp: 93–109, in M. Lynch. (Ed.), *The Arab uprisings explained: new contentious politics in the Middle East.* New York: Columbia University Press.

Lynch, M. (2011) 'The big think behind the Arab spring', *Foreign Policy*, December.

Lynch, M. (2006) *Voices of the new Arab public: Iraq, Al-Jazeera and Middle East politics today.* New York: Columbia University Press.

Lynch, M. (2005) 'Assessing the democratizing power of satellite TV', *Transnational Broadcasting Studies*, 14, Spring/Summer.

Mair, J. (2011) 'Reporter or provocateur? The "Arab Spring" and Al Jazeera's "CNN Moment"', pp: 172–183, in J. Mair & R L. Keeble, (Eds.), *Mirage in the desert? Reporting the 'Arab spring'.* Bury St Edmonds, Suffolk: Abramis Academic Publishing.

Miles, H. (2011) 'The Al Jazeera effect', *Foreign Policy*, February 8.

Miles, H. (2005) *Al-Jazeera: the inside story of the Arab news channel that is challenging the West.* New York: Grove Press.

Mir, M. (2011) 'Was Al Jazeera English's coverage of the 2011 Egyptian revolution "campaign journalism"?', pp: 160–171, in J. Mair & R. L. Keeble, (Eds.), *Mirage in the desert? Reporting the 'Arab spring'.* Bury St Edmonds, Suffolk: Abramis Academic Publishing.

Pintak, L. (2011) *The new Arab journalist: mission and identity in a time of turmoil.* London and New York: I. B. Tauris.

Powers, S. (2012) 'The origins of Al Jazeera English', pp: 5–28, in P. Seib, (Ed.), *Al Jazeera English: global news in a changing world.* New York: Palgrave Macmillan.

Powers, S. M. & Youmans, W. (2012) 'A new purpose for international broadcasting: subsidizing deliberative technologies in non-transitioning states', *Journal of Public Deliberation*, 8(1), Article 13.

Robertson, A. (2013) 'Connecting in crisis: "old" and "new" media and the Arab spring', *The International Journal of Press/Politics*, 18(3): 325–341.

Robertson, A. (2012) 'Narratives of resistance: comparing global news coverage of the Arab spring', *New Global Studies*, 6(2), 1–20.

Rugh, W. A. (2004) *Arab mass media: newspapers, radio, and television in Arab politics.* Westport and London: Praeger.

Sakr, N. (2013) 'Social media, television talk shows, and political change in Egypt', *Television & New Media*, 14(4), 322–337.

Sakr, N. (2007) 'Approaches to exploring media-politics connections in the Arab world', pp: 1–13, in N. Sakr, (Ed.), *Arab media and political renewal: community, legitimacy and public life.* London and New York: I. B. Taurus Publishers.

Sakr, N. (2001) *Satellite realms: transnational television, globalization & the Middle East.* London/New York: I. B. Tauris Publishers.

Sarnelli, V. (2013) 'Tunisia, Egypt and the voices of the revolution in Al Jazeera English', *Journal of Arab & Muslim Media Research*, 6(2/3): 157–176.

Singer, J. B., Hermida, A., Domingo, D., et al. (Eds.) (2011) *Participatory journalism: guarding online gates at online newspapers*. Chichester, UK: Wiley-Blackwell.

Tufecki, Z. & Wilson, C. (2012) 'Social media and the decision to participate in political protest: observations from Tahrir Square', *Journal of Communication*, 62, 363–379.

Usher, N. (2013) 'Al Jazeera English online', *Digital Journalism*, 1(3), 335–351.

Wilson, C. & Dunn, A. (2011) 'Digital media in the Egyptian revolution: descriptive analysis from the Tahrir data sets', *International Journal of Communication*, 5.

Youmans, W. (2014) 'Al Jazeera English's networked journalism during the 2011 Egyptian uprising', pp. 56–78 in S. Bebawi, & D. Bossio (Eds.), *Social media and the politics of reportage*. New York: Palgrave Macmillan.

Zayani, M. & Sahraoui, S. (2007) *The culture of Al Jazeera: inside an Arab media giant*. Jefferson, NC & London: McFarland & Company.

INTERVIEWS

The author conducted research interviews with Al Jazeera innovation department's Mohammed Ziyaad Hassen, Riyaad Minty and Michael Shagoury (AJ+) in Doha, May 2015.

CHAPTER SEVENTEEN

Twitter and the Rolling-News Agenda on Sports Channels

ALAN TOMLINSON

INTRODUCTION

It would be a mistake to underestimate the significance of Twitter, its effects upon cultural consumption and everyday discourse, and its contribution to how rolling, 24-hour news in its constant updating embodies the speeded-up society (Redhead, 2015); Twitter has without doubt accelerated this (Hutchins, 2011). But we must be careful not to ossify the past. The print journalist of the early 1950s faced challenges from radio, and the "Monday morning man" reporting on the Saturday fixture had interpretive challenges in using the material others already had, aware that he "must not be the gramophone on which the same record is played" (Ledbrooke and Turner, 1955: 166).

The traditional press box was no luxury posting, the football writer huddled "with his typewriter in conditions which would never be allowed by any reasonably diligent factory inspector" (Hall and Parkinson, 1974: 14). If such journalists strayed from, say, the club chairman's agenda, the price could be high. Bob Lord, known among the press as "the Khruschev" (USSR president) of Burnley, and a "passionate partisan of his town and his football team" (Hopcraft, 1968: 147), banned a Sunday and daily papers and six individual journalists in a feud over what he saw as media misrepresentation and intrusion. Getting to a player's house? Ringing him up on the telephone "at all hours"? Lord (1963: 116) himself called these "the hole-and-corner methods of the few to which we take objection." Twitter would have turned a football boss like Lord into a Stalinist figure purging the dissidents of both press and players.

Social media are in conception the source of an unprecedented form of inclusion and participation across publics and cultures, and would have been anathema to the management style of Lord, also known more locally as the Burnley Butcher. But they are also immensely lucrative forms of technological development and cultural production, in part fuelled by companies with little ethical concern about business practices and relationships. It is this dual character that makes a balanced evaluation of their socio-cultural contribution a more demanding task than it would seem to be at first sight.

Media professionals may feel generally that it is "perilous to be left behind in a 24-hour news cycle characterized by 'media-herd' behaviour" (Rowe, 2011: 25). At a gathering of international sport journalists in early 2015, the threat of Facebook, Twitter, Instagram and other social media sites—leaders in the rapid growth of digital media—was seen as much more than a passing danger to traditional media: "traditional media like newspapers and magazines…now face the reality of extinction" (Aparicio and Olobulu, 2015: 42). Jerome Cazadieu, editor-in-chief of digital at French paper *L'Equipe*, warned that alternatives to traditional practice must be found, as "if we keep thinking of the same strategies, we can be pushed out of business."

Social media are far from universally welcomed. Former England cricketer Ed Smith generated gasps in the BBC Radio 5 *Breakfast* studio when, in the context of the Pietersen issue discussed in detail in a section below, he responded to a listener's text message damning England's last two cricket captains: "The text was written by an idiot," he said. Reflecting on this, Smith wrote that if every issue is referred to "an online referendum, pandering to the mob…we in the media…are accomplices in infantilisation on a massive scale" (Smith, 2015: 38). When Sky Sports pundit Paul Merson commented that the England football team's Andros Townsend, scorer of an equalizing and face-saving goal against Italy, was lucky to be in the national squad, Townsend retorted on Twitter: "Not bad 4 a player that should be nowhere near the squad ay@PaulMerse?" (Spurling, 2015: 18). The triumph of the trivial? A modern technological version of a democracy's right to free speech? Important instaneity? Take your pick.

But it is beyond dispute or debate that participation and the polyphony of voice, in whatever its specialist sphere of profile, performance or politics, has been immeasurably boosted by the rise of social media, in particular the purportedly egalitarian but potentially tyrannical form of the 140-character utterance of the Tweet. This is particularly true of the popular cultural star or celebrity, but also of the follower or fan. Whether such utterances, the spluttered as well as pithily precise, aphoristically arresting responses and exchanges that they can generate, and the diluted forms of dialogue that they can foster, are the stuff of news is a definitional and territorial question best left to experts on the epistemology of news-making. But social media have without doubt shifted the balance of voice in the communications landscape and the 24-hours news culture. Dedicated sports channels also crowd the field, ranging from single-club television or media operations, to single-sport broadcast by national sports associations. Global players for whom sport is the fulcrum of a running preview, broadcast and review agenda include pioneers and dominant forces Sky/Fox Sports and ESPN, along with channels such as the US's NBC (SC) and the Gulf's beIN sports, out of Doha/Qatar. This book provides in-depth and wide-ranging coverage of 24-hour news channels generally; but the emergence and profile of 24-hour sports channels remains as yet only partially understood and explicated.

In this chapter, I take Twitter as a case through which to offer reflections on the digital transformation of news-sourcing and news-making in the sphere of the popular, in particular sport. Twitter was launched in mid-2006 and by 2015 had more than 500 million users, 300 million of which were reported as active. The scale of the phenomenon is beyond dispute; its significance remains much more elusive, and debatable.

In an earlier piece on the rolling menu of 24/7 sports news, with John Sugden (Sugden and Tomlinson, 2010), we argued, or at least suggested, that a rolling 24-hour sport news provision would inevitably lead to a trivialization of sport writing and reportage; it was also acknowledged that sport chatter (Eco, 1986), gossip and trivia are fundamental to the culture and practice of those who participate in and follow sport. These processes were symbolized in that piece by posing the question of how the expanding forms of such journalism might become preoccupied with questions such as "what might Beckham have had for breakfast?" Half a decade on, with David Beckham no longer playing at the top level, passing the age of 40, and his 2012 role at the London 2012 Summer Olympics Opening Ceremony fading into the mists of contemporary history, perhaps fewer people would be quite so interested in the question; but with the rise of the Twittersphere, perhaps Beckham himself, already with 50 million "likes" on his Facebook page, and having instigated his Twitter account for his 40th birthday in April 2015, would be more likely to join the conversation on what he had, or pass the query on to his wife Victoria whose tweeting engages with the mundanities of everyday celebrity lifestyle more easily than her husband has ever managed; Victoria Beckham boasted 8.83 million Twitter followers in September 2015. But before glancing Twitterwards towards what can only be described as the extraordinary statistics of the phenomenon, a historical perspective on the rise of the celebrity sporting profile provides illuminating context.

IDOLS OF CONSUMPTION

In a brilliant and prescient essay demonstrating the importance of undertaking serious analysis of the popular, Leo Lowenthal wrote of the demise in the print magazines of the US of writing on the lives of men of industry and politics. The essay, borne of Lowenthal's Frankfurt School sensitivities and critical antennae attuned to the emerging consumerism of the country and culture that welcomed him as a refugee from European fascism, was written in 1943 and first published in 1944. In it, Lowenthal noted that the new elite, whose presence was lighting up the pages of the widely popular magazines, comprised movie stars, singers and entertainers, and sportsmen and women. In 1901–02, of 21 biographies published in *The Saturday Evening Post*, "eleven came from the political sphere, seven from the business and professions, and three from entertainment and sport" (p. 112). By 1940–41, the 100 biographies published in a single year in *The Saturday Evening Post* and *Collier's* were dominated by the entertainment category, providing 55% of cases dwarfing the business/professional (20%) and "political life" (25%) biographies (pp. 111–12); close to a half of those classified as "entertainment" were from what we might call the sports industry. In a telling interpretive formula, Lowenthal referred to these emerging personalities, stars and celebrities as idols of consumption, displacing the makers and the rulers of industry and politics, the traditionally respected and sometimes revered idols of production, in the popular consciousness. The essay was prescient because Lowenthal observed from the data one of the key features of modern celebrity politics and culture, its increasing sense that the star could be reached out and touched, could be seen as someone who could have been you, if only you'd had the breaks. The relationship between these idols of consumption and their followers was distant, of course, based on the forms of exclusivity that relative wealth can buy and protect. But in performance, in self-profiling of individual image, the idols of consumption could profit from not just an aura of glamor and a concocted sense of the charismatic, but also from a cultivated informality that could flavor the spheres of consumption in cinema halls, sports stadia and racetracks, and at concerts and performances. Teams of image-managers remain aware of this in the Twitter business, as 21st century idols of consumption monopolize the profile charts of the Tweet Parade.

TWITTER'S TOP 100

There is certainly a representative of the old idols of production in Twitter's Top 100, as Barack Obama's team brings the White House into play in the global game of instant and pithy communication: in September 2015, Obama had more than 63 million followers, in third position behind Katie Perry (almost 75 million) and Justin Bieber (67 million). But it is women and (fewer) young males from the realm of popular music who dominate the follower figures. Sport is not so prominent. Here are the figures for the only sports stars to figure in the Twitter Top 100 in the spring of 2015; they are all men:

Name	Sport	Position	Followers
Cristiano Ronaldo	Football	13th	35,457,090
Kaka	Football	28th	22,640,305
LeBron James	(American) Basketball	33rd	20,984,000
Louis Tomlinson	(American) Football	38th	19,493,106
Neymar Jnr	Football	45th	18,102,774
Amir Kahn	Boxing	82nd	12,501,343
Wayne Rooney	Football	97th	11,181,744

Source: http://twittercounter.com/pages/100, compiled by author consulted May 11th 2015.

Lest this Olympian scale of tweeting be mistaken for some new form of dialogic relationship, Ronaldo follows just 92 other tweeters, and Amir Kahn follows a mere 7 lucky recipients of his musings and accumulated wisdom; at least Wayne Rooney and his team show some reciprocal interest, following 256 other tweet accounts. Kahn may be at the lower end of the Top 100, but with his US-based profile, for a mere 280 tweets he has generated an astonishing number of followers; perhaps there is an *en masse* registration of boxing fans in the Philippines, a fanatical boxing nation that will have followed every move of the boy from Bolton, Lancashire. Kahn also uses his profile in the social media for good cause, linked to his Amir Kahn foundation, and responding to tragedies such as the Syrian migrant/refugee crisis; the shocking picture of three-year old Aylan Kurdi's body washed up on the coast of Turkey at the beginning of September 2015, flopping like an ancient ragdoll in the hands of a hardened Turkish soldier, prompted Kahn into action. A message on social media raised donations filling seven large trucks with long-life food, clothing and bedding bound for the island of Lesbos (Pidd, 2015).

Institutions can also figure high in Twitter's listings; Ronaldo's club Real Madrid has 15,841,533 followers, at 54th position, its great Spanish rival Barcelona in 58th position with 14,941,675 followers. Most notably, six of the seven listed in the table above are footballers, of one code or another, with the one exception of Kahn. Rooney's tweets, viewed between March 30th and May 8th 2015, included Happy Birthday greetings to "my wife, Coleen," and a proud note on seeing his son Kai "pick up his achievement certificate at his assembly"; the Rooney tweets congratulated teammates and avoided controversy, carefully managing the image of an ordinary family man. The tweets were, in all likelihood of course, ghosted; meanwhile, Rooney himself gets on with his own interests, such as in February 2015 boxing in his kitchen with a former teammate, revealed in a leaked video of him sprawled and knocked-out on his kitchen floor. So, of course, tweeting is as much image-management as spontaneous expression or emotional outpouring, and the nature of the tweet can vary as widely as the spoken word. Tweeting sportsmen and women communicate with their followers in a variety of forms, choosing multiple combinations of regularity, opinion-rich utterance, polemical intent, and egotistic posturing. And this has generated a non-stop flow of material for sport journalists and those writing from a range of perspectives about sport.

THE KEVIN PIETERSEN SAGA

Rolling news can take the form of the episodic, the mini-series and the saga (Sugden and Tomlinson, 2010). In England, one sportsman to harness the capacity of Twitter to make headlines is South African-turned-England cricketer Kevin Pietersen. Novick and Steen (2014: 127–28) summarize what they call the KP saga, in which Pietersen, English cricket's "most flamboyant and incautious self-promoter" (Novick and Steen, 2014: 127), used Twitter to announce—or complain about—his omission from the team in 2011. He then went on to text, during the course of a match between England and South Africa, members of the South African team on the characteristics and vulnerabilities of English players. His outstanding score in that match preceded the departure of the then England captain Andrew Strauss: "Social media" took the issue out of the "dressing room and put it into public arena" (Novick and Steen: 128). The KP saga rumbled on for several years with Pietersen himself utilising Twitter, and the launch of his autobiography, to continue sniping at those branches of the cricket establishment and the sports media from whom he did not receive favorable or sympathetic coverage. In May 2015, *The Guardian* summarized things, with Strauss now Director of Cricket in the England set-up:

> According to a tweet from Piers Morgan, who appears to have appointed himself Pietersen's social media cheerleader, Strauss and Harrison were going to inform Pietersen that, despite the incoming ECB chairman, Colin Graves, having indicated the door might not be closed on the batsman's England career, there is in fact no way back into the international fold.

As indeed there wasn't. Pietersen's social media cheerleader and unofficial agent Morgan probably did little more than further alienate the cricketing establishment, and the England team went on to regain the Ashes from the visiting Australian team without the involvement and assistance of the swashbuckling but self-serving South African. "The truth about Kevin is that he is a phenomenal cricketer. But over months and years, trust has eroded between Kevin Pietersen and the ECB. There is a massive trust issue between me and Kevin," Strauss told the BBC (http://www.bbc.co.uk/sport/0/cricket/32703824). Working with cases taken from media reportage of professional sports in the US, Gibbs (2013) identifies four categories of what one might call Twitterthreat, in that they identify the sources of potentially image-damaging coverage and effects. These are the Rookie Reporter, the Team Insider, the Opportunist, and the Imposter. PR professionals at clubs and with responsibility for the profile of the team would, Gibbs notes, be wise to be wary of these sources, so deflecting potential negative consequences. Obviously if your new/social media leak is the egotistic star-turn in the squad, the act by the Team Insider will have huge, potentially detrimental and damaging consequences for spirit, morale and trust, as Andrew Strauss indicated in his firm judgement on the Pietersen case.

SERENA SLAM

As the tennis summer was warming up in May 2015 and the question of Beckham's breakfast choices was at the back of my mind, I wondered how the woman with the most competitive instinct and record in tennis was keeping in with her fan base. Serena Williams has won 20 Grand Slam singles, and has a career trajectory of dramatic falls and comebacks that a thriller writer would envy. At the forthcoming French Open, her 20th Grand Slam, Williams also achieved her second "Serena Slam" by holding all four of the top championship titles at once. In the build-up to this, how was Serena taking to Twitter? On May 10–11, 2015, she tweeted her 4.67 million followers about an Instagram item she'd spotted, of what looked like a straightjacketed Yorkshire terrier dog wriggling manically (perhaps to dog-lovers this was everyday sweet and charming stuff) to escape its bizarre trappings. She'd also posted a photograph of herself ending a doubles encounter a little before that, so sport wasn't off the agenda. And clearly she is communicating also in support of good causes, promoting her swf.org (Serena Williams Foundation): "Living, loving and working to help you." Somebody's been working hard at least. Williams's account has logged 12.9k tweets, in 5 years apparently; that's an average of 11 a day over half a decade.

BARTON: BAD BOY TO CULTURAL PUNDIT

English footballer Joey Barton has had a classically checkered career in English football. As a young starlet with Manchester City, he notoriously poked a fellow player in the eye with a burning cigar; he has criminal convictions for violent assault, and has served a custodial jail sentence. He has played for Newcastle United and Marseilles, and appeared for England. But his celebrity has been framed in the main by the social media. In 2015, he commanded a Twitter following of 3.07 million, who were treated on September 21st to his thoughts on Lord Ashcroft's new unauthorized biography of Prime Minister David Cameron: "Can see the Daily Mail headline now. Our Dave may have rattled a pig but at least he sung the National Anthem. #piggate." Barton has mixed sporting, political, and even philosophical utterances in his outpouring of 16,200 tweets, building and enhancing a reputation as a pundit alongside a declining playing career as he fell into the second tier of English football with Queen's Park Rangers and then, in 2015, Burnley. The tweets on social topics and in philosophically aphoristic style have attracted widespread comment, sport journalist Henry Winter saying that "he's a bright guy trying to deal with his demons" (de Castella, 2015). In 2014, he appeared on the BBC's *Question Time* and has positioned himself as a potential media pundit cum celebrity, in ways unimaginable without his Twitter profile.

SNIPPETS AND COLUMNS

The young English footballer Saido Berahino had a very productive season in the English Premier League in 2014–2015, scoring 14 goals for his club in the English Premier League. This made him a target for bigger clubs than West Bromwich Albion (West Brom), and one of the main foci for press speculation and widespread rumor-mongering in the period up to the closure of football's transfer window, a period in which players can be bought and sold between clubs, before the player's current contract ends and his value in the transfer market tumbles to zero. The closing date of the transfer window in the early days of the 2015–16 season was September 1st, and on the morning of that day, at 9:03 a.m., Berahino tweeted: "Sad how i cant say exactly how the club has treated me but i can officially say i will never play Jeremy Peace." Peace, chairman of West Brom, had been chief negotiator or at least club spokesman in relation to Tottenham Hotspur's bids, peaking at £23 million, for Berahino. The player's tweet generated a torrent of speculation across established and informal social media about player power, strike threats, and club disciplinary processes. But how "official" could this statement be? Where did it mention strike action? What did "never play Jeremy Peace" actually mean (beyond the "for" that most commentators inserted before "Jeremy")? The tweet was retweeted 45,000 times, and deleted on the Thursday night. Meanwhile, one hastily-concocted tweet of 23 words kept innumerable football writers going for several weeks, match reports three weeks later still referencing the "strike" threat as Berahino was scoring for West Brom again, hugging his West Brom manager in celebration, and getting down to his work again, no apology in sight. A particularly striking use of this 23-word pronouncement was in a Marina Hyde column in *The Guardian* (Hyde, 2015: 8). She reminded us that Berahino arrived in England as a 10-year-old refugee from the Burundian civil war, which claimed the life of his father; on arrival, he was placed in a care home until DNA testing could confirm his maternal parentage. Hyde, in her wittily articulate put-down style, expressed herself to be "intensely relaxed about Berahino tweeting something drama-queeny about a 59-year-old multimillionaire club owner." She added that she was equally relaxed about the 22-year-old "taking his mum along" to transfer negotiations, and posting a photo of himself having won a competition to fly in a private jet. Hyde's point is simple: here is a young man whose refugee past has shaped not just his aspirations for a future of his own, but challenged the stereotypical and prejudiced responses of both fans and media. Hyde also reminded us that Berahino has a charitable foundation, and has made a video of his refugee story for the UNHCR (the United Nations Refugee Agency). The story of Berahino's successful post-refugee life is bigger, and more important, Hyde concludes, than a "childish and materialistic" spat with a club owner. What is noteworthy here is that the 23-word tweet, referred to but never quoted, could generate a full 1,000-word (weekly) column. The single tweet can be a gift to a certain kind of writer, particularly one with an eye on the bigger picture, and with a distinctive tone such as Hyde's penchant for the absurd.

It would be easy to reduce sporting discourse to the accumulation of fragments of gossip and opportunistic comment, and to see this as a transformation of a traditional craft. Steen claims that "just as the emphasis once shifted from print to television and thence to the internet, the latest transformation has come courtesy of the Twittering classes" (2013: 218). Seeing "transformation" here may be to overstate the determining influence of Twitter, and to list Twitter alongside print, screen/broadcasting, and the web confuses matters. One feature of the Twittersphere is, though, indisputable, and that is its combination of instaneity with speed of dissemination, and response. After the Premier League match between Chelsea and Arsenal on Sunday, September 20, 2015, Chelsea player Kurt Zouma told beIN sports, referring to team-mate Diego Costa: "Everyone knows Diego, and this guy likes to cheat a lot and put the opponents out of his game, and that happened in the game. He's a real nice guy in the life and we are very proud to have him." The following day, Zouma, or someone, penned the following tweet: "Sorry for any confusion…English is not my first language and I did not mean to accuse anyone of cheating. Simply to say Diego is a player who puts pressure on his opponents and who I have huge respect for" (Fifield, 2015). Were these Zouma's own words? The follow-up tweet is cover-up or *volte face*; agents and PR men and women surrounding the sporting superstar are now on a constant standby to protect the brand of their client, joined by journalistic specialists and citizen commentators alike, desperate to feed the insatiability of the sporting public.

Steen (2013: 218) implies that new freedoms and possibilities are created by the medium:

> Freed from the tyrannical lash of mundane cliché and PR-speak, media-trained footballers have used the new medium to cut out the middleman, bypassing print and television and communicating directly with anyone who has the vaguest interests in their utterances. Journalists, as a consequence, are on perpetual Twitter-watch. This cuts both ways. While what ensues is largely a bonfire of the inanities, exceptions, refreshingly, are on the rise.

He cites as an example, Javi Poves, a former Sporting Gijon footballer who tweeted that the game that had provided him with a good living was "putrid" and "corrupt." How, then, have professionals responded to this sort of expansion of voices and perspective?

THE PROFESSIONAL CHALLENGE

Established, print-trained journalists have had no choice but to respond to the avalanche of commentary generated by new media. Simon McEnnis (2013) has interviewed practising journalists in a study identifying two key themes, the significance of trust as a professional ideology or central principle of practice, and the disruption caused to the core working practices of the profession by the breaking news approach. Journalists note the tension between being in the field, on the beat so to speak, and constructing stories from the torrents of information available through social media (McEnnis, 2013: 428):

> We have to step it up and raise our game. Journalists trawl through social media and put their stories together. That's fine but you shouldn't be too dependent on it. You've got to be out there talking to people all of the time. (*Sun* sport reporter Justin Allen)

Print-based journalists who have cultivated sources over many years cannot afford to breach the trust of those sources (McEnnis, 2013: 428):

> You can't take a gamble. I've spent years building up contacts, whether in boardrooms, press rooms, or at training grounds. I talk to people behind the scenes at clubs. Information is checked and double-checked. I don't go on rumor or supposition. (*Scottish Mail on Sunday* sport reporter Graeme Croser)

Yet there is a kind of compulsion that drives the traditional practitioner to look in on the new media sources, like a neighbor looking in on, though not invited to attend, the newcomers' party next door:

> When you're not covering a game, you find yourself in front of the TV commenting—almost commentating—minute by minute and giving the twittersphere the benefit of our wisdom. (freelance sport journalist Nick Szczepanik) (McEnnis, 2013: 429)

Sky Sports News reporter David Craig recognizes a shift towards the instant reporting of the moment:

> Here at *Sky Sports News,* the big kick is that I am telling you something that you don't know, and now the advent and development of social media can make anyone a journalist. Without doubt it has affected us. I like the live stories and the big news stories, but we're doing them less and less because of new media. (McEnnis, 2013: 430)

Justin Allen of *The Sun* comments too that the "insider" so often cited in news-paper reports may, in the current climate, be tweeting him or herself, and that in the face of the possible reduction of sources, writers—in the tabloid cases for instance—need more than ever to use their creative resources in the production of sport content. The social media can become in this sense a fruitful new source of material:

> You get a lot of great ideas…from what players tweet. Rafael van der Vaart tweeted that his missus Sylvie had a Barbie doll made of her. Straight away my creative juices think… that sounds fun, let's mock up Van der Vaart as Crystal Ken. You need to make the story happen. You wouldn't get that from a citizen journalist, they just wouldn't think like that. (McEnnis, 2013: 431)

It is widely recognized that Twitter has led to "a digital search for scandal, dis-agreements and disclosures that elicit a response from the public and the subjects of the stories" (Hutchins, 2011: 244, quoting J. B. Thompson), and Twitter can feed the voracious appetite of an expanding readership, fueling a trivialization of sport journalism. At the same time, journalists respond creatively to professional challenges, operating across multiple platforms (Hutchins and Rowe, 2012: 126) and as McEnnis (2015) shows in his study of live blogging, they look to respond to the defining new dynamic of speed by balancing old and new skills, recognizing the need to hyper-produce, seeking to build and interact with audiences, and to exploit creative opportunities. As the BBC's Chris Bevan said to McEnnis, on this issue of creativity: "Drive it how you like" (McEnnis, 2015/16).

Conceptions of creativity can of course clash with conventions of established and ethical journalism. Harcup (2007: 46–47) recalls alternatives outside the main-stream, where alternative newspapers could question what any story was about, or ask whether what journalists call the public interest equates with the public good. The Twitter challenge to professionals embraces questions concerning the nature of everyday practice and craftwork, and poses, too, the question of the fundamen-tal nature of journalistic practice.

CONCLUSION: FADING STARS, CROWDED CONSTELLATIONS

It is easy to be searingly critical of the rise of forms of digital consumption and round-the-clock reporting from a particular perspective, of age/longevity for in-stance. French film star Catherine Deneuve, 71-year-old *grande dame* of the movie industry, spoke of this situation in strikingly dismissive tones: "We see a huge amount about people who are very famous, who have millions of followers…and who have done absolutely nothing" (Willsher, 2015: 17).

We hear a great deal, too, from individuals advised by agents and PR machines to sign up for Twitter. It is, as Willsher adds, a little rich if not hypocritical of Deneuve to talk of low profile after a career blending movies and make-up, as the face of Chanel No. 5 and L'Oréal, and promoter of her own name-celebrating perfume, but she is of the generation for whom mass exposure of the image was the holy grail, leaving it up to the fans to gasp, admire and converse among themselves. An intensifying problem for sport reporters and comparable media professionals in the age of permanently rolling pronouncements via the Tweetroute is that Deneuve's "absolutely nothing" is now the routine stuff of the daily grind of screen-scanning. It might be some compensation for the lack of real access that has been created by the feeding of scraps at PR-dominated press conferences, but not much.

The Twittersphere is, literally, worlds (of communication) away from an earlier world of sport writing and sport journalism. But Raymond Williams warned us that the technology does not necessarily transform the ideological meanings conveyed by the cultural form; "we have to reject technological determinism, in all its forms" (Williams, 1974: 130). In a section on sport in *Television: Technology and Cultural Form*, he noted that there is "a large sub-culture of sporting gossip which takes a great deal of television time but which was already basically present in newspapers…television has been a powerful agency of certain trends which were already active in industrial society, rather than a distinctly formative element" (1974: 68). This is a vital point. Twitter mobilizes what was already in the mix— the transfer gossip, the celebrity snippet. Sport journalists have claimed that their primary use of Twitter is for breaking news, but analysis of their Twitter content (Sheffer and Schultz, 2010) has shown rather that tweets feed commentary and opinion. We could add that in sport it provides a seemingly indispensable tool for different interests within the sporting industry to have a say, seek to make or deepen a mark.

In any evaluation of the latest technological innovation we must continue to heed Raymond Williams's warning. Symbolic of the strengths of a traditional, mainstream sport journalism, Geoffrey Green (1953: 202) believed that the "first duty of the Press is to obtain the earliest and most correct intelligence of the events of the time, and instantly, by disclosing them, to make them the common property of the nation." The Press, he added, has a responsibility of disclosure. Twitter, by contrast, has a responsibility for nothing, but a huge and expansive capacity for image-boosting, profile-promoting, chattering, critiquing, questioning, in potentially equal measure. Multimedia professionals combine an awareness of changing means of communication with a creative openness to new ways of conversing and communicating. In the world of sport, and other spheres of the popular, Twitter's major contribution may, though, turn out to be, against the potential to foster dialogue and increase inclusivity, the consolidation of the cultural and communicative base of contemporary idols of consumption.

REFERENCES

Aparicio, M. and Olobulu, T. (2015) 'New media threat to traditional journalism: Newspaper and magazine are struggling to compete at the same level of digital and social Media', *AIPS Magazine*, (January), p. 42.

Boyle, R. and Haynes, R. (2009) *Power Play: Sport, the Media and Popular Culture*, 2nd edition, Edinburgh: Edinburgh University Press.

de Castella, Tom (2011) 'Joey Barton: What's behind his Twitter philosophy', *BBC News Magazine*, 26th August 2011, http://www.bbc.com/news/magazine-14662175, consulted 22nd September 2015.

Eco, U. (1986) 'Sports chatter', in *Travels in Hyper Reality—Essays* (translated from the Italian by William Weaver). San Diego: A Harvest Book/Harcourt Brace & Company, pp. 159–65.

Fifield, D. (2015) 'FA charge Costa with violent conduct after Koscielny clash', *The Guardian (Sport)*, online edition, 22 September.

Gibbs, C. (2013) *Twitter's Impact on Sports Media Relations*, unpublished PhD thesis, Stirling University, downloaded from http://dspace.stir.ac.uk/bitstream/1893/18588/1/GIBBS_Dissertation_Nov_6_2013.pdf, on 12 September 2015.

Green, G. (1953) *Soccer: The World Game—A Popular History*, London: Phoenix House Limited.

Hall, W. and Parkinson, M. (eds.) (1974) *Football Report: An Anthology of Soccer*. Newton Abbot: Sportsmans Book Club.

Harcup, T. (2007) *The Ethical Journalist*. London: Sage Publications.

Hopcraft, A. (1968) *The Football Man: People and Passions in Soccer*. London: Collins.

Hutchins, B. (2011) 'The acceleration of media sport culture: Twitter, telepresence and online messaging', *Information Communication and Society*, 14(2): 237–257.

Hutchins, B. and Rowe, D. (2012) *Sport Beyond Television: The Internet, Digital Media and the Rise of Networked Media Sport*, London and New York: Routledge.

Hyde, M. (2015) 'Berahino's refugee past does not oblige him to conform', *Guardian (Sport)*, 10 September, p. 8.

Ledbrooke, A. and Turner, E. (1955) *Soccer from the Press Box*. London: The Sportsmans Book Club (first published Nicholas Kaye Ltd, 1950).

Lord, B. (1963) *My Fight for Football*. London: Stanley Paul & Co. Ltd.

Lowenthal, Leo (1961) 'The triumph of mass idols' (1943/4), in *Literature, Popular Culture, and Society*. Palo Alto, CA: Pacific Books, pp. 109–40.

McEnnis, S. (2013) 'Raising our game: Effects of citizen journalism on Twitter for professional identity and working practices of British sport journalists', in *International Journal of Sport Communication*, 6: 423–433.

McEnnis, S. (2015) 'Following the action: How live bloggers are reimagining the professional ideology of sports journalism', in *Journalism Practice*, online publication 6 August, Taylor & Francis.

Novick, J. and Steen, R. (2014) 'Texting and Tweeting: How social media has changed news gathering', in A.C. Billings and M. Hardin (eds.), *Routledge Handbook of Sport and New Media*. London: Routledge, pp. 119–129.

Pidd, Helen (2015) 'The Saturday interview: "I'm not like I used to be. I'd rather do charity work now"', *The Guardian*, online edition, 19 September.

Redhead, S. (2015) *Football and Accelerated Culture: This Modern Sporting Life*, London: Routledge.

Rowe, D. (2011) *Global Media Sport: Flows, Forms and Futures*. London: Bloomsbury Academic.

Sheffer, M. L. & Schultz, B. (2010) 'Paradigm shift or passing fad? Twitter and sports journalism', *International Journal of Sport Communication*, 3(4): 472–484.

Steen, R. (2013) 'Sheepskin coats and nannygoats: The view from the pressbox', in R. Steen, J. Novick and H. Richards (eds.), *The Cambridge Companion to Football*. Cambridge: Cambridge University Press, pp. 217–230.

Sugden, J. and Tomlinson, A. (2010) 'What Beckham had for breakfast: The rolling menu of 24/7 sports news', in S. Cushion and J. Lewis (eds.), *The Rise of 24-hour News Television: Global Perspectives*. Oxford: Peter Lang, pp. 151–166.

Smith, E. (2015) 'Left field: The Pietersen Poll delusion, reading the election, and a tour of Tony Crozier's Barbados', *New Statesman*, 22–28 May: 38.

Spurling, J. (2015) 'Negative energy', *When Saturday Comes*, Issue 340, June, p. 18.

Willsher, K. (2015) 'Stars killed off by the selfie age—Deneuve', *The Guardian*, 11 May, p. 17.

CHAPTER EIGHTEEN

Producing News in the Moment: Video Journalism in an Increasingly Converged 24/7 Media Environment

MARY ANGELA BOCK

"You can't go back." That was one of the things a major newspaper reporter, who was experimenting with video journalism and largely enjoying the endeavor, said was one of his biggest challenges. When he was writing a story, if he had another question, he could always call the source and ask another question. With video, he had to get everything the first time, because returning to the moment was generally impossible. In the world of the 24-hour news cycle, there is simply no time for going back.[1]

His observation gets to the root of what makes video journalism both compelling for the audience and challenging for reporters. Video engages people with material moments in time; it can transport viewers to a scene with powerful emotion. Because it requires physical access to those moments in time—on top of a host of new technical and narrative skills —today's video journalism is disrupting existing practices. This chapter is based on several years' worth of observational visits and interviews with video journalists in England and the United States, and examines the way journalists are using video to advance the 24-hour news cycle even as these new time pressures present multiple challenges and ethical concerns.

WHY VIDEO JOURNALISM?

Video is one of the few bright spots on the financial landscape for the newspaper industry, which can sell pre-roll ads online. The medium is not new to television, of course, but employing video journalists who work *alone* saves money and can allow for more flexibility in the story assignment process. For all institutions, using video is essential to maintain technological credibility with an audience that is using video every day on Vine, SnapChat and Facebook messenger. While institutions work their way through the maze of storytelling styles, delivery modes and mobile opportunities, the question is not whether to incorporate video, only how.

CONSTRUCTIVISM AND GATEKEEPING

The constructivist theory of news-work as a form of manufacturing has long been a useful conceptualization for the ways stories are selected and framed (Bantz, McCorkle, & Baade, 1980; Gans, 1979; Tuchman, 1978). Constructivism asserts that journalism creates a reality for its audience based on ideologically shaped institutions, organizations and human interaction. A helpful source for mapping this process comes from gatekeeping theory and its contemporary elaborations (Bennett, 2004; Berkowitz, 1990; Shoemaker & Reese, 2013; White, 1950). Digital technologies and their new equipment, modalities, materialities and shattered schedules are re-drawing the gatekeeping map and the stories constructed within it.

The influence of multimedia technologies brings more than a few new graphic elements to the page; journalists are confronted by a new media "logic," to use Altheide's (2013; 1979) term, which is forcing changes in the way journalists work and *think*. Multimedia reporting is one of a few categories where jobs are growing; video journalists are in demand. At the same time, much is demanded of them. The instant information age has disrupted the rhythm of traditional newsrooms even as the demands of legacy *modalities* continue. [I use the term legacy modalities rather than media to point out that most institutions are in the process of converging, and only some of their processes and products are tied to traditional modalities (Reich, 2015).] Newspaper writers must still write for a print-oriented deadline while also posting social media updates up until that deadline. Television reporters post social media updates, contribute video packages to a newscast and then post stories to their station's web interface. These legacy deadlines exist within a 24-hour framework in which deadlines are "always" and "never." That is, journalists are expected to post news immediately when it happens—always—and the never-ending workday means there is "never" a break from deadline pressure. This mixture of demands complicates the typical VJ's day.

THE VJ'S PROCESS

To understand how these increased demands on a journalist's time and attention change the stories they tell, it's first necessary to understand how video stories are produced, and in particular, how this process contrasts with other forms of news-gathering. What follows is a description of a day in the life of a VJ who worked for a mid-market television station in the US (She has since moved to another job). Her process was devoted to the needs of a broadcast product, but key elements of that process are common for all video production.

One VJ's Day

During the newsroom's editorial meeting on this particular morning, [NC] was assigned to produce a story about a local school building that had been declared a historic site. The story is considered a "rip" because its basic facts were in the morning newspaper, and the television station will convert it to video. During the meeting, [NC] and her colleagues *pre-conceptualized the story* as an exposition of the way buildings in this city are designated "historic." [NC] does some on-line and telephone research to learn more about the process and decides to add a harder edge to the story by comparing the way this building, in a rich, white neighborhood, will receive such a designation, but a similar building in a poorer, black neighborhood has not.

During this preliminary fact-finding, [NC] explains that she always has to answer the question "what video can I use?" That is, how will she convert the facts of this story to a visual narrative? So in addition to gathering information for this rip, she also makes the phone calls necessary to arrange for interviews and access to the building—to *locate and access the elements* she'll need to construct the story as it is pre-conceptualized. In this case, she'll need interviews with those involved with the historic designation process and access to the buildings.

To record those elements, [NC] must travel to the buildings to *record these elements*. Before heading out she packs her camera gear, reporting essentials like pens and notepads and because she is a television reporter, a kit with hair care and cosmetic accessories. She uses a printed map and a GPS to find her way around the city quickly. Visual journalism is a body-centric activity. Unlike facts, which can be communicated over the phone, images must be created in real time and space. For this reason, [NC] has lobbied her managers to assign VJ stories close to the station so that they can return to the station by 3 p.m. in order to produce their stories in time for the news broadcast. Upon arrival at the school building [NC] uses a tripod to shoot some exteriors before going inside. Once inside, escorted by a school representative, she shoots architectural details inside the building and an interview with a school manager.

When she returns to the station in the early afternoon, she *narrativizes* the elements she's recorded, re-contextualizing them into a linear story structure that combines her voice, clips from the interviews she's collected, and illustrative images of the school buildings. She uses a digital editing system to craft a story that lasts a little more than one minute. When she's finished, she sends it to a large computer server for her managers to approve the work, and she returns to her desk to freshen up her hair and makeup.

At 5 p.m., she *presents* her story during the newscast; another form of re-contextualization. [NC] stands in the studio, where she's introduced by a studio anchor, and presents her own live introduction. Then she returns to her desk to convert them to a web version, which involves writing out every sound bite word-for-word (called "verbatim" in this context, they are merely precise quotations) for web users to read, not watch. She politely ignores a request from one of her managers to convert one of her shots to a still for the web because she does not know the technical steps to do so. "Sometimes I have to say 'something's got to give.'"

When she finishes around 6 p.m., she has worked straight through without a lunch break. Nevertheless, she considers this to have been "a nice, neat day."

The Video Journalism Routine

The five basic steps [NC] took as she moved through her day are common to video journalists:

1. Pre-conceiving the story and its elements
2. Locating & gaining access to elements
3. Recording those elements, thereby de-contextualizing them for future use
4. Narrativizing, or re-contextualizing those elements into story form
5. Presenting the story

While this process is similar to what news writers follow, in that it collects facts and re-contextualizes them into narrative, it is distinguished by its *intrinsic physicality*. This blending of body and camera in time and space change journalistic routines throughout the newsgathering process and in turn affects what is delivered to the audience (Bock, 2011). Space, for instance, is an ultimate limitation, because any story is determined by what can be photographed on a particular day within a geographic boundary. Gaining access to elements is similarly a matter of transportation, geography, and social interaction with those who control access to locations. Recording the elements involves the body in concert with a camera, which means that journalists must work with their muscles as well as their brains, a shift that some find off-putting. Narrativizing elements is both a matter of writing and using material technologies, as is story presentation, which engages a very material page, screen or phone. The digital world is often imagined as a buzz of binary code floating through the atmosphere, but its equipment has mass; it must be carried, charged, cared for and cleaned.

NEW ROUTINES

Digital video journalism therefore, disrupts the entire gatekeeping process. Two disruptions are particularly important for their impact on what is eventually presented to the audience: First, video production requires a tighter pre-conception of the story before journalists even walk out the door, and second, they are required to multitask throughout the process. These two factors influence one another and are related to the physicality and materiality of the video journalism process. Neither of these changes in routine is inherently positive or negative, but they do present ethical challenges to journalistic institutions.

Tighter Pre-Conception

Pre-conceiving of a story is not new; Tuchman (1973) carefully explicated how news conventions turn one particular set of facts into a story and ignores other sets of facts. That is different about the pre-conception for a video journalist is the hard, geographic and temporal limits on their work, which tightens the pre-conception. [HT], who works for a television station and must construct a story every day (three versions of it, in fact), plans out his stories in detail before leaving the station:

> You need to know how you're going to approach it and do it, partly because as a VJ you have so little time to do everything yourself you have to pre-decide how you're going to do it before you can leave the building or the meeting or you won't make your deadline. It's not like the old days when you'd go out and see what was actually happening and then decide how to cover it.

Pre-conception drives the overall process, though it is malleable. For instance, when [LT] is assigned to cover the presidential primary as experienced by people in the Bronx, she first tried to reach Democratic Party leaders in the area. They would not, and could not, be interviewed. [LT] still had an assignment, though, so after consulting with her managers, she decides to base her story on the way local high school students are learning about the primary. She couldn't stick with the original story idea because there was no video to match it. Similarly, during a morning meeting at [TV-MID], a reporter's idea for covering the way the African American community was inspired by Barack Obama's campaign was passed over, because there were no gatherings, no places to go, to record visual scenes of inspired people. The fact that video of "sunrises and weirdoes" was already shot and in-house at a BBC newsroom guaranteed coverage of the annual solstice coverage at Stonehenge.

A tighter pre-conception is essential for VJs if they are to meet multiple daily deadlines and multitask while on the job. They must know what to shoot before they start; it's not possible, in today's news factory, to just watch as a situation unfolds:

[DT]: When I first started doing it I was going out there and just shooting ridiculous amounts of video, you know shoot an hour of video for a three minute piece. I just didn't understand the process. I'm getting a whole lot better at it now. I'm understanding a lot more than I did, that capturing video is a lot different than just taking the pictures or even putting together a slide show.

The tighter pre-conception may be a sanity saver for journalists who are working to deliver material in the era of never/always deadlines. It is also a way to cope with competing mental and technological demands from digitization: that is, the need to switch attention from word to images, from grabbing stills to shooting video to copy editing, all in one day's work.

Multitasking

Whether working for legacy print or television organizations, the participants in this research consider multitasking to be one of the challenges of their current job, with varying degrees of acceptance. [LA] is a sports photographer whose responsibilities now include shooting video and stills from the same event. She says time management is the key:

You have to be able to determine when you want to do stills and when you want to do video because you cannot do both at the same time, unless you have a tri-pod, which football fields don't allow. So you have to determine, 'I'm going to do stills right now, and then for this next play I'm going to shoot video.' And you just have to deal with it.

[CD] copes with the competing camera needs by *literally* shooting some stories with both hands—one holding a DSLR camera for stills, the other operating a video camera. Such athletic feats are not common, and hardly safe, but the logical outgrowth of an environment in which no camera (yet) is perfect for all situations at all times.

A number of VJs complain about the difficulty of trying to interview a person while also shooting the interview, monitoring the audio and camera position while also listening to the interviewee and asking intelligent questions. A number of VJs identified the problem not as a matter of "too much" to think about but of *competing* cognitive paths, or as [CD] put it "I have to have a different brain to do video." Similar complaints invoked the idea of "left-brain/right-brain" or "word-picture" conflicts in their heads. [MM] worries that he'll lose an important shot from breaking news when he's set up for his own standups. [FJ] accepts this reality as a form of compromise when she has to shoot her own standups and cutaways:

I use the viewfinder. I flip it around to where I can kind of see where I am in the shot, if he's in there, OK. Then I just hit the record button. I really try not to look at viewfinder and look at the camera lens itself. That's a little bit of a distraction sometimes if you're not used to that. That's how I do it. Aim for myself and then it looks like I'm in there, OK, and then hit record. I'm used to it at this point.

A newspaper photographer with more than 20-years' experience with still images, [CD] struggles against video's temporal modality, because he is more accustomed to seeking what emotional, or as Cartier-Bresson (1952) put it, a "decisive" moment. Speaking of the assignment he covered on the day of observation, he noted, "You want interesting pictures, and I don't think I got any interesting pictures today. That's just me."

Summary

It is no surprise that work routines are changing in the wake of visual journalism's inherent physicality and the digital environment's always/never deadlines. Interviews and observations in a variety of newsrooms indicate that journalists are finding ways to cope with these pressures, by more tightly conceptualizing and planning their stories and by multitasking. These responses do not have a neutral impact on the news product, however; they have implications for the normative role of journalism and its democratic responsibilities.

DISCUSSION

A smartphone or modern video camera is certainly smaller and lighter than tape-based TV cameras of the past, but when added to a full kit with lights, tripod, batteries and other accessories, a gear bag can get heavy quickly. Add to this the physical demands of shooting: the better shots are made by crawling around on the ground, climbing fences or running to the action. One research participant said he lost 14 pounds after becoming a video journalist. A former television producer who once managed network crews of two or three people, says "Holy bananas," it's hard, even though he considers his work a vast improvement over his previous position.

Few members of the audience will worry about complaints from journalists about having to wear more comfortable shoes to work or put in more physically demanding days. The audience might be more worried, though, about the impact of the current economic, technological and organizational demands that are tightening pre-conceptions and forcing increased multitasking. Anderson (2011) has raised concerns about the connection between news decisions and instant metrics, and Reich (Reich & Godler, 2014) has pointed out that the demanding environment is leading to less fact-checking. This research raises additional concerns, namely that multitasking and tight preconceptions of stories may cause journalists to rely more heavily on sources who control geographic access to story elements, to favor softer, more visually pleasing "one-stop" feature stories, and to further bifurcate an already fragmented menu of news offerings between click bait and long-form stories.

Keepers of the (Literal) Gates

Without physical access to the interiors of the historic school building, it's likely that [NC] would have pursued a different story on the day she was observed. This was a ripped feature, not a breaking news story, and if school officials had denied access to the building, [NC] would have not had the shots necessary to produce her story. This was a light "good news" story that served to enhance the district's reputation, a consideration that was no doubt part of the district's decision to grant her—and her camera—access.

Political handlers have long known the power of the visual for public influence. Visual display has historically been part of what elevates royalty; once the camera was invented it took little time for elites to harness its power. Abraham Lincoln credited his presidential win in part to Matthew Brady's flattering portrait (Morrow, 2007; Savas, 2008). Politicians quickly learned that the cumbersome nature of the earliest film cameras meant that staged events were attractive to the photographic press (Barnouw, 1974). Controlling access to physical space is one way to control the stories journalists cover, and VJs must constantly work cooperatively with representatives of the institutions they cover.

Many news scholars have observed that the source-journalist relationship is a matter of interdependence (Cook, 2005; Sigal, 1973; Strentz, 1977). "Giving them what they need," is the way one government official put it. Whether by releasing images they themselves have produced (as with video news releases), controlling journalists' geographical access to locations, or carefully orchestrating and camera placement at events, image managers are able to control the story (Bock & Araiza, 2014; Gandy Jr., 1982). Sometimes their efforts may seem banal, as with the sports team PR representative who says his rules simply keep photographers safe from speeding football players. Other efforts are more blatantly manipulative, as when the Pentagon was found to have vetted the work of journalists applying to be embeds in the war in Iraq and Afghanistan, using research to favor those reporters whose work was categorized as sympathetic to the military (Reed, 2009).

An additional tension that exists between camera-bodies and intermediaries concerns the degree of corporeal autonomy *within* a space. Video journalists must contend daily with formal and informal attempts to control their freedom of movement. Controls on physical autonomy might be overt, by making permission to enter may be contingent upon remaining in one assigned space, or it might be more subtle, as when a public relations representative escorts a photographer throughout a visit. In the early stages of political campaigns, for instance, photographers experienced in campaign coverage report that regulations are loose, so much so that one network photographer says he's been asked where he thinks his best shot might be. But those same video photographers also say that as campaigns progress, as security tightens and the subjects' stakes go higher, camera placement regulation becomes highly regimented. Photographers learn their place hours before an event when they report for a security sweep, the process by which police officers inspect their equipment by hand and with trained dogs. Once in place, they're locked in, which means that the staging set in place by the candidate is the staging delivered to the audience:

[VF] …for the president you can't move…you have to stay in your spot…pretty much for the press, it's risers. You're stuck with the tripod on a riser.

[WB] Usually they want you to stay in one position, it's not like you can move around a lot…and that can be a pain. If you want to kind of have a little bit of movement usually they confine you to one space one area. You go and come as they tell you so you know they pretty much dictate everything to you.

Finally, for journalists who are under pressure, the temptation to rely on sources for help is considerable. "She made it all possible," said one VJ about the PR representative of a local festival, who arranged interviews and access for a full series of features. A former law enforcement official was cognizant of the number of crews a story might garner, and scheduled his news conferences in a way that allowed TV stations to be there for an arrest and subsequent interviews with police. A VJ for a mid-market newspaper often relied on sources not only for access, but also for narration—he would hand them the mic to record an audio track for his features.

Features

Because they take longer to shoot and require interesting people, scenes and activities, feature stories are a more natural match for video journalism than issue-based, political process stories. There's even a derogatory nickname for meeting stories that use too many shots of participants: "BOPSA", as in "bunch of people sitting around." [NC] did her best to add a sharper edge to her story about a school's historic designation, but the story was already a day old when her editor ripped it out of the paper.

Many of the "day of" the stories produced by the newspaper VJs observed for this research were feature stories, human interests pieces about clowns, an amusement park, or golfing tips. A content analysis (Bock, 2014) of the video stories posted to newspaper sites suggested that as these legacy newsrooms move toward multimedia, VJs are relying on stories that "hold still" long enough to shoot well— and provide easy, colorful images. When asked "what makes a good video story," [CD] started by talking about the visual elements, not the facts of a story:

> Something that people are interested in. Something that grabs your attention like a guy getting bucked off a bull. I don't know. Starting with a GoPro video of a bull in your face, that's kind of interesting, that pulls you in. Now why do we care about these guys? I don't know. That's really hard to comprehend.

Not all VJs are shooting daily features. For legacy TV stations, employing one person to shoot, write edit and post stories is a cost effective way of covering breaking news. [EJ] has this role working for a TV organization's bureau in the southern US:

> I can shoot everything. In a week, I will be shooting a spring breaker story where the girl who fell off the seventh floor balcony and died. The next day, talk about the alcohol sales and the concerts that are going on. Then come back and talk about the Matamoros violence across the border and three deaths, or shootings that just happened. Then do a live story like our mosquito problem where I go talk to neighbors. I'm literally everywhere. I shoot anything and everything.

Studies of TV news continue to find, of course, that the bulk of coverage focuses on fires, accidents and flashy stories that are visually appealing (Barnhurst & Steele, 1997; Chiricos, Padgett, & Gertz, 2000; Kaniss, 1991; Lipschultz & Hilt, 2002). Interestingly, though newspaper journalists have traditionally eschewed this sort of coverage, this category of video may be what builds their audience in today's market.

Spot Clips and Docu-Projects

Years after experimenting with solo-practiced video journalism, some newsrooms at the BBC cut back: quality stories took too long to shoot and breaking news suffered. As one journalist [FD] put it: "You have to cover all stories, even ones that are picture challenged." Today, the BBC training program continues to hold up the subject-driven, observational model as an ideal while at the same time accepting that the temporal demands of a daily broadcast impede the development of compelling, intimate observational stories. "They want it both ways," says a leader in the multimedia training division of the managers, who want a VJ to be able to shoot something quickly for the daily show, but also with the pathos and depth of an observational documentary.

During a 2015 visit to a major market newspaper that is incorporating video into its daily operations, newsroom managers were trying to have it both ways with using a different model. One set of "breaking news" journalists used their smartphones and video editing apps to cover the sort of spot news once considered the low-class territory of TV news: fires, accidents and police scenes. The goal of maintaining higher standards, or "click bait with a conscience" remains, but the reality is that today's audience is interested in viewing compelling videos, even when they are not beautifully shot or produced. As one supervisor [WL] put it:

> Yesterday we had this video of a police exploding 10 tons of confiscated fireworks. There's just no way to write about 10 tons of confiscated fireworks exploding all at once. You just have to see it. So that was a case where a video itself drove traffic. Occasionally we'll have a zoo animal doing something funny. That video drives traffic on its own. Those are kind of like, quirky things, but usually they're supplementing traffic. It's a value add. Readers are coming to read the story and we've also got this video of it.

Metrics can tell editors what the audience clicks and how long they'll stay. One organization found that the audience nearly disappears about thirty seconds into a video, even a beautifully, carefully produced feature. The take-away? Such well-produced pieces might not be necessary day to day. Convergence for legacy media may bring an end to the traditional 90-second video package, and instead inspire a system of quick clips for breaking news and more elaborately produced videos in long-form stories. [WL]'s organization is moving in that direction, with quick clips from spot news on the social media feed, and long-term, full-page multimedia projects devoted to team-based investigations.

SUMMARY

This chapter relies on several years of research observing multiple newsrooms with more than a hundred video and visual journalists, to trace the impact of today's 24-hour digital news environment on storytelling. The nature of video journalism amid this high-speed, high-pressure system is changing the traditional gatekeeping map for news. Because their work is intrinsically physical and requires geographic access to news events, video journalists must rely on favors from the very people they cover for location access. They are also more apt to cover lighter stories that can be better visualized and have a longer shelf-life. Finally, the pull of the two digital deadlines, "always and never," is in turn bifurcating story treatment along those lines, with quick clips uploaded for spot coverage, and longer, more extensive documentary treatment for long-form multimedia projects.

These trends are not merely interesting in the academic sense; they have very real implications for the normative role of journalism in democracy. The degree to which authorities can control location access—and therefore the news agenda—should be of particular concern to journalists who hope to maintain independence and integrity. Balancing the need to gain access to visual elements with the obligation to cover institutions with cool detachment is one challenge facing news organizations as they embrace video journalism. Using video to cover features is completely logical, but presents the danger of further diluting the overall news product. Finally, as much as longer-form multimedia projects can inform the public in creative, interactive and useful new ways, the other side of that bifurcation, quick clip coverage of fires and accidents, runs the risk of providing mere snacks to an audience that occasionally needs serious nutrition. As they struggle to maintain relevance and economic stability in the face of the ferocious winds of today's technology, it is essential that journalists hang on tightly to their social responsibilities.

NOTE

1   Portions of this chapter are based on Bock, M. A. (2012). *Video Journalism: Beyond the One-Man Band* (New York: Peter Lang) and are used with permission.

REFERENCES

Altheide, D. L. (2013). 'Media Logic, Social Control, and Fear', *Communication Theory*, Vol. *23*(3), 223–238. http://doi.org/10.1111/comt.12017

Altheide, D. L., & Snow, R. P. (1979). *Media Logic*. Beverly Hills: Sage.

Anderson, C. W. (2011). 'Between Creative and Quantified Audiences: Web Metrics and Changing Patterns of Newswork in Local US Newsrooms', *Journalism*, Vol. *12*(5), 550–566. http://doi.org/10.1177/1464884911402451

Bantz, C. R., McCorkle, S., & Baade, R. (1980). 'The News Factor', *Communication Research*, Vol. *7*(1), 45–68.

Barnhurst, K. G., & Steele, C. A. (1997). 'Image-Bite News: The Visual Coverage of Elections on U.S. Television, 1968–1992', *The Harvard International Journal of Press/Politics*, Vol. *2*(1), 40–58. http://doi.org/10.1177/1081180X97002001005

Barnouw, E. (1974). *Documentary: A History of the Non-Fiction Film* (Vol. 1983). Oxford: Oxford University Press.

Bennett, W. L. (2004). 'Gatekeeping and Press-Government Relations: A Multigated Model of News Construction'. In L. L. Kaid (Ed.), *Handbook of Political Communication Research* (pp. 283–314). Mahwah, NJ: Lawrence Erlbaum Associates.

Berkowitz, D. (1990). 'Refining the Gatekeeping Metaphor for Local Television News', *Journal of Broadcasting & Electronic Media*, Vol. *34*(1), 55–68.

Bock, M. A. (2011). 'You Really, Truly, Have to "Be There": Video Journalism as a Social and Material Construction', *Journalism and Mass Communication Quarterly*, Vol. *88*(4), 705–718.

Bock, M. A., & Araiza, J. A. (2014). 'Facing the Death Penalty While Facing the Cameras', *Journalism Practice*, Vol. *0*(0), 1–18. http://doi.org/10.1080/17512786.2014.964496

Cartier-Bresson, H. (1952). *The Decisive Moment*. New York: Simon & Schuster.

Chiricos, T., Padgett, K., & Gertz, M. (2000). 'Fear, TV News, and the Reality of Crime', *Criminology*, Vol. *38*(3), 755–786. http://doi.org/10.1111/j.1745-9125.2000.tb00905.x

Cook, T. E. (2005). *Governing the News: The News Media as a Political Institution*. Chicago: University of Chicago Press.

Gandy Jr., O. H. (1982). *Beyond Agenda Setting: Information Subsidies and Public Policy*. Norwood, NJ: Ablex.

Gans, H. (1979). *Deciding What's News: A Study of CBS Evening News, NBC Nightly News, Newsweek and Time, 25th Anniversary Edition*. New York, NY: Pantheon Books.

Kaniss, P. (1991). *Making Local News*. Chicago: University of Chicago Press.

Lipschultz, J. H., & Hilt, M. L. (2002). *Crime and Local Television News : Dramatic, Breaking, and Live From the Scene*. Mahwah, NJ: L. Earlbaum Associates. Retrieved from http://ezproxy.lib.utexas.edu/login?url=http://search.ebscohost.com/login.aspx?direct=true&db=nlebk&AN=79419&site=ehost-live

Morrow, K. (2007). 'The Birth of Photojournalism', *Civil War Times*, Vol. *46*(7), 40–46.

Reed, C. (2009). Journalists, Recent Work Examined Before Embeds. Retrieved from http://www.stripes.com

Reich, Z. (2015). 'Comparing News Reporting Across Print, Radio, Television and Online', *Journalism Studies*, Vol. *0*(0), 1–21. http://doi.org/10.1080/1461670X.2015.1006898

Reich, Z., & Godler, Y. (2014). 'A Time of Uncertainty: The Effects of Reporters' Time Schedule on Their Work', *Journalism Studies*, Vol. *15*(5), 607–618.

Savas, T. P. (2008). *Brady's Civil War Journal: Photographing the War 1861–1865*. New York: Skyhorse Publications.

Shoemaker, P. J., & Reese, S. D. (2013). *Mediating the Message in the 21st Century: A Media Sociology Perspective*. Hoboken, NJ: Taylor and Francis.

Sigal, L. (1973). *Reporters & Officials: The Organization and Politics of Newsmaking*. Lexington, MA: D.C. Heath & Co.

Strentz, H. (1977). *News Reporters & News Sources: What Happens before the Story is Written*. Ames, IA: Iowa State University Press.

Tuchman, G. (1978). *Making News: A Study in the Construction of Reality*. London: The Free Press.

White, D. M. (1950). 'The Gatekeeper'. *Journalism Quarterly*, Vol. *27*(4), 383–390.

National Contexts and Journalistic Challenges

The International Newsgathering Challenge for Public Service Australian and Canadian 24/7 TV Channels

COLLEEN MURRELL

ABC Australia and CBC-Radio Canada are two old and venerable public service broadcasters (PSB) that have managed the difficult transition into the digital age: they are battered and bruised, but somehow still standing. The two countries and their media systems have a lot of the same characteristics and challenges: the ABC (founded in 1932) and CBC-Radio Canada (founded in 1936) both have to service relatively small and scattered populations across vast, geographic distances. Both companies do some broadcasting in Indigenous languages, and CBC has both English and French speaking services. The two companies also both have roots in the BBC "Reithian" tradition of being called upon to "educate, inform and entertain." However, in terms of funding they have followed very different paths, and neither service is currently operating within a healthy financial framework. The ABC is a PSB that is almost entirely funded by government, with its TV channels going out advertisement-free. But as we shall see below, this money is now being cut back drastically without warning. CBC, on the other hand, is financed by a mixture of the government (48%), advertising (24%), subscriber fees (13%) and financing/other sales (15%), according to its financial report (CBC, 2014: 4). It operates in the congested airwaves of North America with a lot of competition for viewers across Canada and the US.

Both companies have 24/7 TV channels—ABC News 24 and CBC News Network—and they showcase some of the most vibrant and exciting journalism to be found within the organisations. But recent cuts pose challenges to the stations in their ability to carry out extensive international newsgathering, which is expensive in time and resources. On ABC News 24's online news streaming site, which is free to all Australians to view, the mission statement argues its global reach:

> ABC News 24 channel features continuous, commercial-free coverage of major breaking stories in Australia and around the world. (ABC, 2015)

At CBC's News Network, a commitment to international newsgathering is also made on its online landing page:

> Every day, every hour, across Canada and around the world, stories are breaking. CBC News Network is on top of those stories and more, showcasing the best of CBC journalism covering stories with speed, but adding context and meaning along the way. CBC News Network is also the destination for original journalism, with added depth from CBC News bureaux across the country and around the world. (CBC, 2015)

In seeking to understand the international newsgathering priorities and finances of the ABC and CBC, I interviewed thirteen senior managers from TV, radio and online in 2012–2013. Following the severe cuts to newsgathering in both companies (particularly in 2014), I conducted follow-up interviews with four senior people from these organisations in 2014–2015 to find out what changes had had to be implemented in terms of production and newsgathering to the 24/7 TV operations. I also checked programme content against the interviews: during a selected period in 2014, I monitored the main evening news bulletins on ABC News 24, ABC1's Victoria bulletin, CBC's News Network and CBC's *The National*. I wanted to find out how much international news was running on the bulletins, and whether or not the 24/7 channels were running more of it than the traditional evening bulletins. I also wanted to discover how much of it was generated by international news agencies, or by the companies' foreign correspondents—either in situ at the datelines of breaking stories or from the new, bigger hub bureaux. And finally, I was looking for evidence of innovative and/or cheaper ways of conducting international newsgathering.

SOME ABC PARTS FORCED TO RUN ON THE WHIFF OF AN OILY RAG[1]

In Australia, the ABC was late to the 24-hour TV news scene, launching its service 21 years after CBC's, in July 2010. Springing to life in the wake of the Global Financial Crisis, its competitor (Sky News) and its detractors *inside* the ABC claimed it would be a profligate and unnecessary service that would leach money away from other sections of the news and current affairs divisions (Murrell, 2013: 84). The ABC is almost entirely funded by the government, to the tune of $1077.2

million in 2013–14, pre-cuts. It also received $158.6 million from commercial sales of programmes, etc. (ABC Annual Report, 2014: 133). News 24 is freely available to Australian citizens via television and the internet, although it is geo-blocked outside the country. The station broadcasts Australian content from 6 a.m. to midnight each day, with the night hours showing different BBC programmes (*BBC World News*, *BBC Focus on Africa*, *BBC Outside Source*) and *Al-Jazeera Newshour*. During the day, the ABC's content mostly consists of news programming, with some of it made directly for the 24-hour station, such as the rolling news hours and *The World*, a one-hour international news program that goes out in the evening. The rest of the programming consists of simulcasts or rebroadcasts of programming made for the main domestic, free-to-air ABC channels—including *Insiders* (political chat show), *Landline* (country issues), *Parliamentary Question Time* and *Q&A* (a weekly current affairs audience-based programme).

When I first interviewed the ABC managers in 2012–13, the organisation was in the middle of changing some of its work practices. According to Steven Alward,[2] the Head of International News, the Newsgathering Project directed by Venture Consultants was designed to stop newsgathering being done separately in radio, television and online silos, and was promoting instead a "story-driven agenda" (2012). He said that part of the reasoning behind this was due to the creation of News 24 and the competing demands that the 24/7 station was putting on ABC systems. In October 2012 when I interviewed Gaven Morris,[2] Manager, ABC News Content, he was upbeat about the renewed government contract for Australia Network, the international television service that was then broadcasting into the Asia Pacific region. He believed this service had great international content spinoffs for News 24, as "we carry some of the programmes produced by the 'Asia Pacific News Centre' (APNC) which is a dedicated international production team in Melbourne that produces content for the Australia Network." After the cuts in 2014 put an end to the Australia Network and made many posts redundant from the APNC staff, he noted:

> *That* resulted in broad cuts to our foreign output across television and radio and the cessation of a number of dedicated international reporting roles. We reduced reporting resources in New Delhi and Jakarta and some Australian based international reporting roles like two Pacific correspondents (offset by an Australian-based VJ reporting role) and an Asia business correspondent. (2015)

But even back in 2012, Morris was previewing some of the ideas that are now in the process of being implemented: for example, in backing up traditional foreign correspondents with, "a network of people that you might have on a set number of days a year who are already in situ and can supplement your coverage by making your news gathering model more agile." According to Steven Alward, agility is a necessary by-product of the fact that politicians do not care sufficiently about international newsgathering. He said, "Governments are not inclined to give money to news operations. They're inclined to give money to an organisation like the ABC for *regional* projects and talking to their constituents" (2012).

ABC journalists spent the weeks before Christmas 2014 swimming anxiously in "shark pools" (Knott, 2014) while undergoing a digital skills audit to find out which of them would still have a job in 2015, a process some commentators likened to "The Hunger Games" (Patty, 2014). Forced by a conservative government into accepting $254 million in unexpected cuts to the budget over five years, the ABC's Managing Director Mark Scott appears to have used the opportunity to tweak the workforce to bring it more in line with his own digital vision. Of the 400 staff to be cut during 2014–15, Scott announced: "100 people would go from News and Current Affairs to fund a $20 million digital investment programme and 70 new digital jobs" (Davies, 2014). His mission to take the ABC's share of the national digital audience from 25 to 40 per cent (ibid, 2014) has led to much debate about the future of serious journalism versus "content" on the public broadcaster. This is the second round of cuts within one year for the ABC, with 72 jobs having gone from its International division (APNC) back in July 2014, following the government's cancellation of the Australia Network TV service (Massola, 2014). The latest cuts to people and resources will make serious inroads into the company's international newsgathering. According to departing presenter Quentin Dempster's "Submission to the Senate Select Committee into the Abbott Government's Budget Cuts" (2014), one of the key planks of the restructure, scaling back the number of local producers and moving from a correspondent to a video journalist (VJ) model, has "serious health and safety implications." The ABC in February 2015 said it would not provide "divisional breakdowns" of the job losses incurred, but did list how the bureau structures would change:

- The creation of hub bureaus in Washington and London which would both have three correspondents, two camera/edit positions and two producers.
- The creation of hub bureaus in Jakarta and Beijing which would have two correspondents, two camera/edit positions and two producers.
- The Auckland bureau would be shut down and the money would go towards a new VJ post in Beirut, with one correspondent, one camera/edit position and one locally-hired producer.
- The correspondents in Tokyo, New Delhi, Jerusalem and Bangkok would lose their separate offices and become "home-based" VJ jobs, with the reporters working with a local producer.
- The savings from this restructure would be put into increasing the assignments budget for travelling reporters, building a global network of freelance contributors (some of whom would be former ABC employees) and creating a new role of Chief Foreign Correspondent, to be employed via an internal recruitment process. (ABC News Anonymous,[2] 2015).

CBC: DEATH BY A MILLION CUTS[3]

When I first interviewed CBC managers back in 2012, they were already quite used to a regime of constant cuts, but the particular people whose work involved News Network were still relatively optimistic. Greg Reaume,[4] the Managing Editor of CBC News Coverage, said that CBC News as a whole was then (2012) involved in putting more emphasis into *domestic* news coverage, as this had been previously cut back and was perceived to be a problem. He said, "I don't think that was ever meant to be a sign that we're pulling out of international coverage or anything like that. But there's no question we have to try and cover the world more economically than we have in the past, because the costs are just so prohibitive" (2012). The English-speaking news operation at CBC had already had to shut bureaux (such as Moscow and Nairobi) and, apart from their bureaux in the UK, the US, Israel and a shared correspondent with the French CBC service in Beijing, they were already relying quite heavily on a team of freelancers and agreements with other news organisations for ad hoc TV stories. Reaume said: "We quite regularly air reports by BBC reporters from different places. And you know there's even an arrangement where they will basically do a report for the CBC with a CBC sign-off. Essentially it's the same piece they may have filed for one of the BBC programmes, but it's kind of tailored to us and they do us the courtesy of giving us a CBC sign-off" (2012). On News Network, the executive producer Mark Ross,[4] made creative savings where he could. In the Vancouver newsroom, there was no separate presenter studio, but rather a roving presenter who was followed around the newsroom by a single steady-cam operator. Presenter Ian Hanomansing would drop by and talk to reporters to catch up about breaking news. One of the reporters, Teresa Tang, ran an Asia Desk, from where she would chat to Hanomansing about stories in the news across the Asia Pacific region.

The 2014 fiscal year was a particular *annus horribilis* for CBC, with over 1,000 jobs terminated, according to the Canadian Media Guild (Kane, 2014a). The loss of NHL broadcast rights, a significant cut in government funding and reduced revenue from advertising led to CBC's woes (Huffington Post, 2014). In response, CBC's president Hubert Lacroix laid the groundwork for downsizing the organisation by 25% by a 2020 deadline. Like his counterpart at the ABC, Lacroix has also announced a "digital first" strategy. According to Toby Wong (2014), this equals "shifting priorities away from television and radio over the next five years. That means producing original and exclusive content for mobile devices." Still in Lacroix's sights is "legacy" programming like the award-winning documentary unit, and CBC's portfolio of domestic and foreign real estate. Wong believes that CBC "is at a crucial tipping point that some fear will lead to the eventual dismantlement of the broadcaster" (ibid). In March 2015, there was more bad news when

CBC announced a further 144 local TV jobs would be lost in the English service and 100 news jobs in the French-speaking Société Radio-Canada (Levinson-King, 2015). This move to cut jobs in local services springs from a decision in December 2014 to cut back local TV bulletins from 90 minutes to 60 or 30 minutes (Kane, 2014b). At the same time, Jennifer McGuire, editor-in-chief of CBC News, said the company would be hiring "80 new digital positions over the next year as it grows its mobile offerings" (Levinson-King, 2015). How deep further cuts will be to the corporation are still to be debated. In July 2015, a senate committee report into CBC/Radio-Canada made 22 recommendations, including that it "explore alternative funding models and additional ways to generate revenue to minimize the Corporation's dependence on government appropriations."

In terms of international newsgathering, Greg Reaume insists that the discretionary budget has not changed, but has been protected: "What has changed is the amount we are spending on operational or infrastructure costs" (Reaume, 2014). He went on to list the breakdown of cuts to international news:

- Local hires and administrative staff for the overseas bureaus have been "practically eliminated" and the logistical/accounting work has been shifted to Toronto.
- CBC has cut back on office space in London, Washington and New York.
- CBC has replaced "retiring, one-dimensional technical staff with multi-skilled individuals, providing us with more flexibility and value for money."
- The closure of the Rio de Janeiro office, leaving open the option that it could re-open later, in advance of the Olympic Games.
- The setting up of temporary bureaus, for example establishing a pop-up bureau in Ghana for a few months to cover the Ebola outbreak in West Africa and possibly opening one in Istanbul to cover ISIS-related stories.
- The continuation of the relationship with the stable of freelancers in Paris, Madrid, Rome, Lagos, Kiev plus those from the Global Radio News (GRN) network.
- More "live hits" from locations with different individuals (not necessarily reporters) via Skype, Dejero or spreecasting.
- VJ positions up and running in Beijing, Jerusalem, New York, Los Angeles and London.
- A move towards covering stories like Syria via analysis, drawing on the expertise of returned Middle East correspondents, now based in Toronto.

Against this background of cuts and restructures across the organisations, the English-speaking 24/7 TV news services at both CBC and ABC have not been targeted specifically for redundancies. In Canada, News Network (previously called Newsworld) has been broadcasting on cable since 1989. In its 2014–15 Second

Quarter Financial Report (p. 14), CBC noted the network had 11.3 million cable subscribers and accounted for 1.5% of the total national audience share of television viewers. News Network receives its money from commercial advertising and cable and online subscriber fees. Although it is not meant to receive money from the government's funds to the free-to-air CBC, it does receive a cross-subsidy from being able to use the CBC's reporters and correspondents in its programming. In terms of its online presence, the streaming service is only available via subscription. Given this reliance on subscriber fees, it would seem unlikely that the network would close down its television broadcasting in the near future in order to just appear online, unless it were to generate vast amounts of cash through online subscriptions. At its present rate of $6.95 a month (March 2015), this would seem unlikely. CBC's News Network programming is a mixture of news and current affairs. The rolling news hours are mostly made specifically for the station, and the current affairs programmes are simulcast or rebroadcast from other offerings from across the CBC network—such as *Fifth Estate*, *Doc Zone*, *The Nature of Things*, and *Power and Politics*. CBC runs its own news bulletins throughout the night hours, or replays *The National*, having switched from relying on news programmes rebroadcast from BBC World back in 2009.

INTERNATIONAL NEWS CONTENT ON THE 24/7 CHANNELS

As a small, indicative research study, in December 2014 I compared two weeks' worth of a set evening news bulletin on ABC News 24 with the main evening news bulletins on the ABC's free-to-air TV station. At the same time I compared two weeks' worth of a main CBC News Network hour of evening news, with the main evening news hour *The National*, which is made by the domestic CBC channel. I wanted to find out if the 24/7 news operations showed *more* or *less* international news than the domestic evening "legacy programming" and I also wanted to discover if the 24/7 stations were using different approaches to covering the main international stories of the day. In Table 1, you will find data about the programmes that were monitored from 8–19 December 2014. The data is broken up into the following categories:

(A)   This category reveals the average minutes (mins) per bulletin monitored of *all* international stories covered. This includes video taken in from abroad (agency material broadcast as RVOs[5]), correspondent packages and two-ways;[6] *plus* set-pieces put together in-house that deal with international stories by using graphics or presenter "chats" or interviews with experts or reporters.

(B)   This category shows the average minutes per bulletin monitored of *only* the material sent in from abroad—agency material used as short RVOs and correspondent TV packages or two-ways, i.e. there is no in-house generated "chat" included.

(C)   The data in this category represent the total number of international TV news packages or two-ways by foreign correspondents.

(D)   This category contains data about the number of international news packages or two-ways from the companies' correspondents *in situ* in the country of the news story.

(E)   This category shows the total number of international news packages or two-ways voiced and sent by correspondents *not* from the dateline of the story but from *hub bureaux*.

(F)   The data in this category show the total number of packages from *other companies'* foreign correspondents.

Table 1: International News Content Comparison of ABC & CBC News Bulletins.

Programme	(A) Int'l News mins bulletin (average)	(B) Int'l mins ex overseas per bulletin (average)	(C) ABC or CBC total FC stories or 2-ways	(D) ABC or CBC FC total stories in situ	(E) ABC or CBC FC total not in situ	(F) FC total stories from other channel
ABC News 24	9'30"	8'30"	18	11	7	3
ABC Victoria 7pm[7]	4'30"	4'30"	13	8	5	0
CBC News Network	21'30"	13'00"	13	12	1	2
CBC The National	9'00"	7'30"	18	12	6	1

Source: Author's monitoring of bulletins across two weeks of programming (Mon-Fri) starting 8 December 2014. Int'l = International. Mins = minutes. NB minutes are rounded off to nearest half-minute. Programmes: ABC News 24; ABC Victoria 7PM[7] (30 minutes only); CBC News Network; CBC *The National*.

DATA: ABC NEWS 24

The data for the 24/7 channels shows some interesting results regarding the amount of international news coverage, and examples of innovation shown. (A) On ABC News 24 overall, international news stories accounted for, on average, 9 minutes 30 seconds (9'30") per hour of rolling news programmes that are free of advertising. The interviews, "chats" or graphics sequence "explainers" generated *in-house* only added, on average, one minute per day to the (B) international news material that was generated *from abroad* (8'30"). Therefore, one can see, that there is very little material being created in-house on News 24 that relates to international news, with the bulk of the international news coverage coming from either the bureaux or the agencies. (A) Over the two weeks, there were only three set-piece stories/interviews about international news that were generated in-house. The first was a report on the Russian economy by the finance correspondent, the second was a presenter interview with a Cuban expert from Adelaide and the third was a presenter interview with a Russian expert from Canberra. The bulletin with the most international coverage from abroad had 13 minutes and the one with the lowest had no international stories at all—on the day of the "Sydney siege," when a gunman held hostages in a café in the CBD. Included in these statistics are "reader voice-overs" (RVOs[5]). These are short clips from international news agency video or from other sources, which are voiced by the studio presenter. The amount of international coverage at 9'30" was not that different to the 4'30" that ran on the Victorian bulletin, given that this bulletin is only half as long, at 30 minutes. (There are no nationwide main evening TV bulletins in Australia.)

(C) In terms of voiced packages from ABC correspondents abroad, there were 18 that ran across the ten bulletins monitored on News 24, as opposed to 13 that ran on the Victorian bulletin, showing that the latter leant more towards traditional correspondent packages in its shorter bulletins. (D) Of the 18 foreign correspondent packages on ABC News 24, 11 were voiced by correspondents at the dateline of the story, while (E) seven packages were voiced by foreign correspondents who did *not* have their "boots on the ground" in the country of the story, but were instead in a nearby hub bureau. For example, two Russian topics ("Ruble" and "Putin") were covered three times by a reporter in London; "Pakistan School Shooting" was voiced twice by a reporter in London; and "Ebola" and "Child Soldiers" were covered using pictures from the Central African Republic, but they were voiced by the Africa correspondent from Nairobi. (F) On top of the 18 voiced ABC packages, there were three more international packages, from the BBC. Two were by BBC correspondents in the Philippines, and one was a wrap from the UK about Syrian jihadists. Regarding the 13 foreign correspondent packages that ran on ABC1 (Victoria), five packages were voiced from hub bureaux and all of the packages were from ABC reporters.

DATA: CBC NEWS NETWORK

The rolling news hours on CBC's News Network have around 15 minutes of advertisements. Interestingly, there is a bigger difference on CBC News Network than was found on ABC News 24, between (A) the number of minutes dedicated to international news stories *overall*, and (B) the number of minutes that were *generated abroad*. In category (A), on average there was a total of 21'30" of international news stories per bulletin, including 8'30" that was created in-house. Examples of in-house generation were: the use of graphics and maps to show the location of embassies in Cairo to cover the "Canadian Embassy Closure" story; chats between the presenter and the studio-based breaking news reporter about US police shootings; graphics sequences about global reactions to a Canadian jihadi story, with the breaking news reporter standing in front of a large screen; a presenter chat with a domestic news reporter about an American football story from Oklahoma; a presenter chat with a former CBC Middle East correspondent about the shooting in Sydney; and an interview with a former air traffic controller in Vancouver to discuss air traffic delays in Heathrow. (B) There were also a lot of international stories on this network that were generated from video that was sent from abroad, with an average of 13 minutes per bulletin. This mostly consisted of agency video, cut into short RVOs.

(C) In terms of packages or two-ways with CBC foreign correspondents, there were 13 over the two weeks, or 1.3 on average per bulletin. (D) Of these 13 stories, 12 were by CBC reporters at the dateline. However, it should be noted that four of these 12 were by an Australian-based freelance reporter, Peter Hadfield, who was hired to cover the Sydney Siege and the "Family Shooting" story in Cairns. (E) There was only one story—"Pakistan Shooting"—that was voiced from a hub bureau in London. (F) Alongside the CBC reports there were two stories by journalists from other organisations. One bulletin ran an Australian Channel 9 reporter two-way on the Sydney Siege, and another ran a BBC story, "Pakistan Shooting," that was filmed in situ in Peshawar.

In contrast, *The National*, on the free-to-air CBC channel, had less than half the amount of international news programming of News Network, with 9 minutes on average per bulletin. In its hour-long bulletins (also with approximately 15 minutes of ad breaks) it had 18 international reports or two-ways, of which 12 were voiced by CBC reporters at the dateline of the story. Six of the 18 *were not* from the dateline of the story but were voiced from hub bureaus (some of which were nowhere near the action). The most striking aspect of this was in the stories by London-based correspondent Nahlah Ayed. Of her six stories, only one ("Heathrow Plane Delays") covered a story from her location in London. The other five were "Syrian Refugees," the "School Shooting" in Peshawar, "Ebola" in Sierra

Leone, the "Nobel Prize" ceremony in Oslo, and even the "Sydney Siege" from half-way around the world. *The National* ran one package from a reporter working for another organisation—the BBC's Mishal Hussein on "School Shooting" from Peshawar.

CONCLUSION

From the data, it seems clear that CBC News Network is covering a substantial amount of international news in its rolling news hours. There is an average of 21'30" out of 45 minutes of bulletin (when you discount advertisements) that means that almost half of the coverage is internationally inspired. Around 8'30" is generated each day through having the breaking news reporter stand in front of a large screen in the newsroom and run through a timeline, a graphics sequence or a video sequence. The reporter's commentary is often then followed up with a chat to a presenter, reporter or expert either in the newsroom or via an external studio or Skype-type interview. News Network shows a marked preference for this more innovative method of breaking news updates, in combination with set piece foreign correspondent packages. Interestingly, it did not rely so much on getting foreign correspondents from regional hubs to voice pieces on stories from elsewhere. It was *The National*, CBC's evening "programme of record" that seemed to prefer its international news told through the mouths of its foreign correspondents, even if sometimes they were nowhere near the action—for example, when a London-based reporter voiced a story about Sydney. Todd Spencer,[5] Executive Director CBC News Network, said that in setting up the station, he was convinced there were new ways to approach international stories:

> We do international news all day on News Network without correspondents. We can look at local websites, we can phone and do Skype with somebody who's just been to the protest. There are other ways of doing it....To me it doesn't rule out the need for foreign correspondents, but it certainly cannot be the only way that you tell international news, and we shouldn't go around saying: "We can't tell international news if we don't have correspondents in 20 different locations." (2012)

Essentially, CBC News Network is making the most of the human and technical resources they have—reporters in the "newsroom-cum-studio" who are willing and able to discuss foreign affairs on camera, live. The Vancouver newsroom, which hosts several hours a day of rolling news, is a lively place which has largely dispensed with some of the structures of formal news programming. There is a sense of fluidity and of stories moving, and the editors appear to be willing to run breaking news from abroad, even if they do not have the large numbers of personnel overseas to cover it in the manner in which it is traditionally done. The programme works around its staff with flexibility, and when Teresa Tang left to pursue further study in 2013, the programme folded the Asia Desk. When she later returned, she morphed into a breaking news reporter.

Over at ABC News 24, there is a lot less international coverage happening on a daily basis: with only 9'30" out of an hour-long, advertisement-free news bulletin. This means it is not doing more international coverage than the domestic Victorian bulletin with which it is compared. This partly reflects strengthened domestic coverage that has greatly improved since the station's inception. News 24 is using the ABC's 56 offices and studios across the country to regularly take in live press conferences and breaking news updates on Australian news. However, it is not particularly innovative, nor does it generate much that is new *in-house*, in terms of international news. It does have a slot called *#NX News Exchange* within the bulletin, which usually runs at around seven to eight minutes and which takes in what is trending on social media. Nonetheless, this is not a specific international news slot, with the stories depending to a large extent on domestic entertainment and celebrity. The foreign correspondent package remains the staple way that international news makes its way onto the station, however, in the weeks monitored, there were not that many stories in evidence from the foreign correspondents. Back in 2013, I discovered that some correspondent packages were played on this service for up to 21 hours (Murrell, 2013: 87). Although I did not test for this in 2014, it is possible that this approach has not greatly changed.

At the ABC, the hub bureau model that is meant to be rolled out in 2015, is already in evidence in the data, with correspondents from London and Washington regularly putting together packages from agency footage about stories in their wider region. Of the 18 foreign correspondent packages running over the ten bulletins, seven came out of London and five from Washington. In moving to make these bureaux into formal hubs, it is likely that even more emphasis will be given to these two locations. The cuts to the Australia Network and the APNC in 2014, followed by the closure of the New Zealand bureau in 2015, mean that the Pacific region will now be covered less. Of the 18 international stories that ran on ABC News 24, seven concerned Asia. There were one each from Tokyo, Beijing and Kuala Lumpur, two from Pakistan (done from London) and two from the Philippines on a typhoon (done by the BBC). During this period there were no stories from the Pacific Islands. In February 2015, "Cyclone Pam" caused havoc in Vanuatu and the ABC had to scramble reporters from various locations to Port Vila to cover the crisis. In the event, it did really well—relying on the former Bangkok correspondent Zoe Daniel and Radio Australia's Port Moresby correspondent Liam Cochrane. They both had the necessary skills that meant that they were not parachuting into a crisis story, about which they had no experience. This would seem to be the blueprint for coverage of these kinds of stories for the future, with both the ABC and CBC claiming that their discretionary newsgathering budgets have been respectively increased or preserved for reactive coverage, whilst savings are made through infrastructure and bureau downsizing. The "digital first" strategies of both organisations have yet to fully play out, along with the longer-term consequences for the 24/7 television stations.

NOTES

1. Title of an article on ABC cuts. Cuthbertson, D. (2014, November 25). Some ABC parts forced to run on the whiff of an oily rag. *Sydney Morning Herald.* Retrieved from http://www.smh.com.au/entertainment/art-and-design/some-abc-parts-forced-to-run-on-the-whiff-of-an-oily-rag-20141125-11tm3g.html (Accessed 5 April, 2015).

2. ABC Interviews related to 24/7 TV station:
   - Steven Alward, ABC Head of International News. Interviewed 12 December 2012.
   - ABC News Anonymous. Interviewed 29 January and 26 February 2015.
   - Gaven Morris, Manager ABC News Content. Interviewed 7 December 2012 and 19 November 2014.

3. Title of news headline (2015, March 27): CBC: Death by a Million Cuts. *ArtsJournal.* Retrieved from https://www.artsjournal.com/2015/03/cbc-death-by-a-million-cuts.html (Accessed 5 April, 2015)

4. CBC Interviews related to 24/7 TV station:
   - Greg Reaume, Managing Editor of CBC News Coverage. Interviewed 10 October 2012 and 12 November 2014.
   - Mark Ross, Executive Producer CBC News Network. Interviewed 17 October 2012 and 12 November 2014.
   - Todd Spencer, Executive Director CBC News Network. Interviewed 11 October 2012.

5. Reader voice-overs (RVOs) are when the presenter reads the copy live, over short video clips and are usually comprised of agency video. RVOs are sometimes called "underlay" or "OOVs," meaning the presenter is "out of vision."

6. Two-ways are interviews between the presenter or reporter and a person (often a reporter) from another venue.

7. There are no set-piece national evening bulletins on ABC1. Instead the evening bulletin is a 30-minute state bulletin.
   Colleen Murrell's data gathering in Canada was made possible due to a research grant from the Journalism Education & Research Association of Australia (JERAA).

REFERENCES

ABC Annual Report 2014. (2014). Retrieved from http://about.abc.net.au/wp-content/uploads/2014/12/ABCAnnualReport2014Accessible.pdf (Accessed 24 March, 2015).

ABC News 24 Online landing page. (2015). Retrieved from http://www.abc.net.au/tv/epg/#/channel/ABCN (Accessed 24 March, 2015).

CBC News Network Online landing page. (2015). Retrieved from http://www.cbc.ca/news2/networkstream/ (Accessed 23 March 2015).

CBC-Radio Canada Second Quarter Financial Report 2014–15. (2015). Retrieved from http://www.cbc.radio-canada.ca/_files/cbcrc/documents/financial-reports/q2-report-2014-2015.pdf (Accessed 9 February 2015).

Davies, A. (2014, December 13). ABC cuts put the focus squarely on digital. *Sydney Morning Herald.* Retrieved from http://www.smh.com.au/business/media-and-marketing/abc-cuts-put-the-focus-squarely-on-digital-20141211-124a75.html (Accessed 13 January, 2015).

Dempster, Q. (2014, December 12). Senate Select Committee into the Abbott Government's Budget Cuts—Submission 53. Retrieved from http://www.rssfeeds.aph.gov.au/DocumentStore.ashx?id=ffa277ed-2fd0-4562-9806-7f2ed23485ed&subId=302271 (Accessed 13 January, 2015).

Huffington Post. (2014, October 4). CBC Cuts: More Than 650 Jobs To Be Lost Over Two Years. *Huffington Post Canada.* Retrieved from http://www.huffingtonpost.ca/2014/04/10/cbc-job-cuts_n_5126698.html? (Accessed 24 March, 2015).

Kane, L. (2014a, November 13). Radio-Canada staff refuse award from Hubert Lacroix in protest of job cuts. *The Canadian Press.* Retrieved from http://www.theglobeandmail.com/news/national/radio-canada-staff-refuse-cbc-presidents-award-to-protest-job-cuts/article21570205/ (Accessed 13 January, 2015).

Kane, L. (2014b, December 11). CBC shortens local TV newscasts to 60 or 30 minutes. *The Canadian Press.* Retrieved from http://www.thestar.com/entertainment/television/2014/12/11/cbc_shortens_local_tv_newscasts_to_60_or_30_minutes.html (Accessed 13 January, 2015).

Knott, M. (2014, December 5). ABC star Sarah Ferguson attacks broadcaster's management over digital investment. *Sydney Morning Herald.* Retrieved from http://www.smh.com.au/federal-politics/political-news/abc-star-sarah-ferguson-attacks-broadcasters-management-over-digital-investment-20141205-120oy7.html (Accessed 13 January, 2015).

Levinson-King, R. (2015, March 28). CBC eliminates 144 more jobs from English Services in new round of cuts. *The Star.* Retrieved from http://www.thestar.com/news/canada/2015/03/26/cbc-eliminates-144-more-jobs-in-new-round-of-cuts.html (Accessed 24 March, 2015).

Maclellan, N. (2014, July 22). The Gutting of Radio Australia. *Inside Story.* Retrieved at http://insidestory.org.au/the-gutting-of-radio-australia (Accessed 14 January, 2015).

Massola, J. (2014, November 12). ABC and SBS facing up to 500 job cuts with tens of millions in pay-outs. *Sydney Morning Herald.* Retrieved at http://www.smh.com.au/federal-politics/political-news/abc-and-sbs-facing-up-to-500-job-cuts-with-tens-of-millions-in-payouts-20141112-11kvmh.html (Accessed 13 January, 2015).

Murrell, C. (2013). ABC News 24 and BBC World: a study of limited resources and challenging newsgathering. *Australian Journalism Review.* Vol 35(1), 83–96.

Patty, A. (2014, December 4). "ABC staff reject 'cruel and inhumane' approach to redundancies". *Sydney Morning Herald.* Retrieved at http://www.smh.com.au/nsw/abc-staff-reject-cruel-and-inhumane-approach-to-redundancies-20141204-11zvtd.html (Accessed 13 January, 2015).

Time for Change: The CBC/Radio-Canada in the Twenty-first Century: Report of the Standing Senate Committee on Transport and Communications. (2015, July). Retrieved at http://www.parl.gc.ca/Content/SEN/Committee/412/trcm/rms/14jul15/Home-e.htm (Accessed 4 August, 2015).

Wong, T. (2014, June 26). "CBC to lose up to 1,500 more jobs". *Thestar.com* Retrieved at http://www.thestar.com/entertainment/television/2014/06/26/strategic_plan_cuts_hubert_lacroix.html (Accessed 13 January, 2015).

CHAPTER TWENTY

Anti-Social Media: Watching, Hearing and Talking about Politics in US Cable News Channels

JESSE HOLCOMB

In 2011, the Pew Research Center asked the American public a question that revealed something interesting about the perception of the 24-hour news channels that had become a flash point in political discourse by that time. The survey asked, "what comes to mind when you think about news organizations?" Unprompted, 63% mentioned the name of a cable news outlet. More than any other news brand, respondents named CNN (43%) and Fox News Channel (39%). By way of contrast, little more than a third (36%) mentioned a broadcast outlet such as NBC. Just 5% named a major national newspaper. MSNBC, while mentioned by a modest 12%, still came up more than the *New York Times* (mentioned by 4%) (Pew Research Center, 2012).

Cable news at the time had an air of ubiquity. But in the years since that survey was fielded, social platforms and applications have gone from being supplemental sources of information to central ones. That, along with a host of other competitors for the attention of news audiences, raises questions about the role of cable news as a major broker in American political discourse. But was it ever thus?

Consider that in 2011 only a little over 3 million people in the US were watching one of the three main 24-hour news channels during the key prime-time hours of 8:00 p.m. to 11:00 p.m., according to Nielsen Media Research data. Nearly seven times that amount—22.5 million people on average—were watching one of the three evening news broadcasts aired on the major networks (NBC, CBS and ABC). And while not an indication of time spent reading, the Editor and Publisher Yearbook estimated nearly 50 million households subscribed to a daily paper that year (Pew Research Center, 2013). In other words, cable news was clearly more on the mind than it was on the screen.

Why did cable news brand recognition outpace its actual viewership as measured by Nielsen? Cable news personalities had a way of becoming part of the news cycle themselves. Here's what was going on among cable talent around the time the Pew survey was fielded: In November 2010, MSNBC's Keith Olbermann was suspended from his network for making campaign contributions to Democratic political candidates (Aujla, 2010). Fox News analyst Juan Williams was fired by his other employer, NPR, for making remarks that were deemed to be offensive to Muslims (Williams made the remarks on Fox) (Folkenflik, 2010). And on CNN, another controversial figure was hired—this time it was Eliot Spitzer, the disgraced former governor of New York (TVNewser, 2010).

Regardless of why cable's salience may or may not be overstated, what was true then remains true today: 24-hour news outlets are a destination for news junkies, the politics-obsessed and partisans. It's possible that this describes the outer limits of the cable news audience, too. Yes, cable still occasionally owns a breaking news story as it once did, but in many ways TV news channels have largely ceded that ground to other digital and social media.

ALL NEWS IS SOCIAL

All mediated news events are, on some level, social events. Past research has found that most people follow the news because they enjoy being able to discuss it with their friends, families and others in their lives (Pew Research Center, 2010). More recently, survey research exploring how younger adults engage with news underscores this finding: more than half of so-called millennials (adults age 18–34) get news because they like to talk to others about it (53%) and to feel connected to their community (45%) (Media Insight Project, 2015).

Cable television can, and has, served as a touchstone for large news events—the Gulf War catapulted CNN into a new kind of limelight. The World Trade Center attacks of September 11, 2001, provided a ratings boost to all three major US news channels (two of which, Fox and MSNBC, had only launched about five years earlier). In 2001, the prime-time median viewership for the three channels combined increased by 32%, according to Nielsen Media Research data provided to the Pew Research Center. In 2002, viewership increased by another 41%, to about 2.4 million during the hours of 8 p.m. to 11 p.m. Other sensational fare, from the O.J. Simpson trial in the mid-1990s to a stranded cruise ship in 2013, became office chatter and pop culture vernacular in no small part because of cable's single-minded devotion to those stories. Whether we were watching or—rather more likely—not, cable news has time and again elevated the violent, the banal, the pathological and the histrionic to a level of social currency. You may have heard of "Balloon Boy," but that doesn't mean you actually watched the spectacle unfold on CNN (Gold, 2009).

Increasingly, of course, cable became the place for political discourse. By 2009, following the landmark 2008 presidential campaign, the cable news audience for the three main channels had reached a median of 3.9 million viewers in prime time—a figure that would prove to be its peak to date, as the audience for a certain kind of live television began to decline and its audience continued to age. In the years to follow, the audience would decline to around 3 million on any given night. Among the main three news channels, Fox would continue to be the leader, with a median of 1.7 million viewers per evening in 2014. CNN and MSNBC lost more—the former with 495,000 viewers, and the latter, 568,000.

The rise of political talk on cable TV coincided with a slow, ideological sorting of the audience for 24-hour news. As CNN's less-political fare became its distinguishing feature, the original cable news network's audience began to experience significant declines. The channel, which had built its brand on being a first responder to breaking news, was not only competing with Fox and MSNBC—it was also competing with Facebook and Twitter.

POLARIZED DISCOURSE

The American electorate is more divided along partisan lines than at any point in the past two decades, according to the findings of a landmark survey of 10,000 US adults in 2014 (Pew Research Center, 2014). Although the most consistently liberal or conservative members of the public make up a minority of the total US adult population, the share has doubled between 1994 and 2014 from 10% to 21%. The two main parties are now more sorted along ideological lines than in recent memory, too. Along with polarization and sorting has come what researchers called antipathy: the share of Republicans expressing "very unfavorable" opinions of the Democratic Party has spiked in the past 20 years. The same is true of Democrats and their attitudes toward Republicans. In addition, consistent conservatives (63%) and to a somewhat lesser extent, consistent liberals (49%)—exist in "ideological siloes" at a significantly higher rate than the public at large.

To what extent does this milieu reflect the polarized discourse that has emerged on cable, particularly since the turn of the 21st century, which ushered in the politicized and ideological iteration of 24-hour TV news? Drawing on Pew Research studies of news content on cable TV, a picture emerges of a media platform that is heavily immersed in politics, particularly on Fox and MSNBC. The tenor of that political discourse looks strikingly different depending on the channel. And, increasingly, the political messages delivered on those networks are conveyed through opinion, argument and commentary, rather than conventional news reporting.

One study examining coverage of the 2012 US presidential campaign found that Fox and MSNBC offered starkly different views of the race. On MSNBC, fully 71% of the coverage of the Republican candidate, Mitt Romney, was negative in tone, while just 3% was positive. Meanwhile, on Fox, 46% of its coverage of Democratic incumbent Barack Obama was negative, while just 6% was positive (Pew Research Center, 2012). An earlier study of the 2008 race found a similar pattern, though the gulf between the two channels was less pronounced. In other words, the tone of coverage on Fox and MSNBC toward the leading candidates diverged over time, to the point at which a typical viewer of one channel would be receiving a message even more dissonant than that of another viewer of a different channel. The data also suggested that these two channels were more interested in attacking the "opponent" than in covering the "home team": Fox spent more time covering Obama than Romney, and MSNBC spent more time covering Romney than Obama.

And another content analysis found that fully 85% of MSNBC programming studied during a sample period in 2012 was opinion-based, rather than reported (Fox coverage leaned more toward opinion as well, though not nearly to the same extent) (Pew Research Center, 2013). A platform as polarized as this one serves as a natural destination for a segment of a polarized American public. A 2014 survey of US adults found that those whom the study's authors classify as "consistent conservatives" and "consistent liberals" gravitate toward their own distinctive clusters of media sources when it comes to news and information about government and politics (Pew Research Center, 2014).

Most notable, consistent conservatives are "tightly clustered around a single news source," with nearly half (47%) naming Fox as their primary source for news about government and politics. While MSNBC did not serve the same function for consistent liberals, nevertheless this group was more than twice as likely to name the "Lean Forward" network as its primary source for news about these subjects (12%) than the public overall (4%).

This, then, is cable: a source of news and information for a relatively narrow band of the American public. That narrow band tends to overlap heavily with those on the right and the left. And the messages that emerge from the networks themselves, particularly the right-leaning Fox and left-leaning MSNBC, reinforce the values of the consistent conservatives and liberals who form the core of their audience. With breaking news fast becoming a social media asset, cable news' raison d'etre is disappearing, leaving behind a niche market for groups of people who would probably prefer not to speak with each other. But what else do social media platforms inherit from cable?

SOCIAL MEDIA: POLITICS IN PUBLIC?

For many, Facebook, Twitter and even YouTube are now venues for learning about what has happened in the world. About half of all US Facebook users say they at least sometimes see news on the site, a portion that amounts to about three in 10 US adults. Roughly one in 10 US adults get news on Twitter and YouTube (Pew Research Center, 2013). While these social platforms are emerging as important sources for news, they are not generally viewed as news destinations for large majorities. Take Facebook, for instance, where more than three-quarters of users who get news on the site say they were visiting Facebook for other reasons altogether (Pew Research Center, 2013).

Although the breaking news experience on social media may be relatively incidental, the sheer number of social media users offers profound implications for any legacy news institution that hopes to compete for the attention of those users. Perhaps in tacit recognition of this evolution, CNN and its sibling network HLN have moved in some ways further from the breaking news model and toward a looser, more evergreen documentary and reality television-style approach to content.

But what of political messaging and discussion on social media, either of the kind broadcast by Fox and MSNBC, or in some other iteration? Has the polarized cable TV experience been appropriated by the social web? Here, the data are not yet entirely clear. Some research has found evidence that the ideological divisions that came to mark television news have a home in social media as well. A study of the networks clusters that form on Twitter found that, when it comes to discussion about political topics, polarized "crowds" often tend to form. The study's authors found that "frequently these are recognizably liberal or conservative groups," and that their members seldom interact with each other (Pew Research Center, 2014). Still, however, Twitter users represent a small segment of the total population; in the United States, just 23% of all online adults (Pew Research Center, 2014). Not unlike the core cable news audience, participants in polarized Twitter networks are a unique subpopulation.

Due to the closed nature of Facebook's API (application program interface), comparable data based on observed user patterns is rare. But self-reported survey data finds some ways in which Facebook norms around political news depart from those on Twitter. Early on, research on the difference between the two platforms suggested that news on Facebook is driven more by family and friends (70%) than is news on Twitter (36%) (Pew Research Center, 2012). By contrast, a greater portion of Twitter users said they were getting news from journalists and those outside their personal networks, and were more likely to hear about news they wouldn't have learned about elsewhere.

Though Pew Research Center's media polarization survey work was fielded several years later, one could trace a line: among the groups that Pew Research calls "consistent conservatives" and "consistent liberals," the ones who come across news about politics and government on Facebook are more likely to see views that reflect their own. Among conservatives, 47% say the political views on Facebook they see are either almost or always are similar to theirs. And among liberals, the number is 32% (Pew Research Center, 2014). And yet, the same survey found, of all US adults who ever see posts about politics and government on Facebook (not just the ideologically polarized ones), many are exposed to a wide range of views on that site. Just 2% said that the political posts they see are always or nearly always in line with their own views, and 21% said that that this was the case most of the time. Meanwhile, the bulk of this population (62%) said that the political posts they see on Facebook mirror their own just some of the time (Pew Research Center, 2014).

These latter findings are significant because they represent a far larger portion of the US online population than those pertaining to the polarized public. And other research supports the notion that social media can operate as a network of "weak ties," exposing users to a wide range of ideas and perspectives, rather than simply producing an echo chamber (Barbera, 2014). In the end, the takeaway may be that digital and specifically social media may simply be another arena in which pre-established social behaviors play out (this is a point that internet scholar danah boyd has long argued, and that Ethan Zuckerman discussed at length in his 2013 book, *Rewire: Digital Cosmopolitans in an Age of Connection*). Just as with analog media, where a small cohort of ideologically aligned news consumers gravitate to partisan cable TV programming, so it may be on the social web, where information streams and conversations are as homogenous as one chooses to make them.

Nevertheless, it is notable that the basic functions of 24-hour-news—breaking new information and serving as a platform for partisan, political discussion—are not only replicated on social platforms, but are being engaged there by a growing share of the public.

RELEVANCE, REVENUE, AND THE INCENTIVE TO INNOVATE (OR NOT)

Where will cable news brands find themselves in tomorrow's media landscape, especially when it comes to public political discourse? Outlets like CNN, Fox News Channel and MSNBC may find themselves in something of a crisis on several fronts.

Mobile devices have become the dominant screen for many. For the first time, in 2013, a majority of US adults owned a smartphone, and as of 2015, that portion is nearly two-thirds (Pew Research Center, 2015). A majority of US adults say they use Facebook, and other platforms are gaining market share, too (Pew Research Center, 2014). Those social media platforms are increasingly integrating more visual content, including video, thus entering television's wheelhouse. Mobile applications like Meerkat and Periscope offer a kind of visual rawness that produced television cannot fully replicate. Further still, other video subscription services such as Netflix are gaining steam, shifting consumers' attention and media purchasing habits.

On top of all that, the audience for cable news programming continues to age, and the news engagement behaviors of the young continue to skew toward digital platforms. These factors create significant pressures for the cable mega brands such as those mentioned above, let alone niche ones like CNBC, Fox Business Network and Al Jazeera. Consider a 2015 Pew Research Center report on the political news habits of the millennial generation in comparison to their elders. The survey of online adults found that fully 61% of online millennials turn to Facebook as a source of news about politics and government, while just 39% of members of the Baby Boomer generation do the same (Pew Research Center, 2015). (Still, CNN ranks second in the list of top sources for millennials at 44%—about the same share as for Gen Xers and Baby Boomers. Fox and MSNBC rank higher as a news source among the older generations.)

In light of this competition, how are cable news brands adjusting their strategies to compete in this new environment? No two are reacting exactly alike, but more than any other, CNN has made the most aggressive efforts. CNN under the leadership of Jeff Zucker has invested in documentary style and nonfiction programming, while cutting production and videography staff and simultaneously ramping up its digital operation (Beaujon, 2014). In other words, CNN was subtly ceding some of its breaking news ownership to other players, except in certain instances where it could "flood the zone" such as events like the missing Malaysian airliner. It is also worth noting that CNN was one of the first TV news operations to develop a mobile app that would stream live video to a smartphone, and has made substantial investments in digital operations to maintain its status as one of the premier digital news destinations.

While CNN's tactics suggest it sees itself as multiplatform, MSNBC and Fox doubled down on the legacy aspect of the medium, at least as it came to be defined in the first decade of the 21st century. The programming format on Fox and MSNBC serves as a tacit recognition that breaking news is not part of their core identity (see above). Talk and political debate fill not just their prime-time lineup, but increasingly, their daytime programming as well. Among the three major US cable news channels, only on CNN was a majority of the content that was studied (54%) classified as factual reporting (Pew Research Center, 2013).

There are still, however, some powerful incentives for cable news channels to stick with their present model, which some of them are tweaking but not abandoning. Despite the fact that audiences are shrinking, revenue for cable news outlets, with few exceptions, is growing, and this includes not just the three major news channels, but also the suite of financial news channels: CNBC, Fox Business Network and Bloomberg TV. Profit margins continue to hold in the 30–50% range.

This solid foundation is built on the fact that advertisers are still accustomed to turning to broadcasters as a platform that holds a valuable mass audience. And while the total number of pay-TV households in the United States peaked around 2012, the number is not dropping as rapidly as some would like to suggest. And cable subscriber fees locked in with providers like Comcast and Time Warner help ensure a second revenue stream for cable news channels beyond advertising. Today, news channels often generate more than half their total annual revenue from these fees.

But emerging consumer habits threaten to disrupt this model. Fewer young consumers are purchasing pay-TV packages to begin with. And the cord-cutter trend—i.e., those subscribers who decide to cancel their cable package—is starting to reveal itself as a significant issue for the industry to be reckoned with. If and when cable news outlets whither on the vine, at least in the format in which we know them today, will the social web replace Fox and MSNBC as the platform for partisan political discourse?

In many ways, Facebook, Twitter, and now Instagram and Snapchat are becoming so ubiquitous that these platforms can encompass the whole range of information diffusion and deliberation. At this point, what happens on a cable news channel still has the capacity to stimulate conversation, or at least commenting and sharing on social media. Who emerges as the opinion-leader of tomorrow's political debates is as much a question relevant to journalistic institutions at large as it is one directed at the 24-hour-news juggernaut of the first decade of the twenty-first century.

WHAT NEXT? SEVERAL POSSIBLE FUTURES FOR US 24-HOUR NEWS OUTLETS

As pressures mount from the social web and mobile technology, what kind of near-term future might 24-hour news brands carve out for themselves?

First, if they can adapt while profit margins remain in double digits, that adaption will need to take place in non-native platforms—on smartphones, on watches, in newsfeeds and ephemeral social environments afforded by startups like

Snapchat and whatever it is that will supplant Snapchat. Already, this is happening with offerings such as Facebook's Instant Articles and Snapchat Discover; news publishers that operate in these spaces have decided they need to be where their audiences are, even if it means diverting energy from the legacy platforms they've historically monopolized. Such an approach may ultimately lead to maintaining relevance while risking revenue. It may also be that their hands are forced. The obvious advantage here lies with television news brands' built-in infrastructure to produce video with high production value.

Second, as the bundled pay-TV economy unravels, networks may take a hard look at their "unfair advantages"—whatever it is they are positioned to do better than anyone else. Especially for a brand such as CNN, but also MSNBC aided by the NBC News division, a global infrastructure of bureaus and personnel offer the opportunity to do live video, at high production value, and to verify the material and place it in context. Furthermore, news networks have certain kinds of access that no amateur videographer could even dream of.

Finally, the news networks—at least the two political ones—may decide to double down on the cult of personality. Call it the Glenn Beck model. Beck, a one-time Fox News Channel star, left the network in 2011 to launch his own subscriber-based network, which, as of 2012, was generating tens of millions of dollars in annual revenue, even in its infancy (Friedersdorf, 2012). Fox may have lost one of its major conservative personalities, but in all likelihood the network could have held onto Beck had it needed him enough at the time. Fox and to some extent MSNBC have built their businesses around the distinctive personalities of its program hosts. In a future where advertising and to some extent, cable fees, begin to shrink for cable networks, program hosts offer an opportunity for networks to reach audiences directly. In that scenario, the network is less of a TV channel, or even a news division, and more of a talent agency. A savvy network might enable its program hosts to develop a direct relationship with their core subscriber audiences by providing production, legal and technical support. But the network's brand takes a back seat to the niche brands being cultivated by its talent.

This final scenario in some ways adheres closest to the spirit of the original cable news innovation: fragmenting and capitalizing on the US news audience. Ted Turner, in launching CNN, made a bet that there would be a measurable, albeit small, audience that is ravenous for news at any hour of the day. Rupert Murdoch anticipated an untapped market of conservative news consumers who felt that most news media were unfair and liberally biased. MSNBC did not necessarily tap into the disaffected analog among liberal news audience; it merely imagined there was a core group of liberal political junkies out there who watched TV.

If we anticipate a not-too-distant future in which the economics of pay-TV are disrupted, the big reveal may ultimately be what much of the research has hinted at if not outright declared: that the core audience for cable news, especially its political programming, is quite small indeed. Its agenda-setting power may very well wane as the marketplace for ideas and discourse becomes even more crowded and flattened. And the broader shift in journalism away from an advertising-supported model toward a subscriber model may relegate these threads of political information and debate, for which cable TV has come to epitomize, further into silos. Such a scenario does not necessarily detract from the social aspect of news. But it does chip away at the ideal of a vast information "commons." The question we are left with is whether such a commons was a full reflection of reality to begin with.

REFERENCES

Aujla, Simmi. (2010) 'Keith Olbermann suspended after donating to Democrats.' Available at http://www.politico.com/news/stories/1110/44734.html Accessed March 29, 2015.

Barberá, Pablo. (2014) 'How social media reduces mass political polarization. Evidence from Germany, Spain, and the U.S.' Available at https://files.nyu.edu/pba220/public/barbera-polarization-social-media.pdf Accessed April 8, 2015.

Beaujon, Andrew. (2014) 'CNN laid off more than 40 journalists at end of 2013.' Available at http://www.poynter.org/news/mediawire/236857/cnn-laid-off-more-than-40-journalists-at-end-of-2013/ Accessed April 8, 2015.

Folkenflik, David. (2010) 'NPR ends Williams' contract after Muslim remarks.' Available at http://www.npr.org/templates/story/story.php?storyId=130712737. Accessed March 29, 2015.

Friedersdorf, Conor. (2012) 'Behind the paywall at Glenn Beck's confounding web TV network.' Available at http://www.theatlantic.com/politics/archive/2012/04/behind-the-paywall-at-glenn-becks-confounding-web-tv-network/256328/ Accessed June 5, 2015.

Gold, Matea. (2009) 'What you're watching.' Available at http://latimesblogs.latimes.com/showtracker/2009/10/cable-news-networks-riveted-by-balloon-boy.html Accessed April 5, 2015.

Media Insight Project. (2015) 'How millennials get news: Inside the habits of America's first digital generation.' Available at http://www.americanpressinstitute.org/publications/reports/survey-research/millennials-news/ Accessed April 4, 2015.

Pew Research Center. (2010) 'Understanding the participatory news consumer.' Available at http://www.pewinternet.org/2010/03/01/understanding-the-participatory-news-consumer/ Accessed April 4, 2015.

Pew Research Center. (2012a) 'State of the news media.' Available at http://www.stateofthemedia.org/2012/cable-cnn-ends-its-ratings-slide-fox-   falls   again/cable-by-the-numbers/ Accessed March 25, 2015.

Pew Research Center. (2012b) 'What Facebook and Twitter mean for news.' Available at http://www.stateofthemedia.org/2012/mobile-devices-and-news-consumption-some-good-signs-for-journalism/what-facebook-and-twitter-mean-for-news/ Accessed April 7, 2015.

Pew Research Center. (2012c) 'Winning the media campaign 2012.' Available at http://www.journalism.org/2012/11/02/winning-media-campaign-2012/ Accessed April 5, 2015.

Pew Research Center. (2013a) 'News use across social media platforms.' Available at http://www.journalism.org/2013/11/14/news-use-across-social-media-platforms/ Accessed April 6, 2015.

Pew Research Center. (2013b) 'State of the news media.' Available at http://www.stateofthemedia.org/2013/newspapers-stabilizing-but-still-threatened/newspapers-by-the-numbers/#audience. Accessed March 25, 2015.

Pew Research Center. (2013c) 'The changing TV news landscape.' Available at http://www.stateofthemedia.org/2013/special-reports-landing-page/the-changing-tv-news-landscape/ Accessed April 5, 2015.

Pew Research Center. (2013d) 'The role of news on Facebook.' Available at http://www.journalism.org/2013/10/24/the-role-of-news-on-facebook-2/ Accessed April 6, 2015.

Pew Research Center. (2014a) 'Mapping Twitter topic networks: From polarized crowds to community clusters.' Available at http://www.pewinternet.org/2014/02/20/mapping-twitter-topic-networks-from-polarized-crowds-to-community-clusters/ Accessed April 7, 2015.

Pew Research Center. (2014) 'Political polarization & media habits.' Available at http://www.journalism.org/2014/10/21/political-polarization-media-habits/ Accessed April 5, 2015.

Pew Research Center. (2014b) 'Political polarization in the American public.' Available at http://www.people-press.org/2014/06/12/political-polarization-in-the-american-public/ Accessed April 5, 2015.

Pew Research Center. (2014c) 'Social networking fact sheet.' Available at http://www.pewinternet.org/fact-sheets/social-networking-fact-sheet. Accessed April 7, 2015.

Pew Research Center. (2015a) 'Millennials and political news: Social media—the local TV for the next generation?' Available at http://www.journalism.org/2015/06/01/millennials-political-news/. Accessed June 5, 2015.

Pew Research Center. (2015b) 'U.S. smartphone use in 2015.' Available at http://www.pewinternet.org/2015/04/01/us-smartphone-use-in-2015/. Accessed April 8, 2015.

TVNewser. (2010) 'Eliot Spitzer, Kathleen Parker hired by CNN.' Available at http://www.adweek.com/tvnewser/eliot-spitzer-kathleen-parker-hired-by-cnn/25164. Accessed April 4, 2015.

CHAPTER TWENTY-ONE

The Evolving Format of US Cable News and the Proliferation of Opinion

ALISON DAGNES

In the cable boom of the 1990s, television became the primary medium by which Americans got their news. During the cable expansion, channels carved out niches for themselves by catering their news coverage to particular partisan audiences. Today, cable remains important as it is one vital component of a much larger and ideologically fragmented media system. In an effort to distinguish itself from the competition, cable news channels became partisans and the partisanship of American cable news now extends from television to radio and the internet in a media system that is ideologically reinforcing. It allows Americans to select the politics of the news they receive as it affords political elites the ability to hone a message without contest or question.

Politicians use cable news now to selectively deliver their politics to a receptive segment of the voting public who willingly adhere to their preferred partisan brand while eschewing conflicting opinion. There are great consequences for this kind of specified news content and distribution, to include increasing polarization and an overall distrust of the media and of politicians. This chapter examines the place of cable news in modern American politics as it looks ahead to what will likely emerge in the constantly-changing media landscape.

THE GROWTH OF CABLE NEWS

When cable first emerged in the 1970s and 1980s it was assumed that the bevy of choices that became available for viewers would lead to a brighter media future.

As early as the mid-1970s, academic inquiries that examined the effects of this technological boom began to flourish. One scholar wrote in 1978: "Cable technology promises to provide the 'television of abundance' with its virtually unlimited channel capacity" (Jeffres, 1978: 167). The consequences of such channel capacity were myriad, and before Americans were even able to access the hundreds of channels that we have today it was clear that once viewers had so many options, they were going to have to make some selections. Wrote Jeffres in 1978: "With the increased number of choices provided by cable TV, consumers are faced with a change in the decision-making situation and increased selectivity would be expected" (Jeffres, 1978). This selectivity occurred and led to a kind of niche programming where specific channels crafted their content to appeal to specific viewers. As people had more options, they became far more selective about what they were watching. Channels catering to sports, entertainment, and news sprung to life which meant that viewers could target their viewing and either consume news by the fistful or reject it entirely. If a viewer sought out the news, he could watch it all day thanks to CNN and the channels that followed. As certain Americans became so-called news junkies, cable TV gave them enough material to satisfy their need for a fix. The days of "broadcasting" were over, and instead a kind of narrowcasting became the norm.

The competition among cable news channels had several important consequences. The first was that delivery and consumption of the news sped up considerably. The old adage "If it bleeds it leads" gave way to a new philosophy about being the first to report. Cable news channels rushed to break news and occasionally they actually *broke* the news in the process by mangling a story so completely it bore little resemblance to the truth. Getting things right seemed to matter less than getting things on the air first, and this overarching goal to capture an audience's attention meant that news providers began to take a very broad view of what was actually news. Technology assisted in this speeding-up process where average citizens could use their cell phones to take pictures, video, and audio recordings to send to cable news outlets. CNN actually had a name for this, "iReport," where anchors went to amateur journalists who had sent in footage from their smartphones. This was not considered quality news, but it allowed cable news networks the ability to show video from an incident, accident, or tragedy rapidly which seemed to matter more than value. This lean towards the instantaneous made viewers want news even more immediately. According to authors Rosenberg and Feldman: "The public's right to know has been supplanted by the public's right to know everything, however fanciful and erroneous, as fast as technology allows" (Rosenberg, 2008: 17). This, according to the authors, leads to a "bias of convenience" where viewers are satisfied by answers that are easiest (fastest) to get.

Additionally, other technological advances have allowed cable media to extend across a variety of formats. Now American cable networks use social media such as Facebook, Twitter, and Snapchat to distribute the news and to solicit viewers. The technological advances have led to what Jenkins, Ford and Green call "spreadable media" where content travels at lightning speed between news organizations, media outlets, and individuals (Jenkins, 2013). The increasing influence of social media adds to the speed imperative for cable news with a new requirement for viewers to consume and then share their content across a variety of platforms. Cable news producers have to integrate their product with internet content or else they will be left behind entirely as people move faster and faster online. The structure of online news is wildly divergent from the norms of cable news and thus, cable news networks have had to adapt to very different constraints as they integrate their content online. Cable news once ruled political media by being the most popular way to get the news. Now, cable must continue to broadcast while it also bends to the requirements of the internet. Cable news channels have to maintain their ideological positioning and presence amidst consistently increasing competition, and they must do so while still maintaining their importance on television. The stakes remain high for cable news networks in this rapidly shifting political media environment.

Finally, cable news networks have found that 24 hours of news is not as interesting as programming that centers entirely on emotion and conflict. Accordingly, pundit shows dominate the cable news line up. Pundits are political "experts" who provide their commentary about the issues of the day and unlike journalism that deals only in fact, punditry is rooted entirely in opinion. Pundit shows feature discussions with guests who opine as well, and the preponderance of opinion on cable news, combined with the demands for speed and multiplatform integration, has meant that the news itself has changed. It is now even shorter and easier to consume, more fiery and interesting to attract viewers, and more loudly opinionated, because opinion is easier to pull off than journalistic fact-finding and noisy arguments are more attractive than depth and discussion.

Cable news has developed into something far different from the original idea of the first Cable News Network (CNN) which was founded in 1980 by American media magnate Ted Turner. For more than a decade, CNN was the dominant cable news channel, and during the 1991 US invasion of Iraq, CNN made a name for itself as the cable channel Americans (and international audiences) could turn to during a crisis in order to get news quickly. But the success of CNN soon brought imitators and by the mid-1990s there were a number of all-news cable channels vying for an audience. Fox News was the real innovator here. In 1996, Fox launched its channel with a programming revolution that would entirely change the rules of cable news: It would be unabashedly and unapologetically political, and it would stack the deck in favor of a politically conservative viewpoint. Rupert Murdoch founded Fox News specifically as an antidote to what he perceived as a leftwing bias within the mainstream media and it became a huge success. Amidst this competition, cable news channels developed their characters: Fox News would be the conservative one, MSNBC the liberal counter-balance to Fox, and CNN would try to remain neutral. C-SPAN is not a news channel but rather a public affairs channel without editing, commentary, or (for that matter) ratings. CNBC focuses on business news, as does the fledgling Fox Business Channel, and HLN covers the headlines. As it has worked out, Fox is much more popular than all of the other cable channels combined. Today, Fox News sits as the epicenter of a conservative media system that includes not only the online components of Fox, but also reaches to like-minded bloggers, talk radio hosts, and politicians who exclusively seek out the rightwing news channel as one very important component of their messaging agenda.

One example helps to illustrate the connectivity of cable to other media modes. The conservative blogger Mickey Kaus had been writing for a right-leaning website called The Daily Caller. When Kaus posted a piece that criticized Fox News for neglecting to cover specific stories, his boss Tucker Carlson took the piece off of the website. According to Kaus, Carlson stated that The Daily Caller would not criticize Fox News since Carlson had a show on the network. In an interview about this with the website Politico, Kaus stated: "It's a larger problem on the right: Everybody is scared of Fox…. Fox is their route to a high-profile public image and in some cases stardom. Just to be on a Fox show is a big deal. And I think that's a problem on the right, Fox's monopoly on star-making power" (Byers, 2015). Conservative politicians seek out appearances on the top-rated news network which gives them a solid national audience. The pundits on Fox News are major celebrities in conservative circles around the country. Popular Fox News show hosts often do double-duty as talk radio hosts or, as the Tucker Carlson example illustrates, political bloggers. This kind of synchronicity makes Fox News in particular a powerful entity and it also affords an ideological reinforcement that then bounces around the varying media forms. Certainly, a story that runs on Fox gets picked up by other conservative outlets, and in return Fox uses material from other conservative media sources in their programming and punditry. The other cable news channels are working hard to emulate the Fox News example without success. But these channels remain important as they jockey for position and maneuver themselves to fit into this fluctuating media scene.

Cable news now plays a rather unique role in political media, but the power of cable cannot be denied, even as people increasingly go online to get their news. As cable news channels link to social media, as bloggers and radio show hosts are tied to cable news channels, as politicians use the media for their own purposes, the entire political media system is inextricably linked together and is trussed by partisanship. Although Fox News is the more successful example, there are patterns of this connectivity on the left as well. The cable news networks are one important part of a fixed political media cluster. As a result, the increasing partisan divide in the US is reiterated in the political media as the polarization of the media serves to exacerbate our partisan differences. Cable news, rife with punditry and ideological brawling, remains an essential American news media mode.

SHIFTS IN CURRENT AMERICAN IDEOLOGY & CURRENT TRENDS IN CABLE, TALK RADIO AND INTERNET NEWS

The development of American political ideology in the past forty years has helped to promulgate the partisanship of the American cable news system, and in return the partisan news media have helped solidify the nation's polarization. Scholar James Stinson has tracked political mood and found that the United States has shifted rightward in the past several decades, with more Americans identifying as conservative since 1952 (Bartels, 2013). This ongoing shift to the right must have been an important factor in Fox News Channel's creation. When Rupert Murdoch sought (and failed) to buy CNN, he and Republican strategist Roger Ailes founded Fox News to cater to a conservative audience, what was argued to be an underserved portion of the American viewing public. In the years that followed the launch of Fox News in 1996, the network grew stronger in the ratings, eventually beating out CNN and MSNBC to be the powerhouse of American cable news. They succeeded in no small part thanks to their deliberately conservative line up of pundits, many of whom used to be elected officials from the Republican Party. For viewers and Fox News personalities, Fox News offered up a straw man for conservatives to rally against: the liberal media elite. Crafting the argument that the network's deliberate right turn was fair and balanced, Fox News helped to reaffirm an ideological argument against the left. In doing so, Fox News helped to both usher in an emerging conservative media network as it sought out a conservative audience from whom it asked ideological adherence and viewing devotion. It was a great argument to make: Watch us because we are the only ones you can trust.

The 2014 Nielsen ratings represent the dominance of Fox News on American cable:

Total Viewers

Fox News	1.779 million	(even)
MSNBC	600,000	(down 8%)
CNN	528,000	(down 8%)
HLN	337,000	(down 16%)

(Kissell, 2014)

These numbers show that Fox News is currently the most popular cable news channel in the United States with ratings that equal more than the other three cable news channels combined. While Fox News ratings increased in 2014, the other three networks lost viewers and so it is not surprising that the also-ran networks are trying to mimic the success of Fox in order to raise their own viewership.

Fox News has done a remarkable job of carefully integrating its place on cable with other media forms, and the network often features appearances by conservative media figures. Talk radio is a fecund field from which Fox can pluck talent, as it leans disproportionately to the right. According to Talker Magazine, the trade publication for the talk radio industry, in 2014 the top ten talk radio hosts included seven conservatives, five of whom had some ties to Fox News either serving as contributors and guests or by hosting shows on the network (Editors, 2014). Fox does not maintain an exclusive hold on conservative communication, as certain rightwing policy activists, politicians, and other talk radio hosts look to what has been termed the "counter-establishment" of conservative messaging (Ball, 2005: 64). This secondary conservative media circle has resulted from the rift in the modern Republican Party between tea party activists on the far right, and establishment Republicans who are not as far out on the wing. This "counter-establishment" consists of bloggers, pundits, and political actors who are critical of the mainstream Republican fixtures on Fox. Yet the popularity of the network means that Republican politicians, regardless of their place on the ideological spectrum, still go to Fox News because it is the dominant force in cable.

Leftwing politicians use different forms of media messaging than their rightwing counterparts. Without a successful Fox News counterpart on the left, many Democratic politicians will appear on MSNBC which boasts a panoply of liberal pundit shows at night, but anchors its morning program in the hands of former Republican congressman Joe Scarborough. Many Democrats also seek out liberal bloggers and websites, use social media and alternative platforms such as satirical fake-news shows to reach constituents more directly. This is where the cable channel Comedy Central comes in.

Political satire is at a zenith in modern America. The massive success in the early 2000s of *The Daily Show with Jon Stewart* led to a number of successful spinoffs and imitations, all of which have contributed to a very rich field of political comedy on television today. These shows trend liberal and this fact, combined with their considerable popularity, has meant that politicians frequently appear on these shows to sell themselves and their policies (Dagnes, 2012). Comedy Central, not necessarily a cable channel one would immediately think of when examining the news, hosts an important "news" program: *The Daily Show*. The show takes aim at hypocritical politicians, flaws in the government, and especially at the news media, and has become more than just comedic, instead extending to sharp social and political criticism. Comedy Central followed the success of *The Daily*

Show by launching *The Colbert Report* in 2005. *The Colbert Report* was a spin-off that took specific comedic aim at political pundit shows, *The O'Reilly Factor* (on Fox News) in particular. Comedian Stephen Colbert created a conservative pundit character modeled after Bill O'Reilly and in doing so mocked not only punditry, but also the rightwing ideological philosophy. *The Colbert Report* was pure satire and used this technique (making an argument by using its opposite) so effectively that sometimes the *Colbert* audience had trouble distinguishing truth from satire (LaMarre, 2009). Like *The Daily Show*, the *Colbert* audience was predominantly liberal, making these two shows popular hosts for leftwing political actors. When the *Report* ended in 2014 after a nine-year run, it was replaced by *The Nightly Show* hosted by Larry Wilmore and thus Comedy Central continued its foray into political humor. Embedded in these shows is sharp political and media criticism, and arguably the greatest value of this type of satire is its ability to take critical aim at those in power. Democratic lawmakers appear on these programs more frequently than their right wing counterparts as the accusations of ideological partiality on the part of Stewart and Colbert sank into the Republican Party and discouraged conservative participation. But many politicians know that these shows are a good way to reach a younger audience, and accordingly even some Republicans appear on the programs as well, although most who do have left office and have books they are promoting (Dagnes, 2012).

Cable television beyond Comedy Central now boasts other popular political comedy shows that take a satirical look at the news, and the popularity of these shows indicates an interest in politics beyond the proper cable news channels. On HBO, *Real Time with Bill Maher* and *Last Week Tonight* with John Oliver are two shows that continue the fake-news criticism and political commentary begun by *The Daily Show*. These shows also skew left in their guest lists, issue advocacy, and (accordingly) audiences. Thus the two wings of the American political spectrum use very different media for their messaging and the audiences of cable news (and faux-news) are varied as well. This partisan divide is not without significance, and is important to examine.

CONSEQUENCES OF PARTISAN CABLE NEWS

The consequences of this partisanship within the American news media are significant. Scholars have found these include the rise of negativity in politics. Smith & Searles (2013) studied the content of opinion shows on a variety of American news outlets, to include cable news channels, during the 2008 presidential election and found that these shows did more attacking of opponents than supporting of allies, and that this opposition had the effect of making viewers less favorable toward an opponent (Smith, 2013). Additionally, Levendusky (2013) found that "partisan media polarize the electorate by taking relatively extreme citizens and making them even more extreme" (Levendusky, 2013: 611). The combination of a rise in pessimism with an increase in polarization discourages political participation and increases cynicism about the American electoral process. These pernicious effects are not enough to stem the tide of polarizing political news on cable TV: as the success of Fox News illustrates, there is too much money to be made in the anger, and as the success of *The Daily Show* demonstrates there is much to be gained in snark. But the social and political costs of this kind of mediated fury are rather substantial.

The overt partisanship of Fox and the imitations that have come as a result of their success have had a multifaceted effect on American news. One important outcome has been that political elites now seek out friendly audiences among politically like-minded news organizations. This allows politicians and political actors to create their own version of reality that can be substantiated across a wide variety of media outlets and modes without critique or fear of oppositional reproach. It encourages hyperbolic talk (because it is more interesting) and hostility, both of which make for good theater but lousy politics. When politicians represent their rivals without contest, because they have selected media outlets that favor their political side, viewers will not actually hear *from* political rivals but will instead hear *about* them. These depictions may or may not be rooted in fact, but are filtered instead through the lens of someone else's opinion. This lack of evaluation and debate is damning in a political system where deliberation is supposed to be the basis for compromise and negotiation.

Additionally, this partisanship has led the viewing public to self-select the ideological flavor of the news they wanted to consume. The theory of partisan selective exposure explains that given a wide variety of news choices, people will pick the political flavor of news with which they agree (Stroud, 2008). The idea of partisan selective exposure is contentious, and some scholars discount it while others find it to be highly relevant in today's hyper-mediated age. The real question lies with whether or not people's selection of partisan media outlets motivates their exposure, or if it is the other way around: that they are exposed to certain media which then informs their opinion (Stroud, 2008). Regardless, the partisanship of cable news influences its viewership, and accordingly Americans now choose the ideological understanding of the issues that already confirms their own beliefs.

Perhaps most tragically, the cable news system has become so loud and angry that it discourages news consumption. As Arceneaux, Johnson and Murphy write, the consequences of partisan news include viewers' increasing antagonism toward the news itself, which they call "oppositional media hostility"; this antagonism encourages Americans to avoid the news entirely (Arceneaux, 2012). When satirical "news" shows are examined, a similar phenomenon has been found. Several scholars who examined *The Daily Show* found that the show increased political distrust and cynicism. Hart and Hartlelius accused Jon Stewart himself of political heresy, writing that he has "engaged in unbridled political cynicism" (Hart, 2007: 263). *The Daily Show* is so successful in criticizing the news media, it breeds contempt for the journalism we have today. This distrust of the news, thanks to cable news and political satire shows, is problematic because a populace that disbelieves the news media is less likely to participate in the political process. According to Ksiazek, Malthouse and Webster, we are almost equally split: half the country can be classified as news "avoiders" and the other half of the country as "news-seekers" (Ksiazek, 2010). These authors found that there was a positive correlation between the "news-seekers" and civic participation, especially when the news was from cable television and news magazines (Ksiazek, 2010). But equally interesting was that the authors found that exposure to cable TV news *also* had an influencing effect on those who were not active in civic organizations and had the ability to increase the levels of participation. This would seem to be a good thing, except that the cable news network shows spent so much time attacking political opponents. Exposure to cable news may inspire action, but the question can be asked: What kind of action does it encourage?

The influence of Fox News on the right and its important place in modern political messaging demonstrates that cable news remains essential in American politics. The success of satirical news shows attests to the popularity of this form of political criticism and illustrates an importance in popular culture. Yet despite these successes, cable itself is in a period of uncertainty as technology continues to transform the way we receive information. The future of cable news is far more tentative than that of political comedy, since people will generally seek out the entertainment of satire. They may not be as readily interested in seeking out the real news. Thus, it is the cable news networks that must work the hardest to adapt to the technological shifts that threaten their future.

FUTURE OF CABLE NEWS

Media technology changes so quickly that it is difficult for the "older" forms to keep up with the revolutions and yet they must sustain themselves in order to stay relevant and to maintain their audience (and thus their financial security). Cable news networks are working hard to stay germane in the modern American political era, and they are primarily focusing on two ways to do this: The first is that they are escalating their partisan positioning among their competitors in order to carve out a place for themselves that is singular; the second is that the networks are moving much of their content online in varying and innovative ways in order to stay current.

The most easily identifiable way that the cable news networks continue to promote an ideological stand can be viewed through their primetime pundit show line-ups. On MSNBC, starting at 5 p.m. (Eastern Time), the pundit shows are hosted entirely by progressives: Ed Schultz, Al Sharpton, Chris Matthews, Chris Hayes, Rachel Maddow, and Lawrence O'Donnell. Each of these shows begins from a distinctively leftwing standpoint and the hosts of these shows discuss the news of the day with guests from this perspective. At Fox News, the primetime lineup includes Brett Baier, Greta Van Susteren, Bill O'Reilly, Megyn Kelly, and Sean Hannity, conservatives all. And at CNN, which is trying (without success) to forge a middle ground, the evening programming consists of shows that aspire to be more thoughtful and less divisive. CNN has forgone the pundit shows for interview shows, which means that their audience rests on the subject matter and not on an existing partisan adherence. It should be reiterated that CNN is a distant third in the ratings behind Fox and MSNBC.

The other way cable networks are working hard to stay pertinent in this fluid environment is by using as much of the new technology that comes their way. Were these networks to stay strictly on cable they would certainly die from inertia, since the only constant in the media landscape is change. At present, cable television is at a tipping point. With so many channels available, viewers are beginning to recognize that they only consume a small portion of their cable line up. As online content increases and as providers of streaming media such as Netflix and Amazon Prime have made content available away from cable service, many Americans are beginning to cut the cable entirely out of their media diet. This means that, increasingly, television programming is available by a form of pay-per-view known as OTT which stands for Over-The-Top. OTT refers to media that is available without a multisystem operator to provide content. HBO, CBS, and ESPN are already in the process of offering OTT content as streaming media only, which would allow their viewers to pay for their programming and nothing else. This allows these providers to reach viewers directly, and it also affords these networks the ability to expand their programming reach. HBO, for example, has partnered with Vice News and will launch a new 30-minute daily newscast (Baysinger, 2015). This means that cable news networks will now have increased competition from channels like HBO, a network that heretofore had primarily been an entertainment network that showed movies and scripted series. HBO's entrance into satirical programming also indicates their willingness to expand their reach beyond scripted entertainment.

If and when the cable news networks opt to stream their content, there are several outcomes that could emerge. The first possibility is that the cable news networks will successfully provide content in an affordable way to viewers who eagerly seek out their particular partisan flavor of the news. This would allow the cable networks to stay on cable as they also stream content to those who do not want to pay for a full line up. Another possibility is that cable news does not attract the kind of streaming audience as an entertainment network, which means their hopes for financial success will dwindle. The last possibility is that, thanks to all the news available online, there will be almost no interest in a streaming news content provider which would mean many people will opt not to get any form of cable news at all if they have to pay separately for it. All three of these possible outcomes will have a dramatic impact on the way Americans get the news, as well as the kind of news they get (or do not get, as the case may be).

Since cable networks are struggling to stay viable, they will undoubtedly continue to seek out social media networking as they try to reach younger audiences and spread their content across a variety of platforms. According to the CEO of Sherablee, a company that analyzes social media, companies that want to brand themselves look to social media with great success:

> "The social media marketplace saw huge shifts in 2014, with many brands moving towards measuring audience impressions, clicks, and thinking cross-platform," Tania Yuki, Shareablee's CEO and founder, said, noting that video was the biggest growth area, up 147 percent in 2014. For example, Twitter Video engagement grew 963 percent, she said. (Goff, 2015)

This presents a great opportunity for cable news, and Fox News has already seized on this. Among the most successful companies to brand themselves across social media, Fox News was 7th on the list with 154 million social media actions (Goff, 2015). A relatively new and interesting political media form is Instagram. The photo-sharing service has become one of the most popular social media platforms and, according to the *Washington Post*, politicians are almost certainly starting to use Instagram in their politicking efforts. This is because Instagram users are predominantly young (53% of American age 18–29 are on Instagram) and are also diverse (38% of African Americans and 34% of Latinos are on Instagram) (Blake, 2015). This demographic alone is reason enough for cable news networks to pay close attention to Instagram. If Fox News' social media success is any indication, cable news networks will continue to look to these platforms to enhance their viewership and expand their reach beyond cable.

Another social media revolution for cable news of late comes from Snapchat, the photo-messaging system that allows users to send short clips of content that immediately disappear once read. Snapchat was once regarded as an online tool used almost solely for the purposes of sexting, but new growth has expanded its usefulness and popularity. Snapchat recently developed a new feature called "Discover" where stories and videos would be available for 24 hours. Eleven news and entertainment providers have partnered with Snapchat to offer content for its "Discover" section, to include CNN, ESPN, National Geographic, Vice, Comedy Central, Yahoo News, the Daily Mail and several others (Stelter, 2015). CNN will provide five news stories with video content per day, which allows the cable news network an opportunity to advertise itself beyond the cable line up. But this partnership has some pitfalls for cable news as well, which is increased competition from non-news networks who will be providing news-like content. Depending upon what Comedy Central chooses to post on Snapchat, it is possible that *Daily Show* and *Nightly Show* clips will vie for the same viewers as those who might peruse CNN. This journalistic expansion beyond the networks that are news-centric will provide one more alternative to the already crowded political media landscape. This amounts to yet another potential problem for the news networks.

CONCLUSIONS

All of these potentials for the future of the news add up to an exciting, if not tricky, time for cable. The partisanship of cable has afforded financial success as it has profound effects on the way politicians communicate and the way Americans understand the news and the world around them. The technological advances that are consistently revolutionizing the media landscape pose serious threats to cable, which now must adapt to new forms and communication innovations as it remains on television. At the same time, the positives of cable as a news provider are myriad since it remains so essential. As a result, cable news will continue to spread beyond its home technology and we will see it in different forms and in different ways in the years to come.

REFERENCES

Arceneaux, K. and Johnson, Murphy, C. (2012). Polarized Political Communication, Oppositional Media Hostility, and Selective Exposure. *Journal of Politics*, Vol 74: 174–186.

Ball, M. (2015). Is the Most Powerful Conservative in America Losing His Edge? *The Atlantic Monthly*, January/ February: 64–68.

Bartels, L. (2013). *Americans Are More Conservative Than They Have Been in Decades.* [Online] Available at: http://www.washingtonpost.com/blogs/monkey-cage/wp/2013/09/30/americans-are-more-conservative-than-they-have-been-in-decades/[Accessed 20 March 2015].

Baysinger, T. (2015). *HBO to Launch Daily Newscast in Expanded Partnership With Vice.* [Online] Available at: http://www.broadcastingcable.com/news/programming/hbo-launch-daily-newscast-expanded-partnership-vice/139150[Accessed 27 March 2015].

Blake, A. (2015). *The political potential of Instagram.* [Online] Available at: http://www.washingtonpost.com/blogs/the-fix/wp/2015/02/04/why-the-two-parties-need-to-figure-out-instagram-now/ [Accessed 22 March 2015].

Byers, D. (2015). *On Media.* [Online] Available at: http://www.politico.com/blogs/media/2015/03/mickey-kaus-quits-daily-caller-after-tucker-carlson-204135.html#.VQiknU444E0.twitter [Accessed 23 March 2015].

Chan, J. (2013). Media Proliferation and Partisan Selective Exposure. *Public Choice*, Vol. 156: 467–490.

Dagnes, A. (2012). *A Conservative Walks into a Bar: The Politics of Political Humor*. New York: Palgrave Macmillan.

Editors (2014). *2015 Talkers Heavy Hundred*. [Online] Available at: http://www.talkers.com/heavy-hundred/ [Accessed 25 March 2015].

Goff, L. J. (2015). *Cable Networks Listed Among Top 25 Social Brands*. [Online] Available at: http://www.multichannel.com/news/social-media/cable-networks-listed-among-top-25-social-brands/387590[Accessed 26 March 2015].

Hart, R. and Hartelius, J. (2007). The Political Sins of Jon Stewart. *Critical Studies in Media Communication*, Vol 24: 263–272.

Jeffres, L. (1978). Cable TV and Viewer Selectivity. *Journal of Broadcasting*, Vol 22: 167–177.

Jenkins, H, Ford, S. and Green, J. (2013). *Spreadable Media: Creating Value and Meaning in a Networked Culture*. New York: NYU Press.

Kissell, R. (2014). *Fox News Dominates Cable News Ratings in 2014; MSNBC Tumbles*. [Online] Available at: http://variety.com/2014/tv/news/fox-news-dominates-cable-news-ratings-in-2014-msnbc-tumbles-1201386523/[Accessed 23 March 2015].

Ksiazek, T., Malthouse, E. and Webster, J (2010). News Seekers and Avoiders: Expanding Patterns of Total News Consumption Across the Media. *Journal of Broadcasting and Electronic Media*, Vol. 54: 1–18.

LaMarre, H., Landreville, K. and Beam, M. (2009). The Irony of Satire: Political Ideology and the Motivation to See What You Want to See. *International Journal of Press/Politics*, Vol. 14: 212–231.

Levendusky, M. (2013). Why Do Partisan Media Polarize Viewers?. *American Journal of Political Science*, Vol. 57: 611–623.

Rosenberg, H. and Feldman, C. (2008). *No Time to Think: The Menace of Media Speed and the 24-Hour News Cycle*. New York: Continuum.

Smith, G. (2013). Fair and Balanced News or a Difference of Opinion? Why Opinion Shows Matter for Media Effects. *Political Science Quarterly*, Vol 66(3): 671–684.

Stelter, B. (2015). *CNN and other media brands come to Snapchat*. [Online] Available at: http://money.cnn.com/2015/01/27/media/snapchat-discover-media-deals/[Accessed 28 March 2015].

Stroud, N. (2008) Media Use and Political Predispositions: Revisiting the Concept of Selective Exposure. *Political Behavior*, Vol. 30(3): 341–366.

CHAPTER TWENTY-TWO

24-Hour News in Australia[1]: Public Service and Private Interests

BRIAN MCNAIR

The story of 24-hour news provision in Australia is one of a competitive dynamic between two organisations: one public service media provider, the Australian Broadcasting Corporation (ABC), which produces News 24, and one commercial organization, News Corporation, which operates and owns the Sky News channel (alongside commercial operators Seven Media Group and Nine Entertainment Co).[2] These rivals attract relatively small proportions of the country's TV viewers, and like other "legacy" platforms must now compete also with an expanding range of online news providers eating into audience share. This essay considers the ways in which News 24 and Sky News seek to distinguish themselves from each other, and also how they are adapting to the enhanced competitiveness of the digitized, networked media environment.

24-HOUR NEWS IN AUSTRALIA: AN OVERVIEW

Sky News was launched in 1996 (Young, 2009), the first domestically produced 24-hour news channel in Australia. ABC launched News 24 in July 2010. Paying subscribers to the Foxtel digital network may access a range of externally produced 24-hour channels, such as Russia Today (RT), CNN and Al Jazeera, but in the Australian public sphere, addressing Australian issues and events, ABC News 24 and Sky News comprise a duopoly of real-time news provision.

These channels have relatively low audience ratings by comparison with Australia's free-to-air TV. Where an edition of prime-time news on commercial

Channel 9 may achieve in excess of one million viewers, for example, News 24 rarely exceeds 1.1% of the total viewing audience on a given week. Sky News typically attracts even fewer viewers, averaging 0.4% of audience share in January 2015. ABC News 24 attracts nearly three times the audience of Sky News, then, the latter characterized by one senior ABC manager as "only ever a niche" (although the same could be said of News 24 itself).[3] One possible explanation for the difference in ratings, other than the power of the ABC news brand against that of Sky, is that ABC News 24 is offered free on Australia's Freeview public digital platform, while access to Sky News requires a Foxtel subscription. ABC can also be accessed on Foxtel, but Sky is absent from Freeview.

These ratings understate the key role of 24-hour news in the Australian media ecology, however. A striking feature of Sky News is that its audience draws heavily on the political class in the nation's capital, Canberra, and that broader elite of policy makers and movers across the country. It is indeed the "go to" channel for the political class, and a key platform for them to communicate their messages on a day-to-day, hour-by-hour basis. Sky News' political editor David Speers remarks that

> We deal very much with national politics, and make that our daily guide. We don't have the biggest audience by a long shot, but we've certainly got a strong niche audience, an influential audience—politicians, lobbyists, business, people who are engaged in the political process.

> When, for example, there's a big political story happening during the day, and politicians know that the people watching Sky are going to matter – say for example there's a leadership showdown—those for or against know that there's really not much point in talking to anyone in the print media, because they won't be publishing until the next day. If they want to influence their caucus colleagues, the one place to do that publicly is Sky, so you do start to get a hell of a lot of phone calls, people wanting to go on the record, and also people talking off the record, to try and comment, to influence the debate and the discussion. Because they know that they will be very closely watched. As journalists and hosts you need to be very mindful of that, that you're being used.[4]

THE AUSTRALIAN NEWS ENVIRONMENT

Australia has one of the richest news cultures in the English-speaking world. Before the impact of digitalization and online journalism began to be felt on print circulations, the country supported a large number of mainly city- and region-based newspapers. Even in the era of declining print, per capita print readership in Australia remains comparatively high, although research conducted by the Reuters Institute in 2015 shows a familiar pattern of audience migration to online and mobile platforms for news (Newman et al., 2015). According to the report, 59% of Australians surveyed in 2015 accessed news at least weekly on a smartphone, 35% on a tablet. The report also shows that 44% of Australians now see online platforms as their main source of news, 35% identify TV as the main source, and only 7% go to print first.

Australia has been in the forefront of the global trend to access news online, therefore, but traditional platforms retain a sizeable audience. There are currently six free-to-air TV channels in Australia, delivered by the public service ABC and SBS, and commercial channels 7, 9, 10 and 11. With digitalization, these organisations have expanded into multimedia channels available on the public Freeview and commercial Foxtel platforms. ABC runs ABC2 and ABC3, for example, and ABC News 24. The commercial providers have catch-up and niche channels on Freeview, such as 7mate, and the platform also hosts National Indigenous Television, a public service channel devoted to indigenous themes.

News and current affairs provision on Australian free-to-air TV is concentrated in the ABC, and programs such as *7.30 Report* and *Four Corners* which regularly deal with political, economic and foreign policy. ABC runs a weekly public access political debate show, *Q&A*. Commercial free-to-air TV, however, has largely abandoned the current affairs space. While there are occasional moments of engagement with key public issues in Australia on commercial TV, such as a *60 Minutes* item on child sexual abuse within the Catholic church (an issue in Australia as in other countries), the majority of current affairs output on the commercial channels addresses stories concerned with consumerism, dramatic crime and human interest themes, and often includes product promotion of items about, for example, weight reduction. One primetime show, *A Current Affair*, exemplifies this move away from rigorous current affairs journalism in the commercial sector, presenting as more akin to advertising than journalism. As in many other countries, what was once a significant commitment by commercial broadcasters to investigative journalism and reportage around public affairs has been squeezed by competitive pressures for audience ratings, leaving the public service media as the core location for this key content.

It is in this context that the role of Sky News in the Australian public sphere has become significant. The channel, though available on subscription-only Foxtel, has become de facto the main space in Australian commercial TV for sustained, well-resourced coverage of domestic news and current affairs. Without its presence the public service ABC would be, not only the single provider of 24-hour news in Australia, but the only broadcast organization providing sustained TV news and current affairs coverage of the quantity and quality one would expect of an advanced capitalist society drawing heavily on the British model for its journalism provision.

The ABC News 24/Sky News duopoly parallels that in the UK, where the BBC and Sky compete in a similar manner if against a very different environmental background. The BBC is bigger and better-resourced than the ABC on a per capita basis, although the latter defines its role as a public service media organisation in very similar terms. Moreover, the performance of Sky News in Australia compares to that of the British equivalent in being perceived somewhat differently than the majority of the parent company's outlets. In Australia, as in the UK, News Corp press journalism is seen as bound up with the ideological proclivities of the Murdoch family, to the extent of verging on the flagrantly propagandistic. But Sky News in both markets has occupied a different space in the cultural terrain—still predominantly right of centre in its editorializing, of which the channel supports a great deal—but nonetheless respected across the ideological and political spectra as a quality provider of real-time news, current affairs and analysis.

THE POLITICAL CONTEXT: NEWS CORP AND THE ABC

The real-time news duopoly of ABC News 24 and Sky News has evolved in the context of ongoing criticism of and hostility towards the ABC by News Corporation press outlets in Australia. This can be seen as part of News Corp's broader assault on public service media in general, which in Australia has a particularly fierce tone. As part of the public service system, ABC News 24 is often criticised in News Corp's flagship national outlet, *The Australian*, as a source of left-wing bias, competitive over-reach, and managerial incompetence. A recurring theme of News Corp coverage of the ABC in recent times has been its alleged over-expansion into journalistic and other spaces which Murdoch and his managers see as rightly theirs (including 24-hour news). A June 2010 piece in the News Corp flagship daily, the *Australian*, cited an OECD report documenting opposition, or "pushback" to European public service media organisations expanding inappropriately into the online arena. In Germany, it was noted with some approval, German public service organisations ARD and ZDF had been subject to new legislation to protect commercial media from public service expansion. In Australia, on the other hand, "the ABC has expanded its online offering without any of the kind of parliamentary debate that has been witnessed in Europe. While Europe is updating the legislative rules around public broadcasting to take account of the age of digital platforms, the ABC Act has had no such renovation." An *Australian* editorial of December 13, 2011—"They are our ABC and SBS"—launched a widespread attack on the scale and cost of the ABC and SBS (Australia's other PSM provider). Noting that the ABC now comprised seven TV channels, the same number on radio, online services and the transnational Australia Network, the editorial called for:

an independent inquiry into the future of public broadcasting. The original rationale behind the national broadcaster—to provide countrywide access to information—either is redundant now or, at least, has changed substantially. Yet the ABC expands its platforms while complaints persist about content quality. It is time for a clear-headed debate about whether its charter has kept pace with the digital age and whether, instead of spreading itself so thinly, it should concentrate on some clearly identified objectives.

Another key theme of News' coverage in this period has been that of "content quality" in ABC journalism. ABC news, including News 24, has been accused of being dominated by a left-of-centre ideological perspective, particularly in key policy areas such as climate change and border protection. This was the theme of an article by the *Australian* in November 2011, which accused the ABC of "campaigning" rather than merely reporting. On climate change, "the ABC's extensive reporting has now been revealed as jaundiced and counter-productive. Developments in the public debate have exposed the national broadcaster's misleading alarmism on global warming and handwringing over border protection. Its hyperbole on these issues has polarized public sentiment, made sensible political discussion more difficult, and created a backlash." As in the UK debate on the future of the BBC, the perception of a certain quality of impartiality or objectivity in its journalism is key to continuing public support for the ABC. Were that perception to weaken, the obstacles to News Corp's colonization of much of the PSM space in Australia would be significantly lower. Consequently, the *Australian* and other News Corp outlets maintain a constant assault on the corporation's journalism, as exemplified by the following editorial:

> Any observer contending that the ABC is a paragon of objectivity does not deserve to be taken seriously. This government-funded organisation habitually derides the views of mainstream Australians on issues such as border protection and climate change, and too often its army of journalists acquiesces to government spin. (*The Australian*, March 3, 2012)

News' Australian newspapers have also attacked the ABC on the grounds of inept governance and management. In June 2015 the debate show *Q&A*—similar to the BBC's *Question Time* in its live studio public participation format—came under attack for allowing a Muslim extremist in the audience to ask a question of the panel regarding proposed new citizenship laws designed to combat home-grown support for Islamic State. The questioner made a follow-up remark deemed offensive by many people, and the episode was read as further evidence of the ABC's mendacity by the Coalition government and News Corp press outlets.[5] Prime Minister Tony Abbott announced a government review of the program's production processes, and demanded that the ABC decide "which side it's on." This incident was a reminder that in Australia, perhaps to an even greater extent than is the case in the UK (where the changed fortunes of News International and News Corps post-phone hacking and Leveson have muted the more strident anti-PSM lobbying which was familiar in the past [Davies, 2014]), the role and rationale of PSM is being challenged at core, with the assertion that commercial operators do the job better than a bloated, protected public body can.

Notwithstanding the war of attrition waged on the ABC by News Corp's Australian press outlets Sky News maintains a distance from such views, at least in its news and analysis. Both organisations display mutual respect and regard in their assessments of the performance of the other. The head of editorial policy at the ABC observes that Sky News is a "competitor," and an important contributor to the Australian public sphere.

> It's good and healthy that there's competition and diversity, and clearly with the Australian market at the moment…you really only have the ABC and the Murdoch bloc, and Sky fits into that; to basically say, we will drive the national discussion, we will be where the debates take place. I respect that about News, and I think in large parts, they do extraordinarily well. They have a few weaknesses and blind spots that do them no credit, but by and large they do a fantastic job, and we do see them as a competitor. But we don't see them as one we have to vanquish.

The senior managers of Sky, conversely, refrain from the kinds of attacks on ABC News that can be found nearly every day in News Corp's press outlets. In the words of chief political editor David Speers, "we're mindful of the ABC as a competitor, but we try not to be driven in any way by what they do. They do what they do quite well, and we try to run our own race."

Both channels, then, respect and value the presence of the other in this limited journalistic space, seeing themselves as occupying competing but complementary segments of a small, but influential market of opinion formers and elites. The ABC, of course, is bound by public service broadcasting rules and guidelines, while Sky News has no such constraint. While it does compete with the ABC in rolling news delivery, therefore, it also has more room for opinion and commentary journalism of the type one would see in a newspaper.

AUSTRALIA'S 24-HOUR NEWS DUOPOLY

During the day, both ABC News 24 and Sky News focus on rolling news coverage, much of it repeated or shared with free-to-air (in the ABC case, News 24 also serves as a space for rebroadcasting of programs which show first on free-to-air TV, such as *Q&A* and *Insiders*). ABC devotes time to the Australian parliament when it is sitting (*Capitol Hill*), and has dedicated slots for sport (*Grandstand*) and business (*The Business*). In the evening, it covers *The World*. Sky also covers sport and business, of course, with a 30-minute sport slot in the evening (*Sportsnight*), and regular items throughout the day on both sport and business topics.

News 24's news coverage, like everywhere in the ABC network, is governed by strict rules of "due impartiality" familiar to public service media all over the world. The extent to which the corporation fulfils that remit is challenged regularly by News Corp media outlets, as noted above, but there is little evidence of party political "bias" in the ABC's journalism that would satisfy an observer not on the Murdoch payroll. When it comes to public service journalism, as BBC managers also appreciate, many actors perceive bias, be it the SNP in the context of the Scottish independence referendum, or Labour and Conservative governments at various times. For the ABC too, accusations of bias regularly emerge from whoever is in power at the time.

What one *can* detect at times is a tendency to default towards what some will regard as a liberal stance on issues such as asylum seekers and climate change policy. For example, while the *Australian* newspaper has frequently cast doubt on the scientific case for anthropogenic climate change, and is widely perceived as a hub of climate change "denialism" in Australia, the ABC accepts the global scientific consensus and presents it as such. If one wishes to argue that acceptance of climate change science is a "left-of-centre" position, then the ABC will of course be regarded as accordingly biased.

That argument is itself, however, a product of bias in the direction of climate change scepticism. As in the UK, the debate about ABC bias is ongoing, and usually driven by the anger of a government, be it left- or right-of-centre which conflates its own positions with that of "the nation." To that extent the issue is irresolvable, because it is rooted in shifting balances of ideological and political power rather than a systematic or quantifiable tendency to favour a particular approach to making sense of political events. A right-of-centre Coalition government sees the ABC as left-biased; a Labor government is likely to see the corporation as right-biased, and so on. Where the two providers differ most obviously is in the relative proportion of output each devotes to opinion and editorial. The ABC must exercise "due impartiality" in its news and current affairs coverage, and is thus constrained in the extent to which it can editorialise. ABC News 24 has two strands devoted to political debate: *The Drum*, transmitted on most weeknights at 6 p.m.; and *Insiders*, shared with ABC1 and going out on a Sunday morning. The latter is the corporation's flagship debate show, where the eponymous "insiders"—comprising senior political journalists and commentators, and chaired by distinguished ABC political correspondent Barrie Cassidy—tackle the issues of the preceding week. This is a conventional public service media program of the normatively preferred type—an ostensibly impartial journalist, chairing what is intended to be an ideologically balanced group of expert commentators, discussing the issues of the day in a manner appropriate to the Australian public as a whole. The opinions of guest commentators, and the impartiality of the organisation, are strictly separated in the presentation of views and information.

Sky News, on the other hand, provides several hours a day of highly opinionated journalism, more akin to talkback radio and tabloid editorialising than the studied neutrality of the ABC. Political editor David Speers characterises the channel's approach to the reportage-commentary balance thus:

> We play to our audience. We treat our daytime programming a little bit differently to our evening programming. During the day we have that political niche watching us, and we do a lot more straight up and down, balanced interviews and reporting analysis. At night-time we'll do more opinionated programming targeted at a broader audience who aren't necessarily glued to us during the day. I think people want strong opinion at times. They can get straight news from a lot of different platforms, and more and more so, but people do like to hear debate about an issue. There's only so much news you can do during the day, and then I think it's good to offer that layer of opinion on top, to offer something that's engaging.

The need to engage has intensified with the expansion of the internet and the proliferation of online and social media outlets, and reinforced Sky's drive to occupy that space in broadcasting terms. Although its "liveness" and scheduling flexibility gives it the opportunity to compete with online sources in delivering breaking news more or less immediately, the need to attract and hold viewers in the face of the ABC's competition has driven the evolution of Sky's more opinionated style. Both channels have also striven to include social media, for example in running live Twitter feeds alongside programs, and using social networks to promote and share content.

Sky News' commentators typically occupy one-hour evening slots, such as *Paul Murray Live*. Murray is a radio journalist who successfully transferred his direct, opinionated, man-of-the-people style to TV. In terms of ideological affiliation, he was as of this writing a supporter of the right-of-centre Coalition government, though he has in the past endorsed Labor and Kevin Rudd. He disputes that he is himself "right wing," as opposed to being a critic of whomever he feels deserves criticism. Murray's style is to begin with the broadcast equivalent of a tabloid newspaper column, in which his personal views on what he regards as the key issue of the day will be articulated. Consistent with a tabloid style, he will be outspoken and provocative—a controversialist pundit, in that sense. At other times, he strives for a more measured tone. He engages his audience in a personal manner, and often conveys an authentic anger at the people and parties he rails against. As a TV performer, Murray is frequently compelling, reputed to work without notes or prepared scripts.

Following his opening monologue, Murray typically chairs a debate involving two or three guests, often journalists working for the *Australian* or other News Corp outlets. On occasion, he has guests who are left-of-centre, but the balance of the show is undoubtedly to the right. As such, debate tends to involve a degree of mutual agreement and shared assumptions which will arguably alienate viewers of a different view. This is a feature of other Sky programs of this kind, such as Chris Kenny's *Viewpoint*. Kenny is a right-of-centre pundit who specializes in condemnation of the ABC, and regularly uses his Sky News slots to attack the corporation. These monologues tend to echo his columns in the *Australian*, reinforcing the judgment that these opinion programs are in essence televisual variants of tabloid punditry. Kenny, like Murray, adopts the debate format, where guests tend to share his own frankly stated ideological biases.

One Sky News program, *Richo + Jones*, differs from the above two examples in presenting one left-of-centre, and one right-of-centre commentator engaged in debate with each other, and with their guests, who may be from either "side of politics." Graham Richardson was a Labor minister in the 1990s, while Alan Jones is a leading and highly influential radio talkback host on Sydney's commercial radio. Jones is a notorious controversialist, loathed by the Australian left for his alleged misogyny among other flaws, but prepared to attack the right when he feels it to be justified. Though on opposite sides of the political divide, Richo and Jones are clearly friends, and an engaging double act in televisual terms.

The ABC's head of editorial policy compares Sky's adoption of high-profile editorialising with the approach of News Corp 24-hour channels in other markets:

> You can trace [Sky] back to what Fox achieved in America, because Fox changed the market for rolling news channels in America. The bottom line is that the business models and the numbers tell you in America that this kind of opinionated coverage works. And Sky's flirting—Murray, Richo, Contrarians, etc.—they have actually moved away in the last several years from what was a very traditional rolling news format to much more opinionated programs; to do the Alan Jones of TV, because all the numbers tell you that that's what drives eyeballs. And to me that's what's driving the strategy, rather than any philosophical considerations. It's just building audiences. The ABC is not in that business, but at the end of the day it's also about what's engaging.

But Sky News is about more than frankly stated opinion, and even in its most extreme editorializing rarely matches the absurdist excess of Fox News. I have previously argued that News Corp's broadcasting outlets differ from print in that they must compete with competitors which may have very different journalistic ethos than the average newspaper. Thus, in the UK, Sky has long been a respected complement to the media environment, winning awards for its journalism as long ago as the 1990s (McNair, 2009). This it achieved alongside the existence of a strong BBC, which kept the standard high for all other news broadcasters. In the US, by contrast, there is only a very small public service system, and thus Fox News is able to indulge in what is reasonably characterized as a televisual analogue of the *National Enquirer*. In Australia, like the UK, Sky entered a broadcast journalism market already defined by the public service ABC, and to compete effectively in that market could not simply ignore audience expectations.

And Sky News managers take this challenge seriously. In the run up to the 2013 general election in Australia, for example, Sky News staged three debates involving the leaders of the main parties. These were conducted with scrupulous fairness, and well-reviewed by media observers. On this occasion, as on others, Sky News can be judged to have been successful in providing Australian audiences with important exposure to political debate. Indeed, in 2013 many commentators expressed the view that Sky's debates had been more effective as political media than the ABC's equivalent efforts. Asked by the author to define Sky News' distinctive role within the Australian public sphere, political editor David Speers responded:

> to provide an authoritative platform, to discuss and analyse policy that's coming out, how this government is performing, to hold them to account, also to keep an eye on and analyse how the Labor opposition is responding to what's happened to them and where they need to go, and how they're going in holding the government to account as well.

> But I'm not naïve enough to think that our role is only to be a very dry policy analysis place; it's also to provide a bit of entertainment as well, if that's the right word, where people can engage, and can be informed, but also entertained by issues that press their buttons. That may often mean getting fired up and angry, it may mean sometimes they're having a bit of a laugh at someone or something, or at us. It is that town square, and yes the ABC try to do that as well, and do it very effectively in some ways. But the diversity of platforms is a really important thing, and Sky's place in that landscape is very important.

ABC News 24, consistent with the public service remit of the corporation overall, is obliged to meet the news and current affairs needs of the Australian population as a whole, to be representative of its socio-demographic diversity and ideological plurality, and to avoid taking sides on political issues. While this leads to occasional controversies such as the appearance of the extreme Islamist Zaky Mallah on the public participation *Q&A* format, for which the ABC management was compelled to apologize, the aspiration is clear. ABC News 24 services the Australian public as a whole, and should be accessible and engaging to them all. Reflecting this goal, the channel includes a number of segments and programs devoted to particular groups, such as farmers and those who live in the country (*Landline*). Dotted throughout the schedule are *Quarters*—short magazine programs devoted to indigenous affairs, the arts and other categories of special interest.

Sky, on the other hand, serves a niche, and in so far as it favors a particular ideological viewpoint, this will tend to be to the right-of-centre. As noted above, the more opinionated programs which make up Sky News' evening schedules adopt the style of a televisual tabloid or radio talkback, often containing provocative, or at least opinionated commentaries and editorials (often targeted at the ABC). In its straight news and current affairs programming, however, Sky News maintains an approach which makes it palatable to a broader range of views, and those viewers interested in politics and public affairs in general. There is a consensus amongst Australia's politicians, who frequently appear on the channel, that it is "straight down the middle." The Sunday morning *Australian Agenda* debate show, for example, is presented by University of Western Australia journalism professor Peter Van Onselen, who also presents a daily *News Hour* during the week. The former provides a platform for politicians and others to articulate their views on issues of the day, as Van Onselen and senior print journalist Paul Kelly engage with guests in a traditionally adversarial, but non-partisan manner. Says Van Onselen:

> I would describe *Australian Agenda* as a pretty straight down the line long form political interview program. I think that our interviews are a lot longer than *Insiders* and they're longer than *7:30 Report* (ABC current affairs strands) as well. I think *Australian Agenda* is the only really long form interview program left. And to the extent that it has an editorial bias it's simply the editorial bias of the day that Paul Kelly will put on his editorial of the top, whatever that might be. And that changes from week to week. I'd like to think that we're not biased one way or the other in our political interviewing. I don't think we are biased.

CONCLUSION

The Australian 24-hour news ecology, then, presents a public-private balance in which both providers occupy distinctive spaces within the political public sphere as a whole. They do so with relatively small shares of the total TV viewing audience, as noted above, and Sky News is not a source of short term profit for News Corp in Australia. As with several of his titles around the world, such as *The Times* in London and the *Australian* newspaper, Rupert Murdoch and his managers have supported a journalistic product which delivers substantial indirect benefits to his company, such as cultural prestige and political influence, rather than short-term surpluses of the sort that can be measured in an annual financial statement. Sky News is a typical News Corp loss leader, in short, its perceived impact and influence on Australian politics and culture outweighing the operating losses it incurs. For that reason, too, its managers embrace the constraints of a broadcast news media environment which has been defined by the ABC, and in which audiences expect a level of "quality" not seen, for example, in the US Fox News equivalent. In so far as Sky News in Australia does broadcast opinion and commentary in styles which verge on the tabloidese, it operates in a distinctive niche which also includes political journalism respected across the ideological and party spectrum.

NOTES

1   The research and writing of this essay was supported by an Australian Research Council Discovery grant, DP130100705, Politics, Media and Democracy in Australia: Public and producer perceptions of the political public sphere.

2 News Corp has recently sought to purchase full ownership of Sky News, but as of this writing had not made an offer acceptable to its partners.

3 Interview conducted by the author with Alan Sunderland, Head of Editorial Policy at the ABC.

4 Interview conducted by the author.

5 The questioner, Zaky Mallah, stated that "The Liberals [the lead party of the right-of-centre Coalition] have just justified to many Australian Muslims in the community tonight to leave and go to Syria and join ISIS because of ministers like him." Quoted in McNair, B., "Don't fear difficult debates on Q&A," The Conversation, June 24 2015. https://theconversation.com/dont-fear-difficult-debates-on-qanda-43773

REFERENCES

Davies, N. (2014) *Hack Attack: how the Truth Caught up with Rupert Murdoch*, London, Chatto and Windus.

McNair, B. (2009) *News and Journalism in the UK*, 5th edition, London, Routledge.

Newman, N., Levy, D., Nielsen, R. (2015) *Reuters Institute Digital News Report 2015: tracking the future of news*, Oxford, Reuters Institute for the Study of Journalism.

Young, S. (2009) 'Sky News Australia: the impact of local 24-hour news on political reporting in Australia', *Journalism Studies*, Vol. 10, No.3: 401–416.

CHAPTER TWENTY-THREE

Where Infotainment Rules: TV News from India

DAYA KISHAN THUSSU

In a complex, globalized world, international news is in the process of transformation, partly as a result of an increasingly mobile and globally networked and digitized communication infrastructure together with the digitization of content, enabling global and instantaneous circulation of television news across continents. While the imbalance in the flow of news—from the media-rich North (and within it a US-UK core) to the South—continues to define global television news, the traditional domination of Western, or specifically American, media is diminishing, and, more importantly and arguably, being challenged (Nordenstreng and Thussu, 2015) by the availability of television news from such diverse countries as Russia, China, Iran and Qatar. In this chapter, I focus on how Indian television news is aiming to reach a global audience.

Indian perspectives on global affairs are accessible abroad via such channels as News 18 India (part of the TV-18 group), Headlines Today, as well as NDTV 24x7 (part of New Delhi Television Group), which has provided serious quality news programming now for a quarter of a century. All three are private networks, while the Indian state broadcaster Doordarshan (DD) remains one of the few major state news networks not available on television screens internationally, at a time when global television news in English has expanded to include inputs from countries where English is not widely used, notably China's CCTV News and Russia's RT. This reflects the importance of television news in a country's external communication strategy in the context of visually-driven global geo-politics. Unlike the other non-Western economic and political powers, however, India's presence in the international news arena is being driven by private not public broadcasters, which belies its traditional role in articulating Southern concerns in international fora—most notably during the 1970s NWICO (New World Information and Communication Order) debates within UNESCO. However, while the all-news channels in India are only just over a decade in existence, even in their early formative years their pro-business agenda was well formed. Are Indian television channels the voice of the global South or an echo of a US-dominated, market-driven broadcasting culture, hostage to Bollywoodized infotainment?

Since the 19th century, the global news sphere has been dominated by what Jeremy Tunstall has called a "US-UK news duopoly"—visible both in state-sponsored channels (such as the Voice of America and its associated global networks), and perhaps more effective, private television news with their formidable economic muscle. US government-sponsored news initiatives have been supported by a thriving and globalized private news media—from television news agencies (Reuters Television and APTN) to international news networks (CNN International, CNBC, to name the prominent ones). One reason for the US domination of global media is that the country has always followed a commercial model for its media industry—a venture in which the successive US governments have been a crucial factor. Broadcasting, both radio and television, had a commercial remit from its very inception. The commercially-driven trio of networks—CBS (Columbia Broadcasting System), NBC (National Broadcasting Corporation) and ABC (American Broadcasting Corporation)—provided both mass entertainment and public information. With the globalisation of the past three decades, the US-inspired commercial model of broadcasting has also been globalized, creating a dynamic media system, challenging state control and widening the public sphere, while at the same time also leading to the concentration of media power among private corporations (Thussu, 2016).

Digital communication technologies in broadcasting and broadband have enabled the exponential growth of multichannel networks and given viewers in many countries the ability to access simultaneously a rich mix of local, national, regional, and international television programming, making the global media landscape multicultural, multilingual and multinational. A key part of this has been the explosion in the number of 24-hour news channels (Rai and Cottle, 2010). Changing not only the global broadcasting ecology but the journalism industry itself, real-time news has evolved to become a force in the social and political spheres, influencing political decision-making and empowering social movements across the globe (Cushion and Lewis, 2010).

THE GROWTH OF NEWS TELEVISION IN INDIA

As elsewhere in the world, the rapid liberalization, deregulation and privatization of media and cultural industries in India, coupled with the increasing availability of digital delivery and distribution mechanisms, have created a new market for 24/7 news. Television in India has grown exponentially in the past decade: from a single public-service, state-run channel, Doordarshan, in 1991 to more than 800 channels, including more than 400 news and information networks, making it home to the world's most competitive news arena, catering to a huge Indian audience, both domestic and diasporic (Jain, 2015). In the late 1990s and early 2000s, India's news television sector saw an extraordinary increase in the number of dedicated news channels, most of which were national, but many international in reach, while some catered to the regional markets. Dedicated news networks now operate in a dozen of the 18 state-recognised languages, several of which have large geo-linguistic constituencies, both within the country and among the 25 million strong Indian diaspora (Kapur, 2010).

Such an extraordinary growth shows how much has changed since the introduction of television in India in 1959 as a means to communicate government policies, public information and state propaganda. Its news coverage rarely rose above what critics rightly derided as "protocol news." The stated aim of Doordarshan (DD) was to educate and inform, though it remained a mouthpiece for the government of the day, reflected especially in the way its information bureaucrats ran news operations. The partial privatization of the airwaves started with the introduction of advertising into the state broadcaster in the 1970s, followed by sponsored programmes, and received a boost as India opened up to transnational media corporations in the 1990s (Mehta, 2008).

The Indian government's deregulation of the state channel, and the resulting commercialization of the news and current affairs sector, introduced competition into domestic news television, just as other national and multinational media companies like Murdoch-owned STAR (Satellite Television Asia Region) TV and Zee TV (India's largest media conglomerate) were trying to set foot in the Indian market. DD lost its audience to rival channels like Star News and Zee News while private channels impressed the audience with high-production values and digital transmission. Unlike DD, the private players recognised the possibility of combining news with show-business and advertising (Butcher, 2003; Thussu, 2007; Chaudhuri, 2010). As elsewhere, Indian broadcasting space has been transformed by what Hallin and Mancini have described as the "triumph of the liberal model" (Hallin and Mancini, 2004: 251) of media, partly "because its global influence has been so great and because neo-liberalism and globalization continue to diffuse liberal media structures and ideas" (ibid: 305).

By 2015, more than 800 digital channels were operating, including some joint ventures with international broadcasters reaching more than 500 million TV viewers (Thomas, 2010; Kohli-Khandekar, 2013; FICCI-KPMG, 2015). This unprecedented growth has been accelerated by an increase in advertising revenue as Western-based media conglomerates tap into the new middle-class market of 300 million with enhanced and demonstrable purchasing power and media-induced aspirations to a consumerist lifestyle (Ganguly-Scrase and Scrase, 2008; FICCI-KPMG, 2015). While television news outlets have proliferated in such a liberalized and privatized new economy, the growing competition for audiences and, crucially, advertising revenue, has intensified. Not dissimilar to trends in the US, the growing commercialization of television news has forced broadcast journalists and television producers in India to recognise the need to make news entertaining. They borrow and adapt ideas from entertainment and adopt an informal style with an emphasis on personalities, storytelling and spectacle (Thussu, 2007; Bhusan, 2013; Jain, 2015). As a recent study notes, "the multitude of news outlets and of delivery platforms have come to exert increasing pressure to be the first—often accompanied by compromises in not only the diversity of original sources, but also the ethics and quality of news offerings" (Parthasarathi and Srinivas, 2012: 141).

As cross-media ownership rules are relaxed, there is a greater trend towards concentration of media power: non-media groups have invested heavily in television and telecommunication. Large corporations are increasing their presence across the different segments of the media and entertainment world, creating new media conglomerates, along the lines of the US model. One prominent example is Reliance, one of India's largest conglomerates, which since 2013 owns Network 18, the company that also operates CNN/IBN, one of India's leading English language news networks (Raman, 2014). In the US, such moves have been reinforced by the takeover of news networks by huge media corporations, whose primary interest is in the entertainment business: notable examples include Viacom-Paramount (CBS News); Disney (ABC News); Time-Warner (CNN) and News Corporation (Fox News). So, too, in India media conglomerates who make profit in the entertainment industry have investments in news networks. Such ownership structures can be reflected in the type of stories that receive prominence in news programmes—for example, about celebrities from the world of entertainment and sport—thus strengthening corporate synergies.

In the process, symbiotic relationships between the news and new forms of current affairs and factual entertainment genres, such as reality TV, have developed, blurring the boundaries between news, documentary and entertainment. Such hybrid programming feeds into and benefits from the 24/7 news cycle: providing a feast of visually arresting, emotionally-charged infotainment which sustains ratings and keeps production costs low. The growing global popularity of such infotainment-driven programming indicates the success of this formula (Thussu, 2007). In India, as one commentator has noted, entertainment and infotainment dominate the "ABC of Media—Advertising, Bollywood and Corporate Power" (Sainath, 2010).

NEWS AS INFOTAINMENT

In a fiercely competitive and crowded market such as India, news networks are under constant pressure to raise their Television Rating Points (TRPs) and maximise advertising revenue by acquiring new programming. There is a tendency to infotainment—to make news entertaining—which in the context of India means drawing on Bollywood or Bollywoodized content. This is now common practice on news networks which regularly broadcast "exclusive" stories about the supernatural and the bizarre as examples of compelling TV.

There is a noticeable change in style and content away from a considered, professional approach to a flashier and visually more dynamic presentation; the emphasis seems to be not on the journalistic skills of news anchors and reporters, but on how they look on camera, with style taking precedence over substance. An informal, entertaining schedule is created to increase the audience base; across the channels, ratings and revenue-delivering programmes—sports, entertainment and lifestyle—have increased, while news and analysis have shown a corresponding decline. Parthasarathi and Srinivas note "strong tendencies toward sameness in the themes and emphasis of news, even in vernacular languages—coupled with a near uniform preference for sensational treatment" (Parthasarathi and Srinivas, 2012: 142).

Is commodification of television news and its growing politicisation expanding or eroding the public discourse in the world's largest democracy? Given the symbolic and semiotic power of television news, it can be argued that infotainment-driven television news is contributing to making India a producer and consumer of commodity capitalism. Given the obsession with Bollywoodized content and its regionalized versions that characterize much of television content, the "public" aspects of news is being undermined (Rajagopal, 2009; Nayar, 2009; Thomas, 2010). The popularization of celebrity-driven and sensationalist news may have made it a more marketable commodity, but this has also debased public discourse. A recent study by two eminent economists notes:

> A lack of serious involvement in the diagnosis of significant injustices and inefficiencies in the economic and social lives of people; and also the absence of high-quality journalism, with some honourable exceptions, about what could enhance the deprived and constrained lives of many—often most—people in the country, even as the media presents a glittering picture of the privileged and the successful. (Drèze and Sen, 2013:7)

Most television channels ignore the issues of rural poverty or of developmental concerns in general, as they rarely translate into ratings or interest advertisers, on whose support the edifice of a commercial television news is based. It has been suggested that in a market-driven economy, the media system "is not only closely linked to the *ideological* dictates of the business-run society, it is also an integral element of the economy" (McChesney, 1999: 281, italics in original).

If excessive marketization has contributed to privileging spectacles of sport and celebrity, the shift from bureaucrats to marketing executives has also influenced the politics of television news. News is increasingly shrill, bipartisan and noisy. During Doordarshan's monopoly of broadcast news, news on television was considered little more than the government's view of the day's events, with only primary definers of news—mostly politicians and other elite groups—dominating the discourse. In terms of presentation and style, the news was boring and bureaucratic: audience interest did not matter, as there was no competition with private television. The new visibility of television news has brought new actors on to the national arena as well as influencing the actions of those being filmed. The way television news covered the conflict in Kargil in 1999—India's first televised war (Thussu, 2002)—and the communal violence in Gujarat in 2002, the first major riots of the 24-hour television age (Jain, 2010), is indicative of the power of visuals to shape the public agenda.

With television news mainly in private control, what happens to the public aspects of broadcasting in a country where, despite strong economic growth, more than 300 million people live in poverty—the largest number of poor people in a single country? This infotainment-driven television news has not yet reached large parts of rural and semi-urban India, where Doordarshan still reigns supreme, despite strong competition from private news networks, especially in cable and satellite homes. In 2015, Doordarshan was reaching more than 500 million viewers, while DD News, the first and the only terrestrial news channel in the country, had the highest reach into television households. Though not as bland as during its monopoly days, DD News still lacks the edge and critical dimension that at least some private channels have earned in the past decade of operation (Rao, 2010). As Cottle and Rai suggest, "In a diverse and plural polity such as India, the communicative structures of television news are particularly important in that they variously enable or disable the public elaboration of conflicting interests and identities" (2008: 77). The growing "linguistic and geographical multitude of news outlets," has contributed to an "absolute increase and widening of reportage on marginal concerns and marginalized people. Regional and local outlets, in particular, have consequently better addressed and amplified their immediate issues, which often went under- or un-addressed in the national media" (Parthasarathi and Srinivas, 2012: 141).

Networks such as NDTV 24x7 have arguably broadened the public discourse in India, bringing on board, for example, questions about environmental protection and right to information. The network has undertaken campaigns to promote particular causes and generated both revenues and more importantly awareness on issues such as protecting India's national animal, the tiger. Such instances may be characterised by what Cottle and Rai have called "the campaigning frame," which "declares the news outlet's stance on a particular issue or cause and typically seeks to galvanize sympathies and support for its intervention, political or otherwise, beyond the world of journalism" (Cottle and Rai, 2008: 83).

INDIAN TV NEWS IN THE GLOBAL MEDIA SPHERE

The growing profile of India on the global scene has been helped by the increasing visibility of its cultural and creative industries, its diaspora, and its media operating in a vibrant and expanding media sphere in one of the world's fastest growing economies (despite the global economic downturn, in 2014, India still posted an economic growth of 6%). Given the size and scale of the Indian television industry and the globalization of Indian businesses, the Indian version of news has potentially an audience base, beyond the diasporic one. "By enriching the content of the coverage and analyses of news," write Drèze and Sen, "the Indian media could certainly be turned into a major asset in the pursuit of justice, equity, and efficiency in democratic India" (Drèze and Sen, 2013: 7).

Unlike those in the Western world, the media and cultural industries in India are growing rapidly: in 2013, the Indian entertainment and media industry was worth $29 billion, with a steady annual growth rate. International investment is increasing in India's media sector, as cross-media ownership rules are relaxed (Thomas, 2010; Kohli-Khandekar, 2013; FICCI/KPMG Report, 2015). At the same time, Indian media companies are also investing outside national territories. In the past two decades, India has become an important source of media products, both indigenous, as well as a production base for transnational—largely US-based—media conglomerates (UNCTAD, 2010).

However, India's communication of its developments to a general global audience remains limited, given the US-UK domination of international news and a woeful lack of visibility of Indian news media in the broadcasting scene. This is ironic for a nation with a highly developed model of journalism and the increasing presence of Indian-born or Indian-origin journalists working for leading global news outlets. Unlike many other developing countries, India has a long tradition of politically engaged journalism, with its roots in the anti-colonial nationalist movement. An intellectual engagement with the wider world is a rich legacy of Indian journalism. Indian democracy has been underpinned by a journalism which has by and large delivered its Fourth Estate function.

India is also one of the world's largest English-language television news markets, as many of its news channels broadcast in English (Athique, 2012; Kohli-Khandekar, 2013; FICCI-KPMG, 2015). Some of these channels—especially those broadcasting in English—have a global reach and ambition. NDTV's flagship channel NDTV 24x7 was available in 2015 in the US (via DirecTV), the UK (BSkyB), the Middle East (Arab Digital Distribution) and southern Africa (Multi-choice Africa). Indian companies are in partnerships with global news players such as CNN-IBN, an English news and current affairs channel, launched in 2005, in association with TV-18 Group. The NDTV Group had strategic ties with NBC, while Times Now, owned by the *Times of India* Group (publisher of the *Times of India*, the world's largest English-language broadsheet daily newspaper in terms of circulation), collaborated with Reuters between 2006–2008 in a joint news operation. The growth of English-language journalism in India should open up possibilities for journalistic opportunities offered by the globalization of Indian media industries.

Paradoxically, Indian journalism and media in general is losing interest in the wider world at a time when Indian industry is increasingly globalizing and international engagement with India is growing from across the globe. As for news networks, NDTV 24x7 is the most widely watched internationally. The absence of Doordarshan in the global media sphere can be ascribed to bureaucratic apathy and inefficiency, though in an age of what Seib has called "real-time diplomacy," the need to take communication seriously has never been greater (Seib, 2012). For private news networks, the need for global expansion is limited, since, in market terms, news has a relatively small audience and therefore meagre advertising revenue.

However, perhaps taking inspiration from China, the Indian government has belatedly woken up to promoting its external broadcasting. An eight-member committee headed by Sam Pitroda, former Advisor to the Prime Minister of India on Public Information Infrastructure and Innovation has recommended that Prasar Bharati, India's Public Sector Broadcaster, should have a "global outreach" (Prasar Bharati, 2014). Its vision is ambitious:

> Create a world-class broadcasting service benchmarked with the best in the world using next-generation opportunities, technologies, business models and strategies. The platform should be designed for new media first and then extended to conventional TV. Outline an effective content strategy for Prasar Bharati's global platforms (TV and Radio) focused on projecting the national view rather than the narrow official viewpoint. (Prasar Bharati, 2014: 15)

Recommending professional and financial autonomy, with latest technological support, the committee suggested that such internationalization would contribute to India's soft power. The objective of this "global outreach strategy" should be to create a strong international presence, using all possible platforms and content to portray the story of emerging India and its vibrant democracy to the world: its cultural diplomacy and soft power and influence opinion about India. Uniquely Indian themes such as Yoga, Ayurveda or Bollywood are obvious areas, but information about Indian business successes and the richness and diversity of the country need prominence.

AN ALTERNATIVE GLOBAL VOICE?

As one of the founding members of the Non-Aligned Movement, India pursued a largely autonomous foreign policy. It was a leading voice during the 1970s' and 1980s' debates within UNESCO about the creation of a New World Information and Communication Order. India was a founding member of the Non-aligned News Agencies Pool, an attempt to encourage South-South news exchange to counter Western information hegemony. In the age of BRICS, coinciding with cracks within the neo-liberal model of US-led capitalism, there is now talk of Non-Alignment 2.0 (Khilnani et al., 2012).

Will the presence of Indian news on the international scene pose challenges to the Western-dominated news agendas? One area where a global Indian news presence could make a difference is in the field of development communication: it was the first country to use television for education through its 1970s Satellite Instructional Television Experiment (SITE) programme. India remains home to the world's largest population of poor people: on every major indices of social progress, it shows abysmally low ranking, despite demonstrating robust economic growth and lifting millions out of poverty in the past two decades. India and Indian media, therefore, have a moral and material imperative to be at the forefront of shaping discourse about how to deploy media and communication tools for poverty alleviation programmes internationally. They can draw on the legacy of Mahatma Gandhi's egalitarian journalism (the iconic leader of India's independence movement edited for most of his political life the weekly newspaper *Young India*, later renamed *Harijan*).

More broadly, Indian journalism evolved within the context of a fight for democracy in the tradition of anti-colonialism, represented by leaders like Gandhi. After independence, a "Third Worldist," anti-imperialist ideology continued to define mainstream media under Nehruvian socialism. Making use of this legacy of articulating the voice of the global South, it can make a contribution to international debates, beyond World Bank-dictated anti-poverty programmes. As digital media and communication become more commercialized, India's could be an important voice in articulating Southern viewpoints and perspectives in global forums like UNESCO, ITU and WIPO on such diverse and contested issues as multiculturalism, intellectual property rights in the digital environment, and the safeguarding of media plurality and indigenous media.

Apart from the globalization of Indian news media, the growing Indian presence within the international non-governmental sector, multilateral bureaucracies, and the development communication field could be harnessed to this end (Tharoor, 2012; Thussu, 2013). Would an Indian media perspective on events in other developing countries be less affected by the colonial mindset? The proliferation of news networks and the multiplicity of other media outlets have contributed to the freedom from government control and arguably democratized public communication. In such a news environment, citizens have access to a wider range of information and journalists can help to give voice to the disadvantaged and seek accountability from politicians and bureaucrats (Rao, 2010; Roy, 2012). From electoral politics to economy, to development issues, television news has a major role in shaping public opinion. Networks like NDTV 24X7 have taken up causes in the public interest—such as rural development, environmental protection, freedom of information, gender equality. Their efforts have at times influenced government policy. There is also a strong and growing tradition of investigative journalism to expose social and political misdemeanours, corruption and criminality.

However, such public-interest journalism is confined to a few notable exceptions. Most of television news is entrenched in a "Bollywoodized" media culture, thriving on entertainment and infotainment-driven programming rooted in a crassly commercial media system (Thussu, 2007; Jain, 2015). By overwhelming public discourse with Bollywoodized content, egalitarian aspects are marginalized in the news media, at a time when more than 300 million people in India remain illiterate and the gap between the rich and the poor is growing, making India one of the world's most unequal societies, despite a democratic political system and impressive economic growth.

The issues that confront the Indian situation—about governance, sustainable development, and pervasive poverty—have striking resonances in many other countries in the global South (Tharoor, 2012; Thussu, 2013). Despite India's gradual integration with the US-led neo-liberal economic system, as a producer and a consumer of commodity capitalism, there is a strong and deeply-entrenched tradition of argumentation and critical conversation in the Indian body politic and in its intellectual life, reflected also in journalistic discourses (Sen, 2005; Rajagopal, 2009). With a well-established tradition of English-language journalism—the vehicle for transnational communication and commerce—Indian journalists are well able to operate in a global media sphere and play a significant role in leading international news outlets. As Indian media globalize, will this critical mass contribute to strengthening the voice of those at the receiving end of the excesses of neo-liberalism, or will India, under a pro-business government of Narendra Modi, play second fiddle to the US media, acting as surrogate to the US-dominated, entertainment-driven news?

REFERENCES

Athique, Adrian (2012) *Indian Media: Global Approaches*. Cambridge: Polity.

Bhusan, Bharat (2013) 'News in Monochrome: Journalism in India', *Index on Censorship*, Vol. 42 (36): 37–42.

Butcher, Melissa (2003) *Transnational Television, Cultural Identity and Change: When STAR Came to India*. New Delhi: Sage.

Chaudhuri, Maitrayee (2010) 'Indian Media and its Transformed Public', *Contributions to Indian Sociology*, Vol. 44(1–2): 57–78.

Cottle, Simon and Rai, Mugdha (2008) 'Television News in India: Mediating Democracy and Difference', *International Communication Gazette*, Vol. 70(1): 76–96.

Cushion, Stephen and Lewis, Justin (eds.) (2010) *The Rise of 24-Hour News Television: Global Perspectives*. New York: Peter Lang.

Drèze, Jean and Sen, Amartya (2013) *An Uncertain Glory: India and its Contradictions*. Princeton, NJ: Princeton University Press.

FICCI-KPMG (2015) *The Stage is Set—FICCI-KPMG Indian Media and Entertainment Industry Report 2015*. Mumbai: KPMG in association with Federation of Indian Chambers of Commerce and Industry.

Ganguly-Scrase, Ruchira and Scrase, Timothy (2008) *Globalization and the Middle Classes in India: The Social and Cultural Impact of Neoliberal Reforms*. London: Routledge.

Hallin, Daniel and Mancini, Paolo (2004) *Comparing Media Systems: Three Models of Media and Politics*. Cambridge: Cambridge University Press.

Jain, Anuja (2010) '"Beaming It Live": 24-hour Television News, the Spectator and the Spectacle of the 2002 Gujarat Carnage', *South Asian Popular Culture*, Vol. 8(2): 163–179.

Jain, Savyasaachi (2015) 'India: Multiple Media Explosions', pp. 145–165, in Nordenstreng, Kaarle and Thussu, Daya Kishan (eds.) *Mapping BRICS Media*. London: Routledge.

Kapur, Davesh (2010) *Diaspora, Development, and Democracy: The Domestic Impact of International Migration from India*. Princeton, NJ: Princeton University Press.

Khilnani, Sunil; Kumar, Rajiv; Mehta, Pratap Bhanu; Menon, Prakash; Nilekani, Nandan; Raghavan, Srinath; Saran, Shyam and Varadarajan, Siddharth (2012) *Nonalignment 2.0: A Foreign and Strategic Policy for India in the Twenty-First Century*, New Delhi: National Defence College and Centre for Policy Research.

Kohli-Khandekar, Vanita (2013) *The Indian Media Business*. Fourth edition. New Delhi: Sage.

McChesney, Robert (1999) *Rich Media, Poor Democracy–Communication Politics in Dubious Times*. Champaign, IL: University of Illinois Press.

Mehta, Nalin (2008) *India on Television: How Satellite News Channels Have Changed the Way We Think and Act*. New Delhi: HarperCollins.

Nayar, Pramod (2009) *Seeing Stars: Spectacle, Society and Celebrity Culture*. New Delhi: Sage.

Nordenstreng, Kaarle and Thussu, Daya Kishan (eds.) (2015) *Mapping BRICS Media*. London: Routledge.

Parthasarathi, Vibodh and Srinivas, Alam (2012) *Mapping Digital Media: India*. A Report by the Open Society Foundations. London: Open Society Foundation.

Prasar Bharati (2014) *Report of the Expert Committee on Prasar Bharati. Vol. I and II*. New Delhi: Government of India: Prasar Bharati.

Rajagopal, Arvind (ed.) (2009) *The Indian Public Sphere: Readings in Media History*. New York: Oxford University Press.

Raman, Anuradha (2014) 'Big ED in the Chair: Reliance's Takeover of Media Group TV18 Raises Serious Concerns of Credibility, Press Freedom', *Outlook*, 14 July.

Rao, Ursula (2010) *News as Culture: Journalistic Practice and the Remaking of Indian Leadership Tradition*. New York: Berghahn Books.

Roy, Srirupa (2011) 'Television News and Democratic Change in India', *Media, Culture & Society*, Vol. 33(5): 761–777.

Sainath, Palagummi (2010) 'How to Feed Your Billionaires', *The Hindu*, April 17.

Schaefer, David and Karan, Kavita (eds.) (2013) *Bollywood and Globalization: The Global Power of Popular Hindi Cinema*. London: Routledge.

Seib, Philip (2012) *Real-Time Diplomacy: Politics and Power in the Social Media Era*. New York: Palgrave Macmillan.

Sen, Amartya (2005) *The Argumentative Indian*. London: Penguin.

Tharoor, Shashi (2012) *Pax Indica: India and the World of the Twenty-first Century*. New Delhi: Penguin.

Thomas, Pradip (2010) *The Political Economy of Communications in India: The Good, the Bad and the Ugly*. New Delhi: Sage.

Thussu, Daya Kishan (2002) 'Managing the Media in an Era of Round-the-Clock News: Notes from India's First Tele-war', *Journalism Studies*, Vol. 3(2): 203–212.

Thussu, Daya Kishan (2007) *News as Entertainment: The Rise of Global Infotainment*. London: Sage.

Thussu, Daya Kishan (2013) *Communicating India's Soft Power: Buddha to Bollywood*. New York: Palgrave/Macmillan.

Thussu, Daya Kishan (2016) *International Communication - Continuity and Change*. Third edition. London: Bloomsbury Academic.

UNCTAD (2010) *Creative Economy Report 2010*. Geneva: United Nations Conference on Trade and Development.

CHAPTER TWENTY-FOUR

CCTV 24-Hour Chinese-Language News: From Offline to Online

YUNYA SONG, YIN LU, TSAN-KUO CHANG

24-hour television news is still a relatively new phenomena rooted in American journalism that has influenced the world's media landscape for many decades. The US domination had been considered unbreakable by some scholars. In 1977, in a rather pessimistic, but perceptive, prediction, Tunstall (1977: 63; emphasis added) said that "a non-American way out of the media box is difficult to discover because it is an American, or Anglo-American, built box. *The only way out is to construct a new box*, and this, with the possible exception of the Chinese, no nation seems keen to do." Tunstall was certainly right to anticipate the rise of a Chinese built TV box—China Central Television (CCTV)—but missed the mark when, to name just a few, Al Jazeera (1996), CCTV-9 (2000), Russia Today (2005), France 24 (2006), Iran's Press TV (2007), Japan's NHK World TV (2009), and Venezuela's TeleSur (2010) have joined the chorus of global TV news since the mid-1990s. Against the backdrop of growing attention to China's rise on the global stage, the institutional transformation of CCTV is particularly significant, both internally and externally. The Chinese government has been developing a systemic approach to establishing regular channels for the projection of soft power, and CCTV is at the heart of this grand initiative.

CCTV launched its 24-hour Chinese-language news channel, CCTV-13 (formerly named CCTV News), on July 1, 2003. An inherent difference exists in the programming between CCTV-13 and the 24-hour English-language television channel CCTV News, formerly known as CCTV-9. While the English news channel caters to English-speaking overseas audience with only 35% first-hand news (CNTV, 2010, December 15), CCTV-13 is the premier source of news and views for viewers from across the Greater China region and that of the Chinese Diaspora, offering them a real sense of roll-up news in China. Within the first two months of its broadcast, the time length of CCTV-13's live broadcasting had already exceeded that of all CCTV channels for the whole of the year 2002 (Zhou, 2007). The rationality of such technical advances lies in its institutional mission and ambition.

CCTV has been the predominant state television broadcaster in mainland China since its inception in 1958. Over the years, CCTV has sought to become the respected source for Chinese news and culture, but is often dismissed as the Chinese government's mouthpiece for propaganda (Hong, Lu & Zou, 2009). Obviously, the state-run CCTV consistently operates as a vehicle for ideological indoctrination, but the game has changed due to globalization and marketization. It has to compete not only with myriad local satellite channels for domestic audiences, but also with foreign channels and programming for international influence. Since the decentralization of Chinese television networks in the 1980s, CCTV has made continuous efforts to reconfigure operational models and diversify the programming in its competition for audience shares (Zhu, 2012). The changing status of CCTV can be best seen in its window on the world. Song and Chang's (2015) longitudinal observation of international news production from 1992 to 2006 pointed towards a remarkable increase both in the timeliness of news and in the amount of soft news. A comparative study of CCTV-4 and Phoenix TV (a Hong Kong-based satellite channel founded in 1996 to provide information and entertainment to Chinese audiences worldwide) found that their news diet switched towards more negative- and fewer positive-tone stories—and the storytelling approach that involved more open-ended interactions was gaining popularity (Wu & Ng, 2011).

However, CCTV's national monopoly has not been shaken because of the official supervision system. Xu's (2013a) analysis of market-oriented television policies in post-WTO China suggested that private and foreign media capital had to seek cooperation with state media players to survive in China's broadcasting market. It has been obligatory for local television stations to carry CCTV's programs, ensuring the state broadcaster's nationwide reach (Lynch, 1999). CCTV is also accorded exclusive coverage rights to major domestic and international events, and has privileged access to central officials, providing inside information and insights on China's political scene (Zhu, 2012). These administrative privileges have granted CCTV unique news resources that allow it to outperform its competitors in an increasingly commercialized media landscape both at home and abroad.

This chapter will demonstrate how the launching and development of CCTV-13 relate to the accommodations and tensions between political control and the market imperative within the broad political economy of state-regulated market expansion in the Chinese media system, focusing on the interplay between the external disturbances and its institutional responses. Similar to other countries, China's broadcasters are also feeling the impact of the growing digital ecosystem on the media landscape. A multitude of digital news and information content across a variety of platforms (e.g., microblogging services and news portals) has become available to the audiences in China. The new media environment challenges traditional television broadcasters who are online to improve their broadcasting and move towards multimedia content and distribution. The emergence of media that cross multiple domains, their subsequent impact on citizen participation and the burgeoning networked public sphere and, along with other media influences in society, have increasingly led to CCTV's adoption of goal-oriented institutional structures and professional practices to cope with a competitive market economy under strict state control.

Using the 24-hour Chinese news channel in China as the locus of analysis, this chapter will put into perspective CCTV's changing programming strategies and recent migration towards multiplatform distribution. Given the highly restrictive nature of the media environment in China, the operation of news channels has been subjected to the risks of policy variability and uncertainty (Zhu, 2012). CCTV-13's multiplatform development may reveal CCTV's strategic attempts to transform the challenges from digital era into its business and political leverage. Its path goes through different stages, with each displaying specific external contests and institutional reactions accordingly.

THE EMBRYONIC STAGE: RESPONDING TO STATE AND MARKET IMPERATIVES

CCTV-13 was established in response to complaints about the Chinese media cover-up of the SARS epidemic in 2003. As a news outlet, such a dramatic institutional response has been well documented in the literature (Zhang, 2007). Audiences have a stronger need for live news broadcasting when there is a natural or man-made crisis. The handling of the epidemic by the Chinese government initiated an intense debate about the role of government and media in a public crisis, which was uncommon in a society that generally discourages open discussion of sensitive subjects and reflected the public's demand for more news and information (Zhu, 2012). CCTV-13 started out as a platform to guide public opinion and disseminate messages from the party and the state. With the slogan "first time, first scene, first need," it aimed to keep up with the world and the times. In its early period, CCTV-13 reported a series of landmark events that demonstrated the growing overall national strength of China, including the human spaceflight missions of Shenzhou-5 and Shenzhou-6, the visits of Taiwanese politicians to mainland China, and the fight against natural phenomena such as typhoons and tsunamis.

For 24-hour rolling news, the channel flow is usually marked by highly structured scheduling such as the hourly cycle. CCTV-13 made its first sortie by developing a programming strategy of broadcasting features and relevant news at fixed time slots (Xie & Jia, 2011). It broadcast 24 rolling news programs of 10 or 30 minutes a day, though the information in its early hourly news programs updated at a slow pace. The rest of the time slots were filled with feature programs and their reruns. The fixed broadcast programming and specialized content could hardly meet the audience demand for a diversity of information. Notably, due to the prearranged division of the 24-hour timeline, the channel could hardly report breaking news in a timely manner. As such, it could hardly fulfill the promise of immediacy—the news "as it happens"— held out by a mature 24-hour news channel. The challenge for rolling news is to accommodate the linear real-world happenings to its cyclical format, ensuring that reporting sticks as closely as possible with the "present" moment.

In 2004, CCTV-13 closed programs with poor ratings and adjusted the airtime of such programs as *Social Record* and *International Review* in order to attract viewers. Meanwhile, the length of hourly news was extended, and similar news feature programs were arranged together in a vertical order. Such arrangements in line with the linear schedule were intended to keep the primetime viewing audience for fringe-time programming (Zhou, 2007). The channel tried to draw the audience to in-depth opinion programming and to cultivate audience loyalty. However, the programming had limitations. Evening features programs were rerun in the morning and afternoon. Top headlines from home and abroad could not update in time across the day. Because of the long production cycle, some evening programs were lacking in timeliness.

Overall, in the early years CCTV-13 was a news channel with a number of non-news programs instead of live news reporting. Its programming was highly predictable as was its form. The live reporting on the channel was mainly confined to planned events rather than breaking news. The reason for the lack of live broadcast is probably more ideological than technical, in that spontaneous coverage is difficult to manage and control and may thus cause unexpected consequences, should something go wrong (Chang, Wang & Chen, 2002).

THE FORMATIVE STAGE: OPTIMIZING THE RESOURCES FOR LINEAR PROGRAMMING

2008 is a landmark year for the live broadcasting of CCTV's news department. The live reports increased dramatically as the biggest snowstorm since 1949 hit southern China. When another major event—the Wenchuan earthquake—struck

Sichuan province on May 12, the government required all Chinese news organizations to use the information relayed by CCTV and Xinhua News Agency, but not to send reporters to the scene at the start (Zhu, 2012). CCTV reacted quickly by setting up four reporting stations on the frontline and dispatching nearly 30 teams that covered the hardest hit areas (Wang, 2008). Given these privileges, CCTV-13 provided exclusive live broadcasts that allowed audiences to catch up on what had been happening on the frontline of relief efforts, earning the Chinese government rare praise from international observers. In addition to making a virtue of its ability to prioritize "showing," CCTV-13 was also applauded for its adoption of the format—"special reports"—that provided in-depth interpretation (Wang, 2008). The emphasis on pictures in live broadcasting often risks lack of hard information. While TV news programs are losing ground in timeliness in their competition with online media, how to reinterpret and reorganize the information becomes a major factor for the channel to gain competitiveness. CCTV-13 tried to find the solution within the linear structure by organizing news events into more well-crafted, coherent stories.

After the Wenchuan earthquake, the channel launched a facelift of its live broadcasting system for breaking news. During the Beijing Olympic Games, it broadcast major events and offered a wide range of Olympic programs so that the audience gained timely information. After the live broadcast of the Olympic Games and the Shenzhou-7 spacecraft launch, CCTV-13 drew on its previous experience of reporting the Wenchuan earthquake and set two time slots for live news: from 8:30 a.m. to 11:30 a.m. and from 1:00 p.m. to 5:00 p.m. in order to deal with breaking news and updates, with a view to delivering the latest news (Zhang & Zou, 2010). These changes are compelling evidence as to how news organizations may rise to the external challenges that require adjustments of institutional practices.

Shortly after the new president took over CCTV in 2009, CCTV-13 was made available in basic television service and no longer required cable subscription (Tan, 2009). The change, which was warmly received by CCTV-13's personnel, allowed many Chinese viewers open access for the first time, and boosted its domestic audience share to 12.5% (Tan, 2009). By acknowledging the news function as a core mission at CCTV, the new president took the pressure of ratings out of the news programming, in a sense bucking two decades of movement toward commercialization (Zhu, 2012). To some extent, CCTV has begun to live up to the expectation of a national TV news network, the size of whose captive audience would be an envy of foreign TV channels.

To enhance CCTV's authority in news coverage, a series of measures were taken to pursue more timely and comprehensive coverage. The biggest makeover was the establishment of a brand new flagship News Center in July 2009, which has continued right up until the present day. The news-making procedure in CCTV used to be criticized for its waste of manpower resources. The launching of the News Center was intended to get rid of the initial mechanism organized around programs, departments and centers by merging all channels' news-gathering and editing resources. It works under a macro-management strategy, with one general editorial department planning topics and coordinating the arrangements of production, editing, and broadcasting (Liang, 2012). Each division has its own sub-desk, and sends representatives to a daily meeting to coordinate its work with other divisions and make detailed reporting arrangements. Efficient, parsimonious production process is associated with an organization's adaptability and flexibility. The operation model of the News Center has greatly reduced redundancies and better leverages shared resources among different channels.

Meanwhile, CCTV made significant breakthroughs in realigning its news-gathering network and information resources (Liang, 2009). Domestically, CCTV first set up emergency reporting sites in eight cities (e.g. Shanghai, Chengdu, and Xi'an) in 2008 and continued its rapid expansion to another ten cities (e.g. Chongqing, Nanjing and Kunming). CCTV's overseas journalists are also under the direct management of News Center, spreading over 19 countries as of 2009. Extensive news-gathering networks at home and abroad make CCTV News Center a formidable news presence. Meanwhile, an information platform for pooling domestic and foreign news resources also took shape. CCTV established an information transmission system with 57 provincial TV stations and 92 news programs on municipal TV channels, as well as an information-sharing mechanism with Xinhua's TV service, Taiwan's ETTV, Hong Kong's TVB, and others (Liang, 2009).

The professionalized organizational structure and international operations have greatly improved the quality of CCTV-13's broadcasting. The previous time slot for live broadcast officially changed its name to *Live News* on August 17, 2009. This program adapted the format of CNN's Newsroom (Ouyang & Wang, 2013). Initially *Live News* was broadcast around the clock, and then the hourly news broadcasts from 1 a.m. to 6 a.m. were taken off. Beginning July 4, 2011, however, *Live News* went on air again between 1 a.m. and 6 a.m. such that 24-hour rolling news was back on track. The official explanation for the restoration of 24-hour news is that late night shows mostly broadcast international news, with a view to building the image of CCTV as an international broadcaster (Fang, 2011). The dynamics of the Chinese state's effort to develop its national image and push its soft power is not only manifest in the medium of English news; CCTV's programming strategies for domestic audiences also resonate with its global ambitions. Up to this point, CCTV-13 has built a mature mechanism comprising a 24-hour news cycle with highly integrated news resources and an information platform, providing regular live-stream and on-the-spot reporting. All sorts of news, ranging from culture, sports, and current affairs to people's livelihoods, have been made available in each time slot and audiences can gain information about their own interests.

THE INTEGRATIVE STAGE: CONVERGENCE AND BROADCASTING TRENDS IN PERSPECTIVE

Among the substantial changes of television in digital media environments is viewers' increasing independence from linear TV programs. The dramatic shift toward online and mobile viewing in China is driving broadcast networks to share viewer attention with thousands of online services. This converged news operation has evolved into a centralized coping strategy in an attempt to make the content across different platforms stand out with a signature brand look. Furthermore, the strategic convergence has allowed the one-to-many model of traditional mass media to incorporate non-linear channels of one-to-one and many-to-many communication on the one hand and to give balanced attention to various media types on the other hand (Cordeiro, 2012). This is currently framed by a set of interrelated elements—branding, information management, and a professional complex system of production and distribution. Multiplatform expansion contributes to the branding efforts of old media organizations both in terms of extended opportunities for content consumption and the potential to develop more effective content performance management based on more refined interactive modalities between suppliers and audiences.

In recent years, CCTV-13 has been striving to take a more web-based and multimedia-oriented initiative. Younger employees, who had grown up with the internet, played an important role in the reform. For example in 2010, as many as seven staff under 35 years old were promoted to executive positions (Wen, 2010, August 30). The main tactic chosen by CCTV has been to deliver content across online and mobile platforms. The country's first national internet TV—China Network Television (CNTV)—was launched by CCTV at the end of 2009, carrying the live streaming of CCTV-13. CCTV-13 has also been actively engaged with various social media platforms such as Weibo, WeChat and mobile apps for different purposes. Although CCTV-13 does not have its own official account in these platforms, it has been able to interact with audiences through the official account of CCTV News Center.

CCTV News Center established its Weibo presence for the timely release of big breaking news. @CCTV News, the official microblog of CCTV News Center, made its public debut on Sina Weibo on November 1, 2012. At the end of 2012, when President Xi visited the poor residents of Hebei province, CCTV reported the news first on its Weibo platform instead of the news channel. This and some other relevant tweets were retweeted more than 5,000 times, and aroused wide attention at home and aboard (Xu, 2013b). Since then, it has gradually become the norm for CCTV to report major events relating to party leaders on its Weibo platform first (Liang, 2013). The Weibo account of CCTV News Center broke a great many domestic and international major news stories, such as the successful take-off and landing of aircraft on China's first aircraft carrier, Bo Xilai's court hearing, and the Beijing-Tibet Railway crash (Yang, 2014a).

However, speed is often not the only factor for news reporting. While @ CCTV News is usually the first to release a breaking news story, further interpretation and discussion would be put up on CCTV-13's television programs and other platforms. In order to meet the needs of timeliness demanded by the Weibo format, television journalists and editors at CCTV took on the responsibility of providing original news for Weibo. More than 70% of messages published on the official Weibo platform of CCTV News Center were self-gathered news (Liang, 2013). By July 19, 2013, @CCTV News had more than 6.8 million followers, ranking it the third most influential Weibo account among all forms of traditional media and the most influential Weibo account among television channels (Liang, 2013). Besides Sina Weibo, CCTV News Center also created a microblog account on Tencent and CNTV as well. These three accounts update the same news at the same time, further expanding the influence of CCTV-13.

CCTV also moved promptly to jump on the WeChat wagon as early as April of 2013. CCTV News is the first media account to attain a record-breaking follower count of 1 million (Zhang, 2014). In contrast to the many-to-many communication mode of Weibo, WeChat delivers information in the one-to-many mode in that it enables news story recommendations to be sent to personal mobile phones with a much higher viewing rate. Hou's (2013) study suggested that breaking news, hot issues, and program promotion were the three types of content that were most likely to be recommended on CCTV's WeChat. Furthermore, the audio function on WeChat enables creation of user-generated content that is personal and interactive. To build audio appeal with audiences, CCTV-13 encouraged their hosts to recommend programs or speak about hot topics in their own voices. An editor of CCTV News Center revealed that those celebrity hosts' monologues inspired a burst of online chatter and greatly boosted audience satisfaction with CCTV news (Hou, 2013).

Opening official accounts on Weibo and WeChat has changed the way news is reported. For one thing, these platforms make disaster reporting more effective and compelling. An exemplary case would be the CCTV News Center's strategy of reporting the 2013 Lushan earthquake (Liang, 2013). Seven minutes after the earthquake struck Lushan, CCTV News Center published the story first on Weibo. Then its WeChat account recommended the story to users, and introduced live broadcasting of relief operations that afternoon, thus breaking the barrier between the medium of television and the medium of the mobile phone. Another exemplary case was the coverage of the 2013 Asiana Airlines crash in which CCTV called on its audiences to leave a reply on WeChat if their friends or family members were on this flight (Hou, 2013). This kind of journalistic practice has incorporated active user participation in the process of live broadcasting. When the traditional TV reporter cannot arrive at the scene, information released by Weibo or WeChat helps meet the public's information needs and also eases public disquiet.

Meanwhile, CCTV's online expansion project also involves the adoption of an app from 2012. On May 1, 2013, the 10th anniversary of CCTV-13, when CCTV announced its news app in cooperation with China's internet giant Sohu, the number of subscribers reached 290,000, breaking the one-day subscription record of Sohu News app (Zhang, 2014). As an extension of television news, CCTV News Center's applications on Weibo, WeChat, and apps are closely connected with its television offerings. By October 3, 2014, there had been more than 90 million users of CCTV News on Weibo, WeChat and apps in total, which means the number of people receiving news from these three platforms has already exceeded the number of people watching *News Simulcast* (Yang, 2014b).

Notably, CCTV's strict editorial control extends into these web platforms (Liang, 2013). CCTV News Center has introduced *Regulations Governing the Operation of Weibo, WeChat and App,* defining the responsibilities of conveying the editorial's standpoint on hot issues and avoiding hyping up sensitive issues. A dual verification system has been imposed to check the accuracy of the facts and the identity of publishers. From the selection of topics to publishing, @CCTV News imposes a three-level review system—editors' preliminary scrutiny, producers' finalization, and proofreading. The reporting of major news stories and sensitive subjects must pass the "gate" of the director on duty.

In addition to the expansion of communication channels, CCTV News Center is assigning teams of reporters and coders to pursue more data-centric stories. A notable example is CCTV's use of big data in news reports. This novel reporting tool is able to satisfy viewer demand for comprehensive coverage of society. During the 2014 Spring Festival holiday period, CCTV utilized data from the search engine Baidu to show domestic migration patterns of people travelling during the holiday, and audiences were able to obtain the latest transportation information during the Spring Festival. After CCTV-13's news program broadcast this segment, the audiences responded warmly on personal social networking sites and the mainstream media hailed it as a landmark in Chinese journalism (Liu, Li, & Zhan, 2014). For the following Qingming Festival, CCTV-13 used the thermodynamic chart function of Baidu Maps, analyzing Baidu LBS big data to fully display the dynamics of population densities in hundreds of cities and thousands of attractions and trade areas. This demonstrated to the Chinese journalism community the value of using Baidu big data to estimate social trends in their reporting. Another notable attempt made by CCTV-13 was the adoption of online opinion mining and visualization techniques in the reporting of China's annual parliamentary sessions, which won acclaim for its increasing emphasis on public concerns and popularization of the big data concept (Chen, 2014, April 11). A CCTV news editor revealed that big data, such as that shown in the Baidu Index, endowed CCTV's reporting with exclusive statistics and reports, which changes news reporting's excessive reliance on experts (Transparent snow ball, 2014, March 27). However, compared to that of Western countries, CCTV's data journalism remains at the initial stage (Cheng & Hu, 2014).

SUMMARY

With the transition from a striped programming structure to a linear one, the CCTV rolling news has successfully fought suggested changes through the necessary stages to achieve its triple goals of political mandate, public service and profit seeking. The formation of a converged, networked media ecosystem has reshaped CCTV as the state institution and its practices of content production, including an emergent role for the audience communities to distribute the content. As such, multiplatform delivery, social curation and real-time engagement with the virtual audience have become the norm in the traditional broadcaster's newsroom. At the core of our analysis is the capacity of these multiple platforms to drive tune-in and extend traditional viewing experiences beyond the television screen. Where television once stood alone autonomously, the flow of content across a range of media and devices (tablet, mobile phone) has linked traditional TV viewing with booming digital consumption. This unravels how converged media are both a reality and an opportunity for more efficient workflows and better integration of media specializations. Our analysis has suggested that the institutional adaptation to the converged news operation requires not only re-versioning of content, but also translating to staffing needs and organizational set-up. CCTV-13's multiplatform approaches go beyond simple cloning (Dailey, Demo & Spillman, 2005), but are explicitly structured as collections of cross-media brands in response to the fast-changing market environment. Differentiation in the area of programming according to the distribution channels is important in order to attract audience attention and to remain competitive.

There is little doubt that the CCTV 24-hour news today is multimedia, multiplatform and convergent in the production and distribution of content. All these, as argued in this chapter, are strengthening the state TV's capacity to leverage the momentum of social conversations and to create feelings of community among audience members, feelings which can only increase CCTV News' brand awareness. However, as our study suggests, with a presence on the internet, CCTV-13 in most cases offers its audiences limited participating possibilities. While CCTV has kept re-balancing its top-down engagement in response to the bottom-up audience demand, the converged newsrooms remain anchored around a centralized superdesk system, with a view to ensuring the effective functioning of its role and responsibilities as a state broadcaster. As it is, CCTV's news practice has yet to be institutionalized in ways that will allow it to become more competitive at the global level.

REFERENCES

Chang, T. K., Wang, J., & Chen, Y. (2002). *China's window on the world: TV news, social knowledge and international spectacles*. Cresskill, NJ: Hampton.

Chen, C. (2014, April 11). 'The coming age of big data television news'. Retrieved from http://tech.caijing.com.cn/2014-04-11/114088088.html

Cheng, L., & Hu, L. (2014). Statistic journalism: The new journalism model in the age of big data. *Today's Mass Media*, Vol. 11: 124–126.

CNTV (2010, December 15). 'A new revision of CCTV English news channel in November'. Retrieved from http://1118.cctv.com/20101215/107704.shtml

Cordeiro, P. (2012). 'Radio becoming r@dio: Convergence, interactivity and broadcasting trends in perspective', *Participations*, Vol. 9(2): 492–510.

Dailey L., Demo, L., & Spillman, M. (2005). 'The convergence continuum: A model for studying collaboration between media newsrooms', *Atlantic Journal of Communication*, Vol. 13(3): 150–168.

Fang, F. (2011, May 7). 'The restart of 24-hour live broadcasting in CCTV news channel'. *Law and Order Evening*. Retrieved from http://www.fawan.com/Article/yl/zy/2011/05/07/114536114908.html

Hong, J., Lu, Y., & Zou, W. (2009). 'CCTV in the reform years: A new model for China's television?' In Y. Zhu & C. Berry (Eds.), *TV China* (pp. 40–55). Bloomington, IN: Indiana University Press.

Hou, H. (2013). 'The interaction research between CCTV News Weibo and WeChat', *China Newspaper Industry*, Vol. 20: 48–49.

Liang, J. (2009). 'Strengthen the key position of news reporting in television station, and raise the communication ability'. Retrieved from http://ad.cctv.com/special/news/20090918/102456.shtml

Liang, J. (2012). 'The efficiency desk: The general situation of command system in CCTV news channel'. *TV Research*, Vol. 8: 27–28.

Liang, J. (2013). '@CCTV News: The new media pioneer of the network integration', *News and Writing*, 2013, Vol. 8: 9–12.

Liu, H., Li, Y., & Zhan, J. (2014). '2014 China television list', *Youth Journalist*, Vol. 36: 12–15.

Lynch, D. C. (1999). *After the propaganda state: Media, politics, and 'thought work' in reformed China*. Palo Alto, CA: Stanford University Press.

Ouyang, Z., & Wang, X. (2013). 'The editing art of rolling news in CCTV news channel', *China Radio & TV Academic Journal*, Vol. 9: 24–26.

Song, Y., & Chang, T. K. (2014). 'A new world of spectacle in the post-cold war era: China's central television and its significant other, 1992–2006, *Public Relations Review*. Retrieved from http://www.sciencedirect.com/science/article/pii/S0363811114001131

Tan, L. (2009). 'The reform of CCTV', *Today's Mass Media*, Vol. 9: 46–51

Transparent snow ball (2014, March 27). Baidu employee index raised news awareness. Message posted to http://tieba.baidu.com/p/2947121855

Tunstall, J. (1977). *The media are American: Anglo-American media in the world*. New York, NY: Columbia University Press.

Wang, X. (2008). 'Race against time', *Chinese Journalist*, Vol. 7: 24–25.

Wen, J. (2010, August 30). Establish a personnel system based on ability competition: An interview of the head of CCTV personnel office. Retrieved from http://blog.sina.com.cn/s/blog_4a15d7660100kyyv.html

Wu, D. D., & Ng, P. (2011). 'Becoming global, remaining local: The discourses of international news reporting by CCTV-4 and Phoenix TV Hong Kong', *Critical Arts: South-North Cultural and Media Studies*, Vol. 25 (1): 73–87.

Xie, H., & Jia, J. (2011). 'Interpret the development of frame of CCTV news programs', *Voice of Screen World*, Vol. 11: 23–24.

Xu, M. (2013a). 'Television reform in the era of globalization: New trends and patterns in post-WTO China', *Telematics and Informatics*, Vol. 30(4): 370–380.

Xu, P. (2013b). 'The counterattack of television media in the age of Weibo: The operation strategy of official Weibo of CCTV News', *Youth Journalist*, 12, 10–12.

Yang, J. (2014a). What to tell through new media. *Chinese Journalist*, 1, 41–42.

Yang, J. (2014b). 'We are waiting for you to embrace the new media'. Retrieved from http://m.news.cntv.cn/2014/11/04/ARTI1415108967005816.shtml

Zhang, X. (2007). 'Breaking news, media coverage and "citizen's right to know"', *Journal of Contemporary China*, Vol. 16: 535–545.

Zhang, Z. (2014). 'The new media strategy and series of creation of CCTV News', *Chinese Journalist*, Vol. 10: 24–27

Zhang, J., & Zou, D. (2010). 'The developing situation, problems, and strategies of the news channel in China', *Journalism Review*, Vol. 6: 62–65.

Zhou, Y. (2007). 'The revision path of CCTV news channel', *Cover and Edit News*, Vol. 4, 18–19.

Zhu, Y. (2012). *Two billion eyes: The story of China Central Television*. New York, NY: The New Press.

Contributors

Mary Angela Bock is a former television journalist who joined the faculty at the University of Texas at Austin in 2012. Her most recent project (with co-authors Shahira Fahmy and Wayne Wanta) is *Visual Communication Theory and Research: A Mass Communication Perspective* (2014, Palgrave). Bock is also the author of *Video Journalism: Beyond the One-Man Band*, and co-editor of *The Content Analysis Reader* with Klaus Krippendorff.

Michael Bromley is interim head of the Department of Journalism at City University London. He was previously head of the School of Journalism and Communication at the University of Queensland, and has also taught at universities in the US. A former journalist, he has published widely on journalism and the news media, and contributed to the first edition of this book. His research interests include journalism education, the practices of journalism and the socio-cultural uses of journalism.

Tsan-kuo Chang is Professor at City University of Hong Kong and Professor Emeritus at University of Minnesota-Twin Cities. He earned his PhD from the University of Texas at Austin and was the recipient of AEJMC's Outstanding Contributions Award in international communication research in 2005. He has published five books and his articles appear in such journals as *Communication Research, International Communication Gazette, International Journal of Press/Politics, International Journal of Public Opinion Research, Journal of Broadcasting & Electronic Media, Journal of Communication, Journal of Health*

Communication, Journalism & Mass Communication Quarterly, New Media & Society, Political Communication, and *Public Opinion Quarterly.*

Simon Cottle is professor and former Head of School (JOMEC), School of Journalism, Media and Cultural Studies at Cardiff University (2013–2015) and formerly Head of the Media and Communications program, University of Melbourne. He is the author of many books on media, globalization and the communication of conflicts, crises and catastrophes. Most recently, these include *Mediatized Conflicts* (2006), *Global Crisis Reporting* (2009), *Transnational Protests and the Media* (ed. with L. Lester, 2011), *Disasters and the Media* (with M. Pantti and K. Wahl-Jorgensen, 2012), *Humanitarianism, Communications and Change* (ed. with G. Cooper, 2015) and *Reporting Dangerously: Journalist Killings, Intimidation and Security* (with R. Sambrook and N. Mosdell, forthcoming). He is Series Editor for the Global Crises and Media Series for Peter Lang Publishing.

Stephen Cushion is a Reader at the Cardiff School of Journalism, Media and Cultural Studies, Cardiff University. He has published numerous journal articles on issues in journalism, news and politics, three sole-authored books—*News and Politics: The Rise of Live and Interpretive Television News* (2015, Routledge), *The Democratic Value of News: Why Public Service Media Matter* (2012, Palgrave) and *Television Journalism* (2012, Sage)—and co-edited (with Justin Lewis) *The Rise of 24-Hour News: Global Perspectives* (2010, Peter Lang).

Alison Dagnes is Professor of Political Science at Shippensburg University in Pennsylvania. She is the author of *Politics on Demand: The Effects of 24-Hour News on American Politics* (2010) and *A Conservative Walks Into a Bar* (2012) which examines political comedy. She has also edited two books on political scandal, and gives frequent talks on political behavior as well as on the modern media. Prior to receiving her doctorate in Political Science from the University of Massachusetts at Amherst, Dr. Dagnes was a producer for C-SPAN in Washington, DC.

Tine Ustad Figenschou, PhD, is a researcher at the Department of Media and Communication, University of Oslo. She is the author of *Al Jazeera and the Global Media Landscape: The South is Talking Back* (Routledge, 2013) and has published widely in international journals such as *Journalism, Journalism Studies, International Journal of Communication, The International Journal of Press/ Politics and Media,* and *Culture & Society.* Beyond satellite news networks, Figenschou's primary research interests include media-elite relations, mediated conflicts and mediatization processes.

Ibrahim Helal was Director of News for the Al Jazeera Arabic channel from 2001–2004 and again from 2011–2015. He has more than 25 years of experience in international media including Egyptian TV, BBC Arabic radio and TV, BBC Media Action, Abu Dhabi TV and the Al Jazeera Network. He has played a leading role in the launch of all Al Jazeera's channels, including Al Jazeera English and its Balkan channel. In 2005, the World Economic Forum chose him as one of their Young Global Leaders.

Jesse Holcomb is associate director of research at the Pew Research Center, where he focuses on journalism, media, and informed communities. Before joining the Pew Research Center, Holcomb served in writing and editorial capacities at the Public Interest Network and *Sojourners* magazine. He received his master's degree from George Washington University's School of Media and Public Affairs and later served as adjunct faculty there. Holcomb is an author of studies on the local news ecosystem, social media, and the US news industry with a focus on cable TV news and digital nonprofit startups.

Peter Horrocks is Vice Chancellor of the Open University. Previously he had a long and distinguished career as a producer, editor and executive in BBC News. Among other roles he was Editor of Newsnight and Panorama, Head of Current Affairs, Head of the BBC Newsroom, where he oversaw the integration of online and TV news, and Director of the World Service. He also edited numerous election and budget special programmes. He won Bafta awards in 1997 and 2005 for his editorship of Newsnight and for the documentary trilogy, *The Power of Nightmares*, with filmmaker Adam Curtis.

Justin Lewis is Professor of Communication at Cardiff School of Journalism, Media and Cultural Studies, and Dean of Research for the College of Arts, Humanities and Social Sciences. His books, since 2000, include *Constructing Public Opinion* (2001, Columbia University Press), *Citizens or Consumers: What the Media Tell Us About Political Participation* (2005, Open University Press), *Shoot First and Ask Questions Later: Media Coverage of the War in Iraq* (2006, Peter Lang), *Climate Change and the Media* (2009, Peter Lang) and *Consumer Capitalism: Media and the Limits to Imagination* (2013, Polity).

Yin Lu is a PhD candidate at the Department of Media and Communication, City University of Hong Kong.

Sean McGuire is Managing Director of Oliver and Ohlbaum Associates, the UK's leading strategy advisors to the media, entertainment and sports industries. He focuses on helping broadcasters, platforms, news organisations, regulators, policy makers and investors around the world meet the challenges (and opportunities) of the digital age. Before joining O&O, he was with LEK

Consulting, Director of Strategic Planning for Telewest Communications plc and was Head of Strategy for BBC News.

Brian McNair is Professor of Journalism, Media and Communication at Queensland University of Technology. He is the author of many books and articles on journalism and media, including *News & Journalism in the UK* (5th edition, 2009), *Cultural Chaos* (2006), and *Journalists in Film* (2010). His work has been translated into fourteen languages including Mandarin, Japanese, Russian, Polish, Greek, Korean, Spanish and Romanian. He is a regular contributor to print, broadcast and online media, including Sky News and ABC News 24 in Australia.

Colleen Murrell is a senior lecturer in journalism at Deakin University in Melbourne, where she chairs and teaches units in radio, television and international news. Her main research interests are transnational broadcasting, international newsgathering and social media. Her book, *Foreign Correspondents and International Newsgathering: The Role of Fixers* was published by Routledge in 2015. In her previous career as a journalist, Colleen worked for the BBC, ITN, TF1, CBC, and ABC Australia and was a founding news editor at Associated Press Television News in London. In 2015–16 she was a visiting senior fellow at LSE.

Michael Peters is CEO of Euronews. Born in 1971 in Flensburg, Germany, of French and German nationality, Michael Peters graduated in 1992 from IAE Lyon III, and achieved a master's degree in Financial Engineering from EM Lyon business school in 1995. Immediately after his studies, Michael Peters started his career at the international auditing firm Arthur Andersen in Lyon, France. He left his position in 1998 to join Euronews as CFO. He was nominated Deputy Managing Director of the international news channel in 2003. In May 2005, at the age of 33, Michael Peters was appointed Managing Director of the media and in 2008 Managing Director of the Executive Board of Euronews S.A. In his role as a dynamic actor of the media's international growth and diversification, Michael Peters negotiated many important deals for Euronews and was appointed Chief Executive Officer and Chairman of the Executive Board in December 2011.

Robert G. Picard is a professor on the staff of the Reuters Institute in the Department of Politics and International Relations at University of Oxford, a fellow of the Royal Society of Arts, and an affiliated fellow of the Information Society Project at Yale University. He is the author and editor of 30 books and has been editor of the *Journal of Media Business Studies* and *The Journal of Media Economics*. Picard received his PhD from the University of Missouri,

Columbia, and has been a fellow at the Shorenstein Center at the John F. Kennedy School of Government at Harvard University.

Mugdha Rai is an MA Course Coordinator at the Media and Communications Program, University of Melbourne, Australia. Her PhD examined media representations of globalization discourses and notions of state sovereignty across the US, Australia and India. She teaches and writes extensively on globalization, transnational media, journalism and democracy.

John Ryley is the Head of Sky News. Over the past 30 years he has worked for all of Britain's news broadcasters, the BBC, ITN, and Sky News, producing news and current affairs programming. He has a track record of leading successful change, challenging convention and championing innovation. He is a board member of Sky News Arabia, The National Council for the Training of Journalists, and the Media Trust. He was educated at Durham University and the Wharton School of Business.

Richard Sambrook is Professor of Journalism at Cardiff University where he oversees the vocational postgraduate training of journalists. He worked at the BBC for thirty years as a producer, editor and manager in BBC News, spending a decade on the board of management as Director of Sport, Director of News and finally Director of Global News and the World Service. He is also a Visiting Fellow at the Reuters Institute for the Study of Journalism at Oxford University where he has published on the future of international news and on impartiality in the digital age.

David Schlesinger is the former Editor-in-Chief of Reuters News, where he ran all aspects of the 3,000-journalist strong international news service. Schlesinger joined Reuters as a correspondent in Hong Kong. He then ran editorial operations in Taiwan, China and Greater China before transferring to New York to serve in turn as Financial Editor, Managing Editor for the Americas, and Executive Vice President and Editor of the Americas. He then took on global roles based in London. Ending his Reuters career as Chairman of Thomson Reuters China, he is now a consultant advising on political risk, strategy and on running media companies.

Mark Scott is Managing Director of the Australian Broadcasting Corporation. Since his appointment in 2006, the ABC has rapidly expanded its range of news services, including the creation of a 24-hour TV news channel, a host of new websites and apps, and established itself in the nation's leadership position on key social media sites. The ABC is Australia's most trusted and respected news organisation. He was previously Editorial Director at the Fairfax

Media Group. He was named an Officer of the Order of Australia in 2011. He tweets at @mscott.

Margarita Simonyan is Editor in Chief of RT, a global TV news network. Launched in 2005, RT encompasses round-the-clock news channels in English, Arabic and Spanish, two documentary channels and Ruptly video news agency. Under Ms. Simonyan's helm, RT won the Monte-Carlo TV Festival award for best 24-hour newscast, became the sole Russian channel with three International Emmy nominations for news and the #1 TV news network on YouTube. Prior to heading up RT, Ms. Simonyan had reported from Chechnya during the Second Chechen Campaign, and North Ossetia during the Beslan School Siege. She has received many journalism awards for her work.

Yunya Song is an Assistant Professor of Journalism and Director of the Applied Communication Research Lab at the School of Communication, Hong Kong Baptist University. She works in the areas of international journalism, computer-mediated communication, and media sociology. Her scholarship straddles English, French and Chinese cultures and media. Her research on journalism and media politics has appeared in, among other journals, *International Journal of Press/Politics, International Communication Gazette, Media, Culture & Society, Public Relations Review*, and *Journalism Studies*.

Daya Kishan Thussu is Professor of International Communication and Co-Director of India Media Centre at the University of Westminster in London. Among his key publications are: *Mapping BRICS Media* (2015, co-edited with Kaarle Nordenstreng, Routledge); *Communicating India's Soft Power: Buddha to Bollywood* (2013, Palgrave/Macmillan); *Media and Terrorism: Global Perspectives* (2012, co-edited with Des Freedman, Sage); *Internationalizing Media Studies* (2009, Routledge); *News as Entertainment: The Rise of Global Infotainment* (2007, Sage); *Media on the Move: Global Flow and Contra-Flow* (2007, Routledge); *International Communication: Continuity and Change, Third Edition* (Bloomsbury, forthcoming); and *Electronic Empires: Global Media and Local Resistance* (1998, Arnold). He is the founder and Managing Editor of the Sage journal *Global Media and Communication*.

Alan Tomlinson is Professor of Leisure Studies in the School of Humanities, College of Arts and Humanities, University of Brighton. He has written, co-authored and edited around three dozen texts on sport, leisure and popular culture, and written for When Saturday Comes, the *Financial Times*, *New Statesman*, Guardian Online, the *New York Times*, *Gulf News*, the *Brighton Evening Argus*, and *Der Tagesspiegel*. He has featured regularly in the broadcast media, and breaking news channels, commenting on the global politics of sport, in particular in relation to issues concerning global governance of sport, at the IOC (Olympics) and at FIFA (football/soccer). He blogs from time to time on his typepad blog.

Ingrid Volkmer is Associate Professor, Media & Communications Program, University of Melbourne. She has published widely in the area of globalization theory and political "public" communication. Among her recent publications is *The Global Public Sphere: Public Communication in the Age of Reflective Interdependence* (2014, Polity).

David Westin is an anchor for Bloomberg News in New York. He served as president of ABC News from 1997 to 2010, which he chronicled in his book, *Exit Interview*. Under Westin's leadership, ABC News earned 11 George Foster Peabody Awards, 13 Alfred I. DuPont Awards, four George Polk Awards, more than 40 News & Documentary Emmys, and more than 40 Edward R. Murrow Awards. Earlier in his career, Westin was President of the ABC Television Network, General Counsel of the parent company, Capital Cities/ABC, and a partner in a major Washington law firm.

Index